# An Introduction to
# Exotic Option Pricing

# CHAPMAN & HALL/CRC
Financial Mathematics Series

## Aims and scope:
The field of financial mathematics forms an ever-expanding slice of the financial sector. This series aims to capture new developments and summarize what is known over the whole spectrum of this field. It will include a broad range of textbooks, reference works and handbooks that are meant to appeal to both academics and practitioners. The inclusion of numerical code and concrete real-world examples is highly encouraged.

## Series Editors

M.A.H. Dempster
*Centre for Financial Research*
*Department of Pure*
*Mathematics and Statistics*
*University of Cambridge*

Dilip B. Madan
*Robert H. Smith School*
*of Business*
*University of Maryland*

Rama Cont
*Center for Financial*
*Engineering*
*Columbia University*
*New York*

## Published Titles

American-Style Derivatives; Valuation and Computation, *Jerome Detemple*

Analysis, Geometry, and Modeling in Finance: Advanced Methods in Option Pricing,
  *Pierre Henry-Labordère*

An Introduction to Exotic Option Pricing, *Peter Buchen*

Credit Risk: Models, Derivatives, and Management, *Niklas Wagner*

Engineering BGM, *Alan Brace*

Financial Modelling with Jump Processes, *Rama Cont and Peter Tankov*

Interest Rate Modeling: Theory and Practice, *Lixin Wu*

Introduction to Credit Risk Modeling, Second Edition, *Christian Bluhm, Ludger Overbeck, and*
  *Christoph Wagner*

Introduction to Stochastic Calculus Applied to Finance, Second Edition,
  *Damien Lamberton and Bernard Lapeyre*

Monte Carlo Methods and Models in Finance and Insurance, *Ralf Korn, Elke Korn,*
  *and Gerald Kroisandt*

Numerical Methods for Finance, *John A. D. Appleby, David C. Edelman, and John J. H. Miller*

Option Valuation: A First Course in Financial Mathematics, *Hugo D. Junghenn*

Portfolio Optimization and Performance Analysis, *Jean-Luc Prigent*

Quantitative Fund Management, *M. A. H. Dempster, Georg Pflug, and Gautam Mitra*

Risk Analysis in Finance and Insurance, Second Edition, *Alexander Melnikov*

Robust Libor Modelling and Pricing of Derivative Products, *John Schoenmakers*

Stochastic Finance: A Numeraire Approach, *Jan Vecer*

Stochastic Financial Models, *Douglas Kennedy*

Structured Credit Portfolio Analysis, Baskets & CDOs, *Christian Bluhm and Ludger Overbeck*

Understanding Risk: The Theory and Practice of Financial Risk Management, *David Murphy*

Unravelling the Credit Crunch, *David Murphy*

Proposals for the series should be submitted to one of the series editors above or directly to:
**CRC Press, Taylor & Francis Group**
4th, Floor, Albert House
1-4 Singer Street
London EC2A 4BQ
UK

Chapman & Hall/CRC FINANCIAL MATHEMATICS SERIES

# An Introduction to Exotic Option Pricing

## Peter Buchen

CRC Press
Taylor & Francis Group
Boca Raton  London  New York

CRC Press is an imprint of the
Taylor & Francis Group, an **informa** business

A CHAPMAN & HALL BOOK

Cover image by Ally Buchen.

CRC Press
Taylor & Francis Group
6000 Broken Sound Parkway NW, Suite 300
Boca Raton, FL 33487-2742

First issued in paperback 2019

© 2012 by Taylor & Francis Group, LLC
CRC Press is an imprint of Taylor & Francis Group, an Informa business

No claim to original U.S. Government works

ISBN-13: 978-1-4200-9100-7 (hbk)
ISBN-13: 978-0-367-38172-1 (pbk)

**Library of Congress Cataloging-in-Publication Data**

Buchen, Peter.
     An introduction to exotic option pricing / Peter Buchen.
         p. cm. -- (Chapman & hall/CRC financial mathematics series)
     Includes bibliographical references and index.
     ISBN 978-1-4200-9100-7 (hardback)
     1. Options (Finance)--Prices. I. Title.

HG6024.A3B83 2012
332.64'53--dc23                                                                2011049467

**Visit the Taylor & Francis Web site at**
**http://www.taylorandfrancis.com**

**and the CRC Press Web site at**
**http://www.crcpress.com**

# Dedication

For my students:

Michael Aked, Sean Anthonisz, Sophie Ashton, William
Bertram, Gordon Browne, Greg Campbell, Daniel Campos,
Sean Carmody, Clare Chapman, Abishek Das, Mike Evans,
Matthew Fernandez, Igor Geninson, Tom Gillespie, Malcolm
Gordon, Stephanie Goulter, Matthew Hardman, Kevin Ho-
Shon, Andrew Jackson, Sally Johnson, James Kelly, Otto Kon-
standatos, Oh Kang Kwon, Tim Kyng, Richard Lawson, Greg
Londish, Michael Lukin, Charles Magri, Hamish Malloch, Hugh
Miller, Takao Ohara, Richard Phillips, Leanne Pithers, Prue
Reid, Stephen Rockwell, Karl Rodolfo, Max Skipper, Jon Tin-
dall, Jonathan Tse, Tony Vassallo, Charlie Wilcox, Martin
Wood, Greg Woodbury, Seung Yang, Hanqing Zhu.

# Contents

# List of Figures

# Symbols and Abbreviations

| | | | |
|---|---|---|---|
| aBm | Arithmetic Brownian motion | IVP | Initial Value Problem |
| ADS | Arrow–Debreu security | itm | In-the-money |
| AFM | Arbitrage Free Measure | ode | Ordinary differential eqn. |
| a.s. | Almost surely | lhs | Left-hand side |
| atm | At-the-money | OTC | Over-The-Counter |
| BS | Black–Scholes | otm | Out-of-the-money |
| ccf | Correlation coefficient | mgf | Moment-generating function |
| cdf | Cumulative density function | MoI | Method of Images |
| EMM | Equivalent martingale measure | MR | Martingale Restriction |
| $\overset{d}{=}$ | Equal in distribution | pax | Pay-at-expiry |
| ESO | Executive stock option | pde | Partial differential equation |
| FBV | Free boundary value (problem) | pdf | Probability density function |
| | | PSR | Principle of Static Replication |
| FK | Feynman–Kac | pv | Present value |
| FTAP | Fundamental Theorem of Asset Pricing | rhs | Right-hand side |
| | | rms | Root mean square |
| FX | Foreign exchange | RNM | Risk Neutral Measure |
| gBm | Geometrical Brownian motion | rv | Random variable |
| GM | Geometric mean | sde | Stochastic differential eqn. |
| GFC | Global Financial Crisis | TBV | Terminal Boundary Value (problem) |
| GST | Gaussian Shift Theorem | | |
| hot | Higher order terms | TV | Terminal Value (problem) |
| iid | Independent and identically distributed | wlog | Without loss of generality |
| | | wrt | With respect to |
| iff | If and only if | ZCB | Zero Coupon Bond |

# Preface

This book is a collection of a large amount of material developed from my teaching, research, and supervision of student projects and PhD theses. It also contains a significant quantity of original unpublished work.

One of my main interests in financial mathematics was to seek elegant methods for pricing derivative securities. Although the literature on derivatives is vast, virtually none outside the academic journals concentrates solely on pricing methods. Where it is considered, details are often glossed over, with comments like: "··· and after a lengthy integration, we arrive at the result," or "··· this partial differential equation can be solved to yield the answer." In my experience, many students, even the mathematically gifted ones, found the subject of pricing anything but the simplest derivatives somewhat unsatisfactory and often quite daunting. One aim of this book is to correct the impression that exotic option pricing is a subject only for technophiles. My plan is to present it in a mathematically elegant and easily understood fashion. To this end:

> I show in this book how to price, in a Black–Scholes economy, the standard exotic options, and a host of non-standard ones as well, without generally performing a single integration, or formally solving a partial differential equation.

How is this to be achieved? In a nutshell, the book devotes a lot of space to developing specialized methods based on no-arbitrage concepts, the Black–Scholes model and the Fundamental Theorem of Asset Pricing. These include the Principal of Static Replication, the Gaussian Shift Theorem and the Method of Images. The last of these, which has been borrowed from theoretical physics, is ideally suited to pricing barrier and lookback options. But don't let this technical stuff deter you from reading further!

While the book is certainly targeted to the mathematically capable reader, it is written in non-technical language, in which theorems and proofs are delivered in heuristic rather than formal mathematical terms. This is the *applied*, as opposed to the *pure* mathematical approach. That is not to say that formal methods are not important — they are. But the technical approach is not the focus of this book. Anyone competent in junior year university calculus should be able to understand this book without difficulty.

The following quote from Freedman [23] recently caught my eye: "*The market crash of 2008 that plunged the world into economic recession from which it is still reeling had many causes. One of them was mathematics.*" Statements such as these should serve to remind the reader that this introductory text is not about risk management, but pricing. The GFC should be fair warning to everyone that no mathematical model yet captures the complexities of real markets.

How is this book different from other books written on exotic options? Most other books concentrate on listing or quoting formulae, computing such formulae, applying them to hedging and risk management or combinations of these. As mentioned above, few devote much space to the actual derivation of prices, and those that do generally follow standard practice by employing integration (or partial differential equation) methods. Given the diversity of exotic options, these integrations can be quite cumbersome and are often very complicated.

This book, by contrast, focuses entirely on pricing exotic options. With very few exceptions, no prices within this book are quoted without derivation. Generally, full details of the calculations are provided. The book contains many ideas and techniques which are perhaps new to the general quantitative finance community.

The book is divided into two parts. The first three chapters contain the necessary financial, mathematical and statistical background for the rest of the book. While much of the topics covered is standard, there is also a significant component of specialized material which might be unfamiliar to the reader.

The ensuing chapters contain the applications to exotic option pricing. They include dual-expiry options such as compound and chooser options, multi-asset rainbow options, barrier and lookback options, Asian options, and much more. Chapter 10 of the book introduces and derives a very powerful formula for pricing a class of multi-asset, multi-period binary options called M-binaries. These include all the standard binary (digital) options which are the basic building blocks for complex exotic options. In a very real sense, these M-binaries represent the limit to which Black–Scholes technology can be pushed.

Some might criticize this book on the grounds that its focus on Black–Scholes pricing is too narrow, or even not relevant in today's rumble-tumble financial world. True, the markets are more complex than can be modeled by a Black–Scholes view of the world. Volatilities are not constant, but highly variable and most likely stochastic. But although some stochastic volatility models have been put forward to explain market behavior, and even price exotic options, none has yet to attain any widespread acceptance. Certainly not to the same extent that the Black–Scholes framework has penetrated the consciousness of practitioners and academics. In any case, stochastic volatility models rarely have closed-form solutions, so option pricing largely becomes an exercise in Monte Carlo simulation. It could be argued that the analytic solutions obtained in the Black–Scholes model, which are so readily produced in this book, can be used as control variates, to reduce the errors of such Monte Carlo simulations.

The book is basically an applied mathematics one, with exotic option pricing as its application area. As such, it is foremost a textbook for advanced university students, typically at the honors and postgraduate levels. Each chapter contains a set of exercise problems to assist in the understanding of the techniques introduced. However, the book should be useful to researchers and practitioners as well, and hopefully not only an addition to their library shelves. The book is not intended to be a complete historical account of exotic option pricing. Some relevant references may unintentionally have been omitted. I take full responsibility for such omissions and for the inevitable typos that always seem to escape detection.

**Acknowledgements**
My interest and fascination with financial mathematics was sparked by my friend and one-time futures trader Bobby Richman in 1985.

This book is dedicated to my many past students, some of whose work has contributed significantly to its content. I take the opportunity here to acknowledge my colleagues who have provided inspiration and collegiality over the years. These include Alan Brace, Carl Chiarella, David Colewell, Jeff Dewynne, David Edelman, Robert Elliott, Volf Frishling, Ben Goldys, Andrew Grant, Michael Kelly, Hugh Luckock, Marek Musiela, Alex Novikov, Eckhard Platen, Marek Rutkowski, Erik Schloegl, Pavel Shevchenko, David Stump, John van der Hoek, and Song-Ping Zhu. Of course, many of my past students have since become my colleagues and personal friends.

Particular thanks also go to David Johnstone and Michael McKenzie who kindly provided the opportunity and resources by which this book was written within the finance discipline of the School of Business, University of Sydney.

Sophie Ashton, Matthew Hardman, Kevin Ho-Shon, Otto Konstandatos, Tim Kyng, Charles Magri, Hamish Malloch, Max Skipper and Tony Vassallo ably assisted with the proofreading. Their generosity of time spent on the task and valuable suggestions are acknowledged with gratitude.

Peter Buchen
Sydney, Australia

# Part I

# Technical Background

# Chapter 1

## Financial Preliminaries

The first three chapters constitute Part I of this book and provide the necessary financial, mathematical and statistical background upon which later chapters on pricing exotic options depend. As an introductory text, the book aims to present a comprehensive treatment of exotic option pricing in the Black–Scholes (BS) framework. Readers who are familiar with this background may skip directly to the applications in Part II and refer back to relevant sections of Part I when necessary.

Part I is divided into three sections: the Financial Preliminaries, the Mathematical Preliminaries and Gaussian Random Variables. The Financial Preliminaries include important concepts such as the no-arbitrage principle, static replication as a pricing tool, the derivation of the Black–Scholes approach to option pricing and a non-technical presentation on the Fundamental Theorem of Asset Pricing.

The chapter on mathematical preliminaries presents many of the technical details, mostly without proof, that underpin modern quantitative finance. These include elements of the stochastic calculus, arithmetic and geometric Brownian motion, martingales and techniques for solving the BS partial differential equation (pde).

The chapter on Gaussian random variables is included because from a statistical viewpoint, the BS model can be expressed almost exclusively in terms of these fundamental quantities. A good working knowledge of Gaussian statistics permits many short-cuts to price evaluation and hence readers are strongly advised to be familiar with the contents of this chapter. In particular, as later chapters will affirm, many exotics can be priced using the Gaussian Shift Theorem, without recourse to any formal integration or pde solving. No attempt is made to present a complete and coherent account of all these topics. Such a task would be just too daunting and other books are available to fill the gaps. Hence it is best to regard these opening chapters as a collection of useful and interesting results which impact on exotic option pricing.

Let us also be clear what this book is not about: it is not concerned with legal, institutional or trading aspects of derivatives, nor is it concerned with the design and risk management of structured products. The book is almost exclusively concerned with the arbitrage free pricing of derivative securities within a standard Black–Scholes framework and its extensions. These derivatives can be highly non-standard and come under the umbrella term: *exotic options*.

This opening chapter on Financial Preliminaries is essentially a review of derivative security generalities, such as no-arbitrage conditions and the two main techniques for pricing derivative securities. These are the Black–Scholes partial differential equation (BS-pde) method and the Equivalent Martingale Measure (EMM) method entailed by the Fundamental Theorem of Asset Pricing (FTAP). Many of the mathematical details that underlie these concepts will be found in the following chapter on Mathematical Preliminaries.

Readers of this book are assumed however, to be familiar with derivative security basics and have a working knowledge of University level calculus and mathematical statistics. Our approach, although mathematical, is nevertheless heuristic rather than technical. The book is an introductory treatise on the subject of exotic option pricing and is targeted to senior undergraduate, post-graduate finance and mathematics students, practitioners and researchers in the field of Quantitative Finance.

---

## 1.1   European Derivative Securities

A derivative security is any contract whose future payoff depends in some way on the price evolution of one or more underlying assets.

Simple European derivatives have a single future payoff date $T$, called the expiration date (expiry date for short) and the payoff depends on the price $X = X(T)$ of a single underlying asset on that date. The term expiry date is treated in this book as synonymous with maturity date, settlement date, and payoff date, though in some contexts these may have subtle differences. For European options, the payoff is a function of $X$ only and we shall write $V(X, T) = F(X)$, where $F(X)$ is called the *payoff function* of the derivative and the function $V$ stands for the value of the derivative.

Generally the underlying asset price evolves according to some stochastic process, which makes $X$ a random variable (rv), and $F(X)$ a function of that random variable. We shall take $F(X)$ to be a measurable function of $X$, which means in practice that given $X$, $F(X)$ is uniquely determined. That is, the payoff depends only on $X$ and not on any other extraneous random variables. Pricing the derivative entails finding the present value (pv), $V(x, t)$

of the derivative before expiry, where $x = X(t)$ denotes the underlying asset price at any time $t < T$. That the price of the derivative $V(x, t)$ depends only on the current asset price $x$, is a consequence the Markov property discussed in the next chapter.

| Vanilla Option Payoffs | |
|---|---|
| *Contract* | *Payoff Function at T* |
| 1. Long Forward | $(X - k)$ |
| 2. Short Forward | $(k - X)$ |
| 3. Call Option | $(X - k)^+$ |
| 4. Put Option | $(k - X)^+$ |

Four well-known examples of European options, together with their payoff functions are listed in the above table. The long/short forward contract is the obligation to buy/sell the underlying asset for $k$ dollars (the settlement price) at date $T$. The call/put option is the right (but not the obligation) to buy/sell the underlying asset for $k$ dollars (the strike price) at date $T$.

The payoffs for the call and put option are written in terms of the plus function, defined by:

$$(x - k)^+ = (x - k)\mathbb{I}(x > k); \qquad (k - x)^+ = (k - x)\mathbb{I}(x < k) \qquad (1.1)$$

and $\mathbb{I}(x > k) = (1$ if $x > k$; 0 otherwise), denotes the usual indicator or step function. The plus function captures the optionality in the call and put contracts. The call is exercised at $T$ only if $x > k$ at expiry, while the put option is exercised at $T$ only if $x < k$.

American options are similar to European options with the additional feature that they may be exercised at any time prior to expiry, as well as on their expiry date $T$. American options are therefore *path dependent* options, while European options are strictly *path independent*. It is well known that pricing American options is a much harder task than pricing their European counterparts. We shall mainly be concerned in this book with pricing European options.

---

## 1.2 Exotic Options

European and American calls and puts are often called *plain vanilla* options and are traded both in organized exchanges and in over-the-counter (OTC) markets.

Exotic options are any derivative securities which are not plain vanilla. Exotics can be classified in several different ways. As for European and American options, they can be either path-dependent, or path-independent. Examples of path-dependent exotic options include barrier lookback and Asian options, all of which are analyzed in detail in later chapters of this book.

Some exotic options are simple portfolios of vanilla options. These are usually referred to as *packages*. An example is a *range-forward* derivative consisting of long position in a European call option and a short position in a European put option. These are the easiest to price, as we shall demonstrate later in this chapter, using the Principle of Static Replication.

There is also a family of exotic options on a single underlying asset, whose payoff depends on the asset price at two future dates $T_1$ and $T_2$ say, with $T_1 < T_2$. We refer to such options as *dual-expiry* options. Examples include compound options such as a call-on-call option, and chooser options where at time $T_1$ the holder chooses either a call option or a put option, both of which expire at time $T_2$. Such dual-expiry options are considered in detail in Chapter 5 of this book.

Other exotic options that have been traded and also appear in the academic literature include derivatives whose payoffs at the single future expiry date $T$ depend on the prices of two distinct, but perhaps correlated, underlying assets. Such derivatives are often referred to as *two-asset rainbow* options or *correlation options*. The best known example is the *exchange option* which gives its holder the right to exchange one asset for another asset at time $T$. Chapter 6 considers several examples of these exotic options.

We shall also study in this book extensions of dual-expiry and two-asset rainbow options, including, in Chapter 10, general multi-period and multi-asset exotics.

---

## 1.3  Binary Options

A particularly important class of exotic options is the family of *binary options*. These options pay out at one or more future dates if and only if some exercise condition is met. If not met, they expire worthless. Binary options are also called *digital options* in the derivatives industry. The best known examples are the cash-or-nothing digital, which we shall refer to as a *bond binary*, and the asset-or-nothing digital, which we shall refer to simply as an *asset binary*. The bond binary pays one dollar or nothing according to whether the exercise condition is met or not. The asset binary pays one unit of the underlying asset if the exercise condition is met, and nothing otherwise.

Binary options are particulary important, as they are often the basic building blocks of more complex exotic options. So if we can price the binary options, we can then price the exotic option on which they are constructed.

In this context, a standard European call option is actually a binary option on a long forward contract. From the previous table of vanilla option payoffs, it is clear that the payoff of the call option is $(X - k)$, that of a long forward, but only if the exercise condition $X > k$ is met. In similar fashion, a standard European put option is seen to be a binary option on a short forward contract with exercise condition $X < k$.

We shall meet many different types of binary options in this book, including dual-expiry binaries and two-asset rainbow binaries. The final chapter is concerned with pricing a general multi-period, multi-asset binary which includes all other binaries studied in this book as special cases.

---

## 1.4   No-Arbitrage

When pricing derivative securities, whether plain vanilla or exotic, we seek the fair price of the derivative at equilibrium. By fair price we mean the *arbitrage free* price. Indeed, the concept of no-arbitrage in derivative pricing is no less important than Newton's laws of motion are for particle dynamics.

In the Black–Scholes framework, the no-arbitrage assumption leads to unique prices for both the underlying asset and derivatives written on this asset. This uniqueness derives from the fact that a Black–Scholes market is complete. In such a market, there exists a set of assets which span the risks in that market. This means the cash flows of any contract over any market asset or group of assets can be replicated by a (usually dynamic) portfolio of the spanning set.

It is generally accepted that real financial markets are incomplete, which has the consequence that derivative prices, though possibly arbitrage free, are not unique. There may exist a continuum of prices, all of which are arbitrage free. Recent advances in Behavioral Finance attempt to price financial assets in an economy which is neither complete, nor arbitrage free. A recent survey of such models can be found in Shefrin [68]. Thus, relative to these wider contexts, the Black–Scholes world is very much an idealized one. Yet, despite the considerable efforts of both academics and practitioners, the Black–Scholes framework continues to be a benchmark against which other pricing models are currently measured.

The assumption of no-arbitrage has a number of important consequences for derivative pricing, three of which we discuss immediately below.

### 1.4.1 The Law of One Price

Perhaps the simplest and best known outcome of the no-arbitrage principle is the *Law of One Price*, which can loosely be stated as follows.

> *If two securities have exactly the same pattern of future cash flows, then the securities must have the same price today.*

The proof of this statement is straightforward. Suppose the two securities are labeled $A$ and $B$ and that their current prices satisfy $V_A > V_B$. Then we could short sell security $A$ and go long in (i.e., buy) security $B$. This strategy gives a positive cash flow now of amount $(V_A - V_B)$ and all future cash flows generated by the two securities exactly cancel. Hence we receive a positive amount now and have zero cash flow for all future states of the world. This is an arbitrage. Similarly, if $V_A < V_B$, we could obtain another arbitrage by reversing the above strategy. That is, we now buy security $A$ and short sell security $B$. In order to avoid either of these arbitrages, there is only one conclusion, namely we must have $V_A = V_B$. $\square$

The simple two security situation above illustrates other important features of general pricing models.

1. It is assumed there are no impediments in the market to short selling the security. Of course in real markets there are often significant restrictions to short selling. In such cases, theoretical arbitrages may develop and persist because there is no practical way to take advantage of them. For example, if there are no short sales allowed, the theoretical price equilibrium may lie outside the set of real market positions.

2. In setting up the arbitrage strategy, there is an immutable rule:

   > *In every arbitrage strategy, buy the relatively under-priced security and short sell the relatively over-priced security.*

   In other words, it can never be optimal in an arbitrage strategy to purchase the (relatively) over-priced or sell the (relatively) under-priced security.

### 1.4.2 The Principle of Static Replication

Another consequence of the no-arbitrage condition is the *Principle of Static Replication* (or PSR), which can be stated as follows.

> *If the future payoff of a derivative security can be expressed as a portfolio of elementary securities, then the price of the derivative must be equal to the price of the replicating portfolio.*

This statement can be made more precise by expressing it mathematically. Let $v_i(X, T)$ for $i = 1, 2, \ldots$ denote the payoff at time $T$ of a set of elementary derivatives and let $v_i(x, t)$ denote their arbitrage-free prices at any time $t < T$ where $X = X_T$ and $x = X_t$ are the underlying asset prices at the designated times $T$ and $t$. Suppose further, the market admits another derivative security whose payoff at time $T$ can be represented as the portfolio

$$V(X, T) = \sum_i \alpha_i v_i(X, T), \qquad (1.2)$$

where $\alpha_i$ are the portfolio weights, which may be positive or negative. The Principle of Static Replication states that the price of the new derivative at any time $t < T$ is given by

$$V(x, t) = \sum_i \alpha_i v_i(x, t). \qquad (1.3)$$

**REMARK 1.1**   The PSR essentially states that the pricing functional, i.e., the rule which determines the price $V(x, t)$ from its payoff $V(X, T)$, must be linear. Observe in particular, that if all the portfolio weights $\alpha_i = 0$, then zero payoff must be worth zero value today, but the converse is not necessarily true. A derivative with the possibility of both positive and negative payoffs at time $T$, may have zero value today, without violating the condition of no-arbitrage.

The Principle of Static Replication gets its name from the following observation. At time $t$ we may replicate the decomposable derivative security with the above portfolio. Once set up, this portfolio may be held to expiry without further adjusting the weights $\alpha_i$. Such a constant weight portfolio is an example of a static portfolio, in contrast to the dynamic portfolios which underlie the Black–Scholes approach to option pricing.

The above principle should not be confused with the idea of *static hedging*, which has become popular in recent times (e.g., see Carr et al. [14]). While there are similarities, the respective contexts are quite different. Static hedging is concerned with decomposing certain exotic options as portfolios of other *traded* securities. There is no requirement in the PSR that the portfolio securities are actually traded: they may be a set of entirely theoretical securities which have never seen the light of a real market. Nevertheless, the PSR is an indispensable tool for pricing certain classes of exotic option, and this is precisely how it shall be used in this book. ☐

## Example 1.1

Consider two elementary derivatives on a stock $X$. The first is an up-type asset binary (an asset-or-nothing digital) with expiry $T$ payoff

$$A_k^+(X,T) = X\,\mathbb{I}(X>k). \qquad (1.4)$$

The second is an up-type bond binary (a cash-or-nothing digital) with expiry payoff

$$B_k^+(X,T) = \mathbb{I}(X>k). \qquad (1.5)$$

The asset binary therefore pays one unit of the underlying asset at time $T$, but only if the asset price is above a given exercise price $k$. On the other hand, the bond binary pays one unit of cash (a dollar say) at time $T$, again only if the asset price is above $k$ at time $T$.

Observe that the bond binary is similar to a zero coupon bond (ZCB) with a one-dollar face value. The main difference is that the bond binary delivers the one dollar face value only if the underlying asset satisfies the exercise condition $X > k$. The ZCB is not really a derivative security at all, because its guaranteed one-dollar payoff is independent of any underlying asset. The bond binary, however, is a genuine derivative security whose payoff has strong dependence on the underlying asset price.

The superscripted plus-signs on $A_k^+$ and $B_k^+$ are not just decoration, but have significance. We shall also consider binaries with the following payoffs

$$A_k^-(X,T) = X\mathbb{I}(X<k) \qquad \text{and} \qquad B_k^-(X,T) = \mathbb{I}(X<k)$$

with an obvious interpretation. These binaries are referred to as down-type binaries, which are in-the-money (itm) when the underlying asset price is below the exercise price.

Suppose for the present, we have priced all these binary options in some asset price model and that for time $t < T$ these prices are given respectively by

$$A_k^\pm(x,t) \qquad \text{and} \qquad B_k^\pm(x,t)$$

where $x = X_t$ is the asset price at time $t$. Now consider a European call option with expiry date $T$ and strike price $k$. The payoff at time $T$ can be expressed as

$$\begin{aligned}
C_k(X,T) &= (X-k)^+ = (X-k)\mathbb{I}(X>k)\\
&= X\mathbb{I}(X>k) - k\mathbb{I}(X>k)\\
&= A_k^+(X,T) - kB_k^+(X,T).
\end{aligned}$$

That is, the European call option payoff is seen to be equivalent to that of the binary option portfolio: long 1 up-type asset binary and short $k$ up-type bond binaries. Hence, by the PSR, the price of the call option today (at time $t < T$) is given by

$$C_k(x,t) = A_k^+(x,t) - kB_k^+(x,t). \tag{1.6}$$

A similar argument leads to the following representation for the European put option price in terms of down-type binaries

$$P_k(x,t) = -A_k^-(x,t) + kB_k^-(x,t). \tag{1.7}$$

Observe that Equations (1.6) and (1.7) are model independent and are therefore valid in any arbitrage-free economy, Black–Scholes or otherwise. Chapter 4 of this book gives more details of the analysis and application of binary options. □

### 1.4.3 Parity Relations

We shall meet throughout this book many different parity relations. Probably the best known parity relation is the European *put-call parity* relation

$$C_k(x,t) - P_k(x,t) = x - ke^{-r(T-t)}; \qquad (t \leq T) \tag{1.8}$$

where $C_k$, $P_k$ denote the present values of a European call and put option, of strike price $k$, expiry date $T$ and $x = X(t)$ is the underlying asset price at the current time $t$.

This parity relation, like all others we shall meet, is just a particular example of the Principle of Static Replication. In the above case, a portfolio of the long call option and the short put option is statically equivalent to a long forward contract. This is readily seen from the identity

$$(X-k)^+ - (k-X)^+ \equiv (X-k)$$
$$\text{or} \quad C_k(X,T) - P_k(X,T) = F_k(X,T).$$

The result (1.8) is established since $(x - ke^{-r\tau})$ is the pv of the forward contract payoff, $F_k(X,T) = (X-k)$.

### Example 1.2

As another example of simple parity relations, we mention here the case of *up-down parity* of the asset and bond binaries (on a non-dividend paying asset) considered in example 1.1. For all $t \leq T$,

$$A_k^+(x,t) + A_k^-(x,t) = x \tag{1.9}$$
$$B_k^+(x,t) + B_k^-(x,t) = e^{-r(T-t)} \tag{1.10}$$

We leave the proofs of these straightforward results to the reader. □

## 1.5 Pricing Methods

Apart from using static replication and parity relations, there are two fundamentally distinct methods for pricing exotic options. The first is the *partial differential equation* method. The second is the EMM or *equivalent martingale measure* method entailed by the *Fundamental Theorem of Asset Pricing*. Both methods have their origins in arbitrage-free pricing technology. However, it should be noted at the outset that the EMM method is a more general method, being valid for a wider range of models than that encompassed by pde methods.

Both methods used in this book are, however, specific to the Black–Scholes model. We often combine the two approaches, in order to simplify the pricing analysis. It is therefore essential that the reader has a good grasp of both these important pricing tools. We review the two methods in some detail in the next few sections.

## 1.6 The Black–Scholes PDE Method

The exposition that follows is somewhat different from that of Black–Scholes [5] and Merton [58] as described in their seminal 1973 papers, updated to reflect modern financial parlance. Let us begin by stating the ten assumptions that underlie the Black–Scholes model.

| | *Black–Scholes model assumptions* |
|---|---|
| 1. | There are no transaction costs or taxes |
| 2. | There are no penalties for short sales |
| 3. | The market operates continuously |
| 4. | The risk-free rate $r$ is a known constant |
| 5. | There is a single traded risky asset of price $X_t$ |
| 6. | The asset pays no dividends |
| 7. | The option payoff at expiry $T$ depends only on $X_T$ |
| 8. | The option is European (i.e., no early exercise) |
| 9. | The market is arbitrage free |
| 10. | $X_t$ follows continuous geometric Brownian motion (gBm) with constant (logarithmic) growth rate $\mu$ and volatility $\sigma$ |

We shall refer to a market which satisfies these assumptions as the *standard Black–Scholes economy*. The old adage that a model is no better than its underlying assumptions certainly holds true for Black–Scholes. However, the model has been found to be extremely robust with respect to the assumptions above. Relaxing one or more certainly leads to changes in outcome, but not in any fundamental way. The most important are of course the last two. We next discuss each assumption in order to bring greater clarity to later developments.

*Assumption-1*: This is the *frictionless market* assumption. Of course in all real markets, transaction costs and taxes are important ingredients which cannot be ignored. However, in many situations, these costs can simply be regarded as add-on costs, above the price of the traded security. Price frictions such as transaction costs and taxes are ignored in the standard BS-model.

*Assumption-2*: Not everyone in the market has the right to short sell assets. Indeed most fund managers are specifically forbidden to short sell assets within their portfolios. In the idealized BS-world such restrictions are relaxed. Indeed the dynamic hedging strategy that underlies the BS methodology would break down if unrestricted short selling were not permitted.

*Assumption-3*: By the very nature of trading on financial markets, all real markets operate in discrete time. However, for a liquid financial asset there may be thousands of trades per day. Since many derivatives have maturity dates measured in months or even years, the assumption of continuous trading may not be too severe. The problem of discrete price changes will then be an issue, only at or very near to the expiry date.

*Assumption-4*: We shall assume mostly throughout this book that interest rates are non-stochastic and time independent. The presence of stochastic interest rates complicates the pricing problem, but not generally in any fundamental way. We show in Section 2.9 how to handle the case of deterministic time varying interest rates.

*Assumption-5*: The simplest derivative securities, such as vanilla call and put options, depend on a single traded underlying security. Early chapters of this book consider such single asset derivatives in detail. Later chapters investigate two-asset and multi-asset derivatives, known as *rainbow options*. For the single asset case, we adopt the notation $X_t$ to denote both the price of the asset at time $t > 0$ and the stochastic process it follows. We also often write $x = X_t$ where $t$ denotes the present time. It is also possible to price options on non-traded assets (e.g., volatility swaps, some interest rate derivatives and so-called real options). Many of these can be analyzed using the BS methodology described in this chapter, but generally extra assumptions, typically on the market price of risk of the non-traded asset, must then be included.

*Assumption-6*: Many equities, particularly those on which calls and puts are written, pay dividends. In practice, the dividend policy of the company issuing the shares can be quite complicated. Indeed, most dividend policies will themselves have a stochastic component, either in the timing of the dividend payout, the amount of the dividend, or both. All such complications are ignored in the standard BS-framework. We investigate the effect of a constant, continuous dividend yield in Section 1.12 toward the end of this chapter.

*Assumption-7*: Vanilla calls and puts are examples of *path-independent* options. Their payoffs depend only on the price of the underlying asset at a fixed expiry date $T$. Classical Black–Scholes is concerned with options of this type. Many exotic options, however, have more complicated payoffs, which of course are the main concern of this book. As mentioned earlier, these include options with multi-period payoffs (e.g., dual expiry options) and also the class of path-dependent options such as barrier and lookback options.

*Assumption-8*: If the option or derivative can only be exercised at its maturity, it is of course a *European option*. Options which can be exercised at any time before maturity are *American options*, while those that can only be exercised at prescribed fixed dates before maturity are termed *Bermudan*. In the standard BS-model, all options are assumed to be European.

*Assumption-9*: As previously stressed, the no-arbitrage assumption is the single most important assumption underlying the BS-methodology of derivative security pricing. A formal definition of arbitrage, in terms of trading strategies, will be presented later in this chapter.

*Assumption-10*: The assumption of geometrical Brownian motion (gBm) for the dynamics of the asset process $X_t$ provides the essential mathematical background to the BS-model and gives it its unique character. Under this model, it is truly remarkable that we are able to obtain analytical, closed-form prices for many extremely complicated exotic options. The model implies that the logarithmic return of the underlying asset is Gaussian, i.e., normally distributed. Unfortunately, assets in real markets rarely exhibit this behavior. Real asset prices and related market indices show far more extreme fluctuations than predicted by Gaussian statistics. It is of course possible, and indeed the topic of much current research, to consider other asset dynamics which address this issue. For example, modeling asset returns by Lévy processes is one such approach currently in vogue. Two recent works that address these models are Schoutens [67] and Kyprianou et al. [50]. But no market model to date has been put forward that has universal or even generally wide support. In lieu of such a model, most practitioners stick with the BS framework or a variant of it with perhaps time-dependent or stochastic volatility. While the subject of stochastic volatility is beyond the scope of this book, we will show how simple European options can be priced in the presence of deterministic time-varying volatility (see Section 2.9).

## 1.7 Derivation of Black–Scholes PDE

Assumption-10 states that the underlying asset price follows geometrical Brownian motion (gBm) with growth rate $\mu$ and volatility $\sigma$. Mathematically, this statement means that the asset price $X_t$ is a stochastic process with dynamics described by the *stochastic differential equation* (sde)

$$dX_t = x(\mu\, dt + \sigma\, dB_t^P). \tag{1.11}$$

Here $x = X_t$ is known at time $t$, and the stochastic price increment $dX_t = X_{t+dt} - X_t$, given $X_t$, is expressed as a sum of two terms: a non-stochastic drift term $\mu x\, dt$ and a stochastic jump term $\sigma x\, dB_t^P$. For small $dt$, the second term dominates, since the increment of Brownian motion

$$dB_t^P = B_{t+dt} - B_t \overset{d}{=} N(0, dt)$$

is Gaussian with standard deviation equal to $\sqrt{dt}$. The superscript $P$ signifies that the Brownian increment is with respect to the real-world or historic measure $P$. Later, when we derive the EMM approach to derivative pricing, we shall see that the real-world measure plays no direct part in the actual price. In fact, a new measure $Q$, equivalent to $P$, called the Equivalent Martingale (or Risk Neutral) Measure is the appropriate one to use.

SDE's of the type described by (1.11) are discussed in the next chapter on Mathematical Preliminaries, so readers who are unfamiliar with such equations may jump ahead to see how to solve them using Itô's Lemma.

Suppose our market consists of two tradeable assets: a $T-$ maturity ZCB (with face value \$1) and the underlying asset $X$. Assuming constant continuously compounded interest rate $r$, the ZCB has pv $P(t,T) = e^{-r(T-t)}$ and satisfies the (non-stochastic) differential equation

$$dP(t,T) = rP(t,T)\, dt; \qquad P(T,T) = 1. \tag{1.12}$$

Now consider the *self-financing* portfolio, initiated at time $t < T$, defined by

$$\mathcal{P}_t = \begin{cases} \alpha_t & \text{units of asset } X \\ \beta_t & \text{units of ZCB} \end{cases}. \tag{1.13}$$

The term self-financing means that over a small time interval $dt$, the portfolio weights $(\alpha_t, \beta_t)$ are held constant. Changes to these weights may be made only in the instant just after $(t+dt)$. From a more technical viewpoint, $(\alpha_t, \beta_t)$ can also be predictable processes which are adapted to the Brownian motion induced by $P$. The value of the portfolio at time $t$ is

$$V_t = \alpha_t\, x + \beta_t\, P(t,T) \tag{1.14}$$

and just before any weights are changed, its value will be

$$V_{t+dt} = \alpha_t\, X_{t+dt} + \beta_t\, P(t+dt, T).\tag{1.15}$$

Subtracting Equations (1.14) from (1.15) leads to

$$\begin{aligned}
dV_t &= V_{t+dt} - V_t\\
&= \alpha_t\, dX_t + \beta_t\, dP(t,T)\\
&= \alpha_t\, dX_t + \beta_t r P(t,T)\, dt.
\end{aligned}\tag{1.16}$$

However, by Itô's Lemma (see Section 2.5), we also have

$$dV_t = V_x\, dX_t + (V_t + \tfrac{1}{2}\sigma^2 x^2 V_{xx})\, dt.\tag{1.17}$$

Here the subscripts on $V = V(x,t)$ denote partial derivatives. Comparing the coefficients of $dX_t$ and $dt$ in (1.16) and (1.17) leads to the identities

$$\alpha_t = V_x \qquad \text{and} \qquad r\beta_t P(t,T) = V_t + \tfrac{1}{2}\sigma^2 x^2 V_{xx}\tag{1.18}$$

But from Equation (1.14), we also have $\beta_t P(t,T) = V - \alpha_t x = V - xV_x$. It follows that

$$\boxed{V_t = rV - rx\, V_x - \tfrac{1}{2}\sigma^2 x^2\, V_{xx}}\tag{1.19}$$

Equation (1.19) is the celebrated Black–Scholes pde.

---

## 1.8   Meaning of the Black–Scholes PDE

The Black–Scholes pde (BS-pde) derived in the previous section determines the price evolution of any simple European derivative security written on the underlying asset $X$. Given the payoff function $V(x,T) = F(x)$ of the derivative, its price $V(x,t)$ for $t < T$, will be uniquely determined by (1.19) subject only to a technical growth condition on $F(x)$.

The BS-pde was derived by replicating the derivative with a portfolio of the underlying asset and a ZCB. That this can always be done (though we have not proved it), is due to the fact that the BS-market is *complete*. However, this replication is performed only over each time interval $(t, t+dt)$. In the limit as $dt \to 0$, this replication process becomes continuous, a process called *dynamic replication* or *dynamic hedging*. This is quite different from the hedging implied by the PSR.

Note that in the original derivation of the BS-pde, the asset and derivative are dynamically hedged to produce a riskless asset. The two approaches are ultimately equivalent.

The prices of simple European options like calls and puts must satisfy the BS-pde, but obviously with different payoff functions. For a call option of strike price $k$, we should take $F(x) = (x-k)^+$, while for a put option we solve the same pde but with payoff function $F(x) = (k-x)^+$.

From a mathematical perspective, the BS-pde is a backward propagating, parabolic or diffusion type of pde with non-constant coefficients, defined on the domain $\{x > 0;\ t \le T\}$. By backward propagating we mean that the solution (derivative price) is known at time $T$ from its given payoff, and the solution at any earlier time is obtained by solving the BS-pde for $t < T$.

We shall refer to pde's of this type as *terminal value* (or TV) problems. It is a simple matter, in fact, to transform the backward propagating BS-pde into a corresponding forward propagating pde, through $\tau = (T - t)$. Thus $\tau$ represents the time *remaining* to expiry and at expiry itself, $\tau = 0$. The TV problem is thus converted by this time transformation into an equivalent *initial value* (or IV) problem for $V(x, \tau)$,

$$V_\tau = -rV + rxV_x + \tfrac{1}{2}\sigma^2 x^2 V_{xx}; \qquad V(x, 0) = F(x) \qquad (1.20)$$

Initial value problems tend to be far more common in engineering, applied mathematical, and theoretical physics applications.

Interestingly, although the underlying structures that led to the BS-pde are stochastic, the final price evolution Equation (1.19) is itself purely deterministic.

Generally, pde's with non-constant coefficients are difficult to solve. However the inhomogeneity in the coefficients in the BS-pde are easily handled. In particular, the pde is readily transformed into the canonical heat equation with constant coefficients. We perform this reduction by two distinct transformations in the next chapter.

Importantly, the BS-pde is seen to be linear, as indeed it must be, in order to satisfy one of the implications of the no-arbitrage condition. Furthermore, the PSR discussed earlier is easily derived from the BS-pde, for if the expiry payoff is $V(x, T) = \alpha F_1(x) + \beta F_2(x)$, then by linearity of the pde, the solution for $t < T$ must satisfy

$$V(x, t) = \alpha V_1(x, t) + \beta V_2(x, t),$$

where $V_1$ and $V_2$ each solve the BS-pde with respective terminal values

$$V_1(x, T) = F_1(x) \qquad \text{and} \qquad V_2(x, T) = F_2(x).$$

This is essentially the Principle of Static Replication.

Readers who are familiar with methods for solving linear pde's will know that a general solution to the TV problem described by Equation (1.19), can be expressed as an integral of the terminal value and the associated *Green's Function* of the pde, over the defining domain. The Green's Function is also called the *fundamental solution* of the pde and is derived for the BS-pde in the next chapter. If $G(x, t; \xi, T)$ denotes this Green's function, then the unique solution of the Black–Scholes TV problem is given by the expression

$$V(x,t) = \int_0^\infty F(\xi)G(x, t; \xi, T)\, d\xi \qquad (1.21)$$

where $F(x) = V(x, T)$ is the terminal value or payoff function of the derivative. The growth condition on $F(x)$ mentioned above is made to ensure the existence of this integral.

**Example 1.3**
Two very simple solutions of the BS-pde correspond to the underlying asset itself and the ZCB with a $1 face value. That is, writing $\tau = (T - t) > 0$,

$$A(x, T) = x \quad \text{gives} \quad A(x, t) = x \qquad (1.22)$$
$$B(x, T) = 1 \quad \text{gives} \quad B(x, t) = e^{-r\tau}. \qquad (1.23)$$

Thus a contract which pays one unit of the underlying asset at time $T$, is worth the current price of the asset today, while a contract which pays one dollar at time $T$ (i.e., a ZCB) has pv $P(t, T) = e^{-r(T-t)}$. The reader may wish to verify that these expressions do in fact satisfy the BS-pde with the given payoff functions. ∐

On a more technical note, no-arbitrage in continuous markets has been more fully analyzed by Delbaen and Schachermayer [18] where they introduce the formal notion of *no free lunch with vanishing risk* (or NFLVR). This idea captures the technical aspects of continuous or dynamic hedging that underlie the basic BS framework.

### 1.8.1 Other Derivatives and the BS-PDE

We have seen above that simple European derivatives on a single asset are priced by solving the BS-pde as a TV problem, with given payoff function $V(x, T) = F(x)$. However, with suitable modification, exotic options in a Black–Scholes economy can also be priced as solutions of the BS-pde.

For example, we shall see that both barrier and lookback options, priced in later chapters of this book, have prices which also satisfy the BS-pde. However, here the price domain is restricted and extra conditions, called *boundary conditions*, must be applied at the domain boundary. We shall refer to such problems as *Terminal Boundary Value* problems, or TBV problems for short. Such problems are obviously more difficult to solve than TV problems. In fact, in Chapter 7 we shall develop specialized methods for solving such problems using the *Method of Images*.

American options, with their early exercise feature can also be priced by the BS-pde. It transpires that these satisfy what is called a *free boundary value* (FBV) problem of the pde. Such problems are well-known in thermodynamics, e.g., the shape of a melting block of ice is determined as the solution of a FBV problem. We shall have very little to say about such FBV problems in this book.

Multi-asset (rainbow) options are derivative securities written on more than one underlying asset. It is relatively straightforward to extend the Black–Scholes framework to include such options. The key idea is to model the asset prices by a vector of correlated geometrical Brownian motions. Not surprisingly, the relevant pde that prices such options turns out to be a multi-variate version of the standard one-dimensional BS-pde. The asset price volatilities and correlation coefficients are the important parameters in this pde. We show how this multi-asset model can be represented in Section 3.10 of this book.

## 1.9 The Fundamental Theorem of Asset Pricing

The discovery of the BS-pde to price derivative securities in an arbitrage free way was an important cornerstone in the field of Quantitative Finance. Another significant advancement was made by Harrison and Pliska [28], who showed how arbitrage free pricing could be represented in terms of the theory of martingales. We present a brief overview of martingales in Section 2.6 of the next chapter.

That martingales appear as quantities in the arbitrage free pricing should not be surprising. Loosely speaking, a martingale is a stochastic variable whose expected future value is equal to today's value.

Consider a gambler (investor), with initial wealth $W_0$, who plays a casino game $n$ times. If the game is fair, the expected future wealth $W_n$ of our gambler should be no different from the initial wealth $W_0$. Thus $W_n$ in a fair game, should be a martingale. In other words, as this simple example illustrates, martingales are closely connected to the notion of a fair game. In financial markets, "fair game" is synonymous "no-arbitrage."

Before developing the theory, let us state the key result of Harrison and Pliska, which is now commonly called the *Fundamental Theorem of Asset Pricing* or FTAP for short. Although the language looks very technical, we shall define all the terms in a heuristic way and attempt to give a clear and simple understanding of this important result.

### THEOREM 1.1 FTAP

*In an arbitrage free market, there exists a probability measure $Q$, with respect to which, the discounted price of any derivative security is a martingale.*

The main ingredient of the FTAP is the new probability measure $Q$. It is known by any of the alternative names:

| | |
|---|---|
| Equivalent Martingale Measure | (EMM) |
| Risk Neutral Measure | (RNM) |
| Arbitrage Free Measure | (AFM) |

The measure $Q$ is said to be *equivalent* to the real-world measure $P$ mentioned in Section 1.7. This is a purely technical statement which means that $P$ and $Q$ share the same null sets. That is, if $A$ is any event with zero probability under measure $P$, then it also has zero probability under measure $Q$, and vice versa.

We shall presently give a derivation of the FTAP which captures the essential approach of Harrison and Pliska [28]. But before we do so, it is possible to get the main idea behind the theorem from a very different direction.

Consider a complete, continuous, arbitrage free market and let $A(x, t; \xi, T)$ denote the price at time $t$, when the underlying asset price is $x$, of a security $\mathcal{A}$ that pays $F(X) = \delta(X - \xi)$ at time $T > t$. Here $\delta(X - \xi)$ denotes the Dirac delta function. Readers unfamiliar with this function might liken this generalized function to an infinite spike located at $X = \xi > 0$, with the unit-intensity property

$$\int_0^\infty \delta(X - \xi) \, d\xi = 1.$$

Security $\mathcal{A}$ is called an *Arrow–Debreu security* (ADS). Financially, one may think of the ADS as a security that pays \$1 if $X = \xi$ at time $T$, and pays nothing otherwise.

That the Arrow–Debreu security exists (in the theoretical sense) at all is a consequence of the completeness assumption, which ensures that $A(x, t; \xi, T)$ is well-defined for all $\xi > 0$. Basically, completeness of the market implies that every payoff (subject only to growth conditions) is attainable. The condition of no arbitrage simply requires $A(x, t; \xi, T) > 0$.

Consider now, an arbitrary derivative security in this market, with time $T$ payoff $F(X)$. We can make use of the substitution property of the Dirac delta function to write, for an arbitrary function $F(X)$,

$$V(X, T) = F(X) = \int_0^\infty F(\xi)\delta(X - \xi)\, d\xi.$$

The integral above can be thought of as a continuous portfolio of Arrow–Debreu securities with portfolio weights $F(\xi)$. Hence, ignoring technical issues, the PSR implies (at least formally) that the price of the derivative is equal to the price of this replicating portfolio, which at $t < T$, is

$$V(x, t) = \int_0^\infty F(\xi)A(x, t; \xi, T)\, d\xi. \tag{1.24}$$

A particular case of this representation is the pv $P(t, T)$ of a ZCB with maturity $T$ payoff $F(\xi) = 1$. Equation (1.24) then reads

$$P(t, T) = \int_0^\infty A(x, t; \xi, T)\, d\xi. \tag{1.25}$$

Now let us define

$$q(x, t; \xi, T) = P^{-1}(t, T)\, A(x, t; \xi, T), \tag{1.26}$$

but under no-arbitrage, $A(x, t; \xi, T) > 0$, which implies

$$q(x, t; \xi, T) > 0 \qquad \text{and} \qquad \int_0^\infty q(x, t; \xi, T)\, d\xi = 1.$$

Hence $q(x, t; \xi, T)$ is a probability density function (pdf).

Equation (1.24) can therefore be written equivalently as

$$V(x, t) = P(t, T) \int_0^\infty F(\xi)q(x, t; \xi, T)\, d\xi.$$

Now in the case of constant interest rates we may write

$$P(t, T) = e^{-r(T-t)} = \frac{D_T}{D_t}, \tag{1.27}$$

where $D_t = e^{-rt}$ denotes the *discount factor* at time $t$.

Let us further define $\tilde{V}(x,t) = D_t V(x,t)$ for any derivative price $V(x,t)$ as the *discounted* derivative price. Then, since $F(\xi) = V(\xi, T)$, the last equation above for $V(x,t)$ can be written as

$$D_t V(x,t) = D_T \int_0^\infty V(\xi, T) q(x,t; \xi, T) \, d\xi.$$

This is equivalent to

$$\tilde{V}(x,t) = \int_0^\infty \tilde{V}(\xi, T) q(x,t; \xi, T) \, d\xi = \mathbb{E}_Q \{\tilde{V}(X,T)\}, \qquad (1.28)$$

since the integral is just the expectation of $\tilde{V}(X,T)$ with respect to the measure $Q$, whose pdf is the function $q(x,t; \xi, T)$ defined by Equation (1.26). Now Equation (1.28) states in words, that the discounted derivative price $\tilde{V}(X,T)$ is a martingale wrt the measure $Q$. This is precisely the statement of the Fundamental Theorem of Asset Pricing, and $Q$ is the corresponding *Equivalent Martingale Measure*.

We have certainly glossed over several technical issues in obtaining this result, but in doing so we have also uncovered some interesting relationships.

First, comparing Equations (1.21) and (1.24) we see that the Green's Function $G(x,t; \xi, T)$ for a derivative pricing model is none other than the Arrow–Debreu price $A(x,t; \xi, T)$. Secondly, Equation (1.26) determines the EMM $Q$ entirely in terms of the Arrow–Debreu price and the ZCB price. Later, we shall see that the FTAP expressed by Equation (1.28) is more than just a martingale statement; it is actually a pricing formula for derivative securities in a very general setting. $\square$

We now turn our attention to deriving the FTAP along the lines of Harrison and Pliska [28]. Again, we shall take a practical and more heuristic approach, rather than a fully technical one, as in fact was adopted by the cited paper. This new approach will elucidate further features of the FTAP and the associated EMM method for pricing derivatives. Most of the mathematical details that follow can be found in the next chapter on Mathematical Preliminaries.

Start from the same self-financing hedge portfolio $\mathcal{P}$ defined in (1.13). Then Equations (1.14) and (1.15) describing the portfolio value at times $t$ and $s = (t + dt)$ are

$$V_t = \alpha_t \, X_t + \beta_t \, P(t, T)$$
$$V_s = \alpha_t \, X_s + \beta_t \, P(s, T).$$

If we multiply these equations respectively by the discount factors $D_t$ and $D_s$ (defined above) and use $P(t,T) = D_T/D_t$, we obtain the pair

$$\tilde{V}_t = \alpha_t \tilde{X}_t + \beta_t D_T \tag{1.29}$$

$$\tilde{V}_s = \alpha_t \tilde{X}_s + \beta_t D_T. \tag{1.30}$$

Now, suppose that the discounted asset price is a $Q$ martingale wrt the filtration[1] $\mathcal{F}_t$. That is,

$$\mathbb{E}_Q\{\tilde{X}_s \,|\, \mathcal{F}_t\} = \tilde{X}_t. \tag{1.31}$$

Then it follows that

$$\mathbb{E}_Q\{\tilde{V}_s \,|\, \mathcal{F}_t\} = \alpha_t \mathbb{E}_Q\{\tilde{X}_s \,|\, \mathcal{F}_t\} + \beta_t D_T$$
$$= \alpha_t \tilde{X}_t + \beta_t D_T = \tilde{V}_t. \tag{1.32}$$

Since $(\alpha_t, \beta_t)$ are adapted to the filtration $\mathcal{F}_t$, they can be taken outside the conditional expectation.

Hence, if $Q$ is a martingale measure for the discounted asset price over $(t,s)$, it is also a martingale measure for the discounted derivative price over $(t,s)$. We now show, by induction, that this result holds true for any finite time interval $(t,T)$, and not just the infinitesimal interval $(t,s)$.

Let $t_n = t + n dt$ and $t_{n+1} = t_n + dt$. The *Tower Law* for conditional expectations (see Section 2.1) states that for any $t < s < u$ and a stochastic process $Z_t$,

$$\mathbb{E}\{Z_u \,|\, \mathcal{F}_t\} = \mathbb{E}\{\mathbb{E}\{Z_u \,|\, \mathcal{F}_s\} \,|\, \mathcal{F}_t\}$$

Choose $s = t_n$, $u = t_{n+1}$ and $Z_t = \tilde{V}_t$ and take expectations wrt measure $Q$. Assume $\mathbb{E}_Q\{\tilde{V}_{t_n} \,|\, \mathcal{F}_t\} = \tilde{V}_t$ is true for some $n > 1$. We have already seen above that it is true for $n = 1$. Then,

$$\mathbb{E}_Q\{\tilde{V}_{t_{n+1}} \,|\, \mathcal{F}_t\} = \mathbb{E}_Q\{\mathbb{E}_Q\{\tilde{V}_{t_{n+1}} \,|\, \mathcal{F}_{t_n}\} \,|\, \mathcal{F}_t\}$$
$$= \mathbb{E}_Q\{\tilde{V}_{t_n} \,|\, \mathcal{F}_t\}$$
$$= \tilde{V}_t$$

The second line above follows from $\mathbb{E}_Q\{\tilde{V}_{t_{n+1}} \,|\, \mathcal{F}_{t_n}\} = \tilde{V}_{t_n}$.

Hence by induction, the result $\mathbb{E}_Q\{\tilde{V}_{t_n} \,|\, \mathcal{F}_t\} = \tilde{V}_t$ is true for all $n > 0$. Now if we take the limit $n \to \infty$ and $dt \to 0$ such that $n dt = T$, then we formally obtain the desired result $\mathbb{E}_Q\{\tilde{V}_T \,|\, \mathcal{F}_t\} = \tilde{V}_t$ for all $t \leq T$. $\quad\square$

We have therefore shown that a martingale measure for discounted asset prices is also a martingale measure for discounted self-financing portfolios $\mathcal{P}$.

---

[1] The meaning of the symbol $\mathcal{F}_t$ is given in the next chapter in Section 2.1.

Suppose further that the market described by the asset $X$ and ZCB is both complete and arbitrage free. Let a derivative security have payoff function $V(X,T) = F(X)$ at time $T$ and let $V(x,t)$ denote its present value for $t < T$, when the asset price is $x$. Then, completeness means that there exists a self-financing strategy $(\alpha_T, \beta_T)$ which replicates the payoff $F(X)$. That is,

$$V(X,T) = \alpha_T X + \beta_T P(T,T) \tag{1.33}$$

since $P(T,T) = 1$.

In general, $(\alpha_T, \beta_T)$ will depend on $X$. No arbitrage means that the price of the derivative at $t$ must be equal to the price of the replicating portfolio. That is,

$$V(x,t) = \alpha_t x + \beta_t P(t,T) \tag{1.34}$$

But we have just demonstrated that (1.34) implies (1.28), which is the mathematical statement of the FTAP. We conclude this section, with an important corollary.

### COROLLARY 1.1
*If the market is complete, the Equivalent Martingale Measure $Q$ is unique.*

We do not offer a proof of this important result, but its truth can be inferred from Equation (1.24) and the uniqueness of the Arrow–Debreu prices $A(x,t;\xi,T)$.

---

## 1.10 The EMM Pricing Method

Now that we have the FTAP we are in a position to state the pricing methodology it implies, and what is commonly called the EMM method. Equation (1.28) is the mathematical statement of the FTAP, and is equivalent to

$$D_t V(x,t) = D_T \mathbb{E}_Q\{F(X) \,|\, \mathcal{F}_t\},$$

where $D_t = e^{-rt}$ and $D_T = e^{-rT}$ are discount factors, and $F(X) = V(X,T)$ is the derivative payoff at time $T$. Thus, we also have the EMM pricing formula

$$\boxed{V(x,t) = e^{-r(T-t)} \mathbb{E}_Q\{F(X) \,|\, \mathcal{F}_t\}} \tag{1.35}$$

The expectation is conditional on all information available up to and including time $t$. Once we know the measure $Q$, the payoff $F(X)$, and how to represent $X$ under the measure $Q$, the derivative price can be found by calculating the expectation in (1.35).

**REMARK 1.2**  The EMM method of derivative pricing is very different from the BS-pde method. We mention three significant differences between the two methods.

1. The BS-pde method is deterministic (non-stochastic), while the EMM method is clearly statistical in nature.

2. The BS-pde depends strongly on the underlying assumption that the asset price follows geometrical Brownian motion. No assumptions on the underlying asset process (beyond that its discounted value is a $Q$ martingale) have been made. Hence, the EMM method is far more general than the BS-pde method, permitting not only geometrical Brownian motion, but other asset processes as well.

3. It is often the case that calculating the expectation in the EMM method is a much simpler task than solving the BS-pde.

It may seem hard to believe, but after we have introduced specialized methods in the ensuing chapters for manipulating the BS-pde and for calculating certain Gaussian expectations, we hardly ever formally solve the BS-pde throughout this book. Nor do we have to calculate an EMM expectation by some complicated integral, even though such integrals are implicitly there.

It should be clear that the pricing formula (1.35) cannot admit an arbitrage. Technically, in the context of the FTAP, an arbitrage can be defined as a self-financing trading strategy such that $V(x,t) < 0$ and $V(X,T) \geq 0$ with $\mathbb{E}_Q\{V(X,T)\} > 0$. This obviously can never occur with (1.35) since the ZCB price $P(t,T) = e^{-r(T-t)}$ is always positive.        □

In order to use the EMM method through (1.35) to price derivative securities, clearly we need some way of determining the martingale measure $Q$. We showed in Section 1.9 that $Q$ is closely related to Arrow–Debreu prices. Unfortunately, these prices are not always easy to find or calculate. In practice, it is more common to find the Arrow–Debreu prices given the martingale measure $Q$, rather than vice versa. However, there is a technique, which we consider next, that determines $Q$ in many cases, including in particular, the arbitrage free measure under either discrete binomial or continuous Black–Scholes asset price dynamics.

## 1.10.1  The Martingale Restriction

We saw in the previous section that if $Q$ is a martingale measure for the discounted asset price, then it will also be a martingale measure for any discounted derivative price. Thus one possible way of finding the EMM $Q$ is to use the asset price equation $D_T \mathbb{E}_Q\{X_T \mid \mathcal{F}_t\} = D_t x$, or in terms of the risk free rate $r$,

$$\boxed{\mathbb{E}_Q\{X_T \mid \mathcal{F}_t\} = x e^{r(T-t)}} \qquad (1.36)$$

Equation (1.36) is called the *Martingale Restriction* (MR) and in many situations is sufficient to determine $Q$ unambiguously.

### Example 1.4

Consider a standard, multiplicative binomial asset price model. In each step, the asset price $x$ say, becomes $ux$ in the up-state and $dx$ in the down-state where $d < 1$ and $u > 1$. Let $\rho = 1 + r$ denote the 1-step future-value factor, where $r$ is now the 1-step risk free interest rate. Let the up-down EMM be denoted by $Q(p, q)$. Then the martingale restriction for this situation will read

$$p(ux) + q(dx) = \rho x$$

which, together with $p + q = 1$, uniquely determines $Q$ through

$$p = \frac{\rho - d}{u - d} \qquad \text{and} \qquad q = \frac{u - \rho}{u - d} \qquad (1.37)$$

provided $d < \rho < u$. This last result is of course the well-known risk-neutral measure (or EMM) for the 1-step binomial model (e.g., see Van der Hoek [75]). Furthermore, the 1-step price of a derivative which pays $(V_u, V_d)$ in the up-down states, will be given by Equation (1.35) as

$$V = \rho^{-1}[pV_u + qV_d] \qquad (1.38)$$

regardless of the expiry payoff $(V_u, V_d)$. $\qquad \qquad \qquad$ ⬚

Notice in this example that we tacitly assumed the EMM was also a binomial measure. Indeed, the martingale restriction will determine $Q$ unambiguously only if we generally make the additional assumption that $Q$ and the real-world measure $P$ belong to the same family of distributions. Still, once the measure $Q$ is determined or specified, option pricing becomes nothing more than calculating an expectation. The next example illustrates the procedure for a discrete time, continuous state-space asset price model.

### Example 1.5

In this example we wish to price at $t = 0$, a European call option of strike price $k$ and expiry date $T$. We suppose that the asset price is currently $S_0$ and at the fixed time $T$, the asset price $X$ is a continuous exponential random variable on $[0, \infty)$.

Assuming that the EMM is also exponential, the corresponding pdf will be

$$q(x) = \lambda^{-1} e^{-x/\lambda}; \qquad x \geq 0$$

where $\lambda$ is some positive parameter.

The martingale restriction states $\mathbb{E}_Q\{X\} = S_0 e^{rT}$, where $r$ is the risk free rate over $[0, T]$. But for the exponential distribution, $\mathbb{E}_Q\{X\} = \lambda$. Hence $q(x)$ is an EMM if and only if $\lambda = S_0 e^{rT}$. The price of the call option is then calculated as follows.

$$
\begin{aligned}
C_0 &= e^{-rT} \mathbb{E}_Q\{(X - k)^+\} \\
&= e^{-rT} \int_0^\infty (x - k)^+ \, q(x) \, dx \\
&= \lambda^{-1} e^{-rT} \int_k^\infty (x - k) e^{-x/\lambda} \, dx \\
&= \lambda e^{-rT - k/\lambda}
\end{aligned}
$$

after performing an integration by parts. Substituting $S_0 e^{rT}$ for $\lambda$, the price of the call option at $t = 0$ is given by

$$C_0 = S_0 \exp\left(\frac{-k e^{-rT}}{S_0}\right). \tag{1.39}$$

This simple expression for the call option satisfies all the standard arbitrage relations for European call options, including,

$$(S_0 - k e^{-rT})^+ \leq C_0 \leq S_0$$

as can easily be verified. Furthermore, it can be checked that $C_0$ satisfies the usual asymptotics

$$\lim_{S_0 \to 0} C_0 = 0 \qquad \text{and} \qquad \lim_{S_0 \to \infty} C_0 = (S_0 - k e^{-rT})$$

for call options as well.                                                      ☐

---

## 1.11   Black–Scholes and the FTAP

We have seen that there are two distinct methods of pricing derivatives in an arbitrage free economy. The first, valid only if the underlying asset follows gBm, gives the price of the derivative as the solution of the BS-pde (1.19). The second, which has wider generality, prices the derivative by (1.35), as a discounted expectation of the derivative payoff with respect to the equivalent martingale measure $Q$. In this section we show how the two approaches are reconciled.

Our starting point is the sde (1.11) for gBm. The solution of this sde is derived in Section 2.5.1, where it is shown that for any $T > t$,

$$X_T = x e^{(\mu - \frac{1}{2}\sigma^2)(T-t) + \sigma B_{T-t}^P}; \qquad x = X_t. \tag{1.40}$$

This representation of the asset price in the real-world measure $P$ is seen to be log-normal, since it is the exponential of a Gaussian. The expected value of the future asset price, given $\mathcal{F}_t$, is

$$\mathbb{E}_P\{X_T \mid \mathcal{F}_t\} = x e^{\mu(T-t)}. \tag{1.41}$$

To find its representation in the arbitrage free measure $Q$, let us first assume that measure $Q$ is also log-normal. Then applying the martingale restriction

$$\mathbb{E}_Q\{X_T \mid \mathcal{F}_t\} = x e^{r(T-t)} \tag{1.42}$$

leads to the conclusion that $Q$ is the EMM if we simply replace $\mu$ by the risk free rate $r$. That is, under $Q$, the asset price has dynamics governed by the sde

$$dX_t = x(r\, dt + \sigma\, dB_t^Q) \tag{1.43}$$

with solution

$$X_T = x e^{(r - \frac{1}{2}\sigma^2)(T-t) + \sigma B_{T-t}^Q}; \qquad x = X_t. \tag{1.44}$$

Now Equations (1.11) and (1.43) are compatible iff

$$dB_t^Q = dB_t^P + \left(\frac{\mu - r}{\sigma}\right) dt. \tag{1.45}$$

That such a relationship exists for two different Brownian motions $B_t^P$ and $B_t^Q$ is a consequence of Girsanov's Theorem (see Section 2.8).

But, since $B_{T-t}^Q \overset{d}{=} N(0, T-t)$, we may write the FTAP (1.35) when applied to Black–Scholes derivative pricing, in the form

$$\boxed{\begin{aligned} V(x,t) &= e^{-r\tau}\, \mathbb{E}_Q\{F(X_T) \mid \mathcal{F}_t\} \\ X_T &= x\, e^{(r - \frac{1}{2}\sigma^2)\tau + \sigma\sqrt{\tau} Z} \end{aligned}} \tag{1.46}$$

where $\tau = (T-t)$ and $Z \overset{d}{=} N(0,1)$ is a standard Gaussian random variable. This is the form of the FTAP we shall use to price derivatives on a single underlying asset in the BS framework.

**REMARK 1.3** In the stochastic representation (1.46), there is only one random variable $Z$, which is Gaussian. One must be aware however, that for each $t < T$, there will be a different $Z$, so a more accurate notation to take this into account would be to write $Z_t$ for this rv. However, $Z$ is a standard Gaussian and has constant (zero) mean and constant (unit) variance, so we are content to stick with $Z$ as an appropriate notation, with the understanding that at each instant of time, a different Gaussian is actually implied. ▯

---

## 1.12 Effect of Dividends

In the derivation of the BS-pde (1.19), we tacitly assumed that the underlying asset was an equity (or stock) which pays no dividends. Let us now consider a dividend paying asset, where dividends are paid continuously at the constant rate $q$. Since the payment of such a dividend stream is equivalent to a continuous liquidation of the firm issuing the stock, the asset price will reflect this by exhibiting, in the real world measure $P$, a reduced growth rate equal to $(\mu - q)$. That is, the asset price will still follow gBm, but with the new sde

$$dX_t = x[(\mu - q)dt + \sigma\, dB_t^P]; \qquad X_t = x \tag{1.47}$$

in place of (1.11).

The expected asset price at time $T$ will then be $\mathbb{E}_P\{X_T \,|\, \mathcal{F}_t\} = xe^{(\mu-q)\tau}$, where $\tau = (T - t)$. The corresponding sde in the EMM will be

$$dX_t = x[(r - q)dt + \sigma\, dB_t^Q]; \qquad X_t = x \tag{1.48}$$

and the corresponding Martingale Restriction becomes

$$\mathbb{E}_Q\{X_T \,|\, \mathcal{F}_t\} = xe^{(r-q)\tau}. \tag{1.49}$$

When a constant, continuous dividend yield of magnitude $q$ is included in the underlying asset, the BS-pde (1.19) is then found to change to

$$\boxed{V_t = rV - (r - q)x\, V_x - \tfrac{1}{2}\sigma^2 x^2\, V_{xx}; \qquad V(x, T) = F(x)} \tag{1.50}$$

(see Q10 in the Exercise Problems at the end of this chapter for a proof). The associated FTAP (1.46) is changed to

$$\boxed{\begin{aligned} V(x,t) &= e^{-r\tau}\, \mathbb{E}_Q\{F(X_T) \,|\, \mathcal{F}_t\} \\ X_T &= x\, e^{((r-q)-\frac{1}{2}\sigma^2)\tau + \sigma\sqrt{\tau}Z} \end{aligned}} \tag{1.51}$$

While the pricing equations for non-dividend and dividend paying assets are different, the following Lemma shows how the derivative prices are related in the BS framework.

**LEMMA 1.1**

*Let $V_0(x,t)$ denote a BS derivative price when there are no dividends paid (case $q = 0$), and let $V_q(x,t)$ denote the corresponding derivative price when continuous dividends are paid (case $q > 0$). Then*

$$\boxed{V_q(x,t) = V_0(xe^{-q\tau}, t); \qquad \tau = (T-t)} \tag{1.52}$$

*In other words, we need only replace the current asset price $x$ in the non-dividend case, by $xe^{-q\tau}$, to obtain the derivative price for the dividend paying case.*

The proof, which depends only on elementary rules of partial differentiation, is left to the Exercise Problems (see Q11) at the end of this chapter.

**Cost-of-Carry Model**

While it is interesting to consider dividends, as above, in their own right, there is an unexpected additional advantage as described in Haug [29] and reproduced below. By appropriate choice of the parameter $q$, it transpires that we can model other financial markets such as futures markets, foreign exchange (FX) markets and commodities markets. This is usually referred to as the *cost-of-carry model*. The cost-of-carry $c$ is defined to be the difference $c = r - q$, where $r$ is risk-free interest rate.

| Market Underlying | Dividend Yield | Cost-of-Carry |
|---|---|---|
| $x$ | $q$ | $c$ |
| Equity price (no dividends) | $0$ | $r$ |
| Market Index (cum dividends) | $q > 0$ | $r - q$ |
| Futures price | $r$ | $0$ |
| Exchange rate $(r = r_d)$ | $r_f$ | $r_d - r_f$ |
| Commodity price | $y_c - y_s$ | $r - y_c + y_s$ |

Thus if we are interested in pricing a derivative security on a futures price, we need only take either of the Equations (1.50) or (1.51) and set $q = r$, the continuously compounded (risk free) interest rate. Similarly, for the FX market, where the underlying is an exchange rate measured as the dollar (or domestic) value of one unit of foreign currency, we take $q = r_f$, the risk free rate in the foreign economy; $r = r_d$ is the corresponding risk free rate in the domestic economy. Options on commodity prices can be obtained by taking $q = y_c - y_s$, where $y_c$ is the convenience yield of the commodity, and $y_s$ is its storage yield.

## 1.13   Summary

This opening chapter laid down the financial foundations we shall be using in later chapters to price exotic options. The mathematical development has so far been kept to a minimum in order to focus on the main ideas and concepts. As mentioned at the start of this chapter, it is best to treat this opening salvo as a collection of the basics, rather than as a comprehensive treatise on the subject of Financial Mathematics. The principal topics covered in this chapter included:

| | |
|---|---|
| Concept of no-arbitrage | Black–Scholes pde |
| The Law of One Price | The FTAP |
| Principle of Static Replication | Equivalent Martingale Measure |
| Asset and bond Binaries | How to include dividends |
| Parity relations | Pricing in other markets |

All of these topics are important in what follows. The next chapter on Mathematical Preliminaries addresses many of the technical issues skipped over in this opening chapter. In particular, since the pricing formula (1.46) depends only on Gaussian random variables, Chapter 3 devotes considerable attention to listing many of their useful properties. Most of these properties will impact directly on our methods for pricing exotic options. □

## Exercise Problems

1. Prove, using arbitrage arguments alone, the following relations for European call and put options of strike price $k$ and expiration date $T$, for all $\tau = (T - t) > 0$

$$(x - ke^{-r\tau})^+ \leq C_k(x,t) \leq x$$
$$(ke^{-r\tau} - x)^+ \leq P_k(x,t) \leq ke^{-r\tau}$$

These relations are often called the *arbitrage boundaries* for European calls and puts.

2. Show that a strike $k$ straddle with time $T$ payoff function $S(X,T) = |X - k|$ can be statically replicated by a long European call option and a long European put option, both of strike price $k$ and expiration date $T$.

3. Consider a market $\mathcal{M}_t$ at two distinct times $t = 0$ (now) and $= T > 0$ (in the future). Suppose $\mathcal{M}_T$ may assume any one of $n$ states indexed by $i \in \Omega = \{1, 2, \ldots, n$. Let $S_i$ denote an Arrow–Debreu security, which pays \$1 at time $T$, if $\mathcal{M}_T$ is in state $i$, and pays zero otherwise. Let $e_i$ denote the time 0, pv of security $S_i$.

(a) Prove that security $\mathcal{X}$, which pays $X_i > 0$ at time $T$ when $\mathcal{M}_T$ is in state $i$, has fair value at $t = 0$ given by

$$X_0 = \sum_\Omega e_i X_i.$$

(b) Hence find the price $B_0$, in this market, of a $T$ maturity ZCB with face-value of \$1, and show that the price at $t = 0$ of a European call option on $\mathcal{X}$, with strike price $K$ and expiry $T$, is given by

$$C_0 = \sum_A e_i X_i - K \sum_A e_i$$

where $A$ is the subset of $\Omega$ corresponding to the exercise states of the option at $t = T$.

(c) Prove that $\pi_i = e_i/B_0$ and $\pi_i^* = e_i X_i/X_0$ are discrete probability measures on the state space associated with $\mathcal{M}_T$.

(d) Show that the call option price above can be represented by the equivalent formula

$$C_0 = X_0 Q_A^* - K B_0 Q_A$$

where $Q_A^*$, $Q_A$ are the probabilities of exercise in the measures $\pi_i$ and $\pi_i^*$ respectively.

4. In relation to the previous question, consider a market with just two states: an up-state and a down-state. Let $B_0 = \rho^{-1}$; $X_0 = x$; $X_i = (ux, dx)$ with $0 < d < \rho < u$. Determine the corresponding Arrow–Debreu security prices $e_i = (e_u, e_d)$ and probability measures $\pi_i = (p, q)$ and $\pi_i^* = (p^*, q^*)$.

Hence price the European call option, assuming that the strike price satisfies $dx < K < ux$.

5. Let $m_k(x, t)$ and $M_k(x, t)$ denote the present values of European derivatives which respectively have payoffs at time $T$, given by

$$m_k(X, T) = \min(X, k) \quad \text{and} \quad M_k(X, T) = \max(X, k).$$

Express the prices $m_k(x, t)$, $M_k(x, t)$ in terms of asset and bond binaries and derive a parity relation connecting $m_k(x, t)$ and $M_k(x, t)$.

6. Show that the EMM, $q(x, t; \xi, T)$ for the standard BS model, satisfies the pde
$$-q_t = rx\, q_x + \tfrac{1}{2}\sigma^2 x^2\, q_{xx}$$
with terminal condition $\lim_{t \to T} q(x, t; \xi, T) = \delta(x - \xi)$.
This pde is known as the Backward Kolmogorov Equation.

7. Write down the FTAP for a European call option price $C_k(x)$ of strike price $k$ and a fixed time $T$ to expiry, on an underlying asset price $X = X_T$ with continuous risk-neutral pdf $q(x; T)$ on $x > 0$. Hence derive the Breeden–Litzenberger Equation
$$q(x; T) = e^{rT}\left[\frac{d^2 C_k(x)}{dk^2}\right]_{k=x}.$$

8. Suppose the underlying asset price at time $T$ has RN-pdf which is uniformly distributed on the finite interval $[0, A]$ for some $A$. If $x$ is the current time zero asset price, show that the price of a European, strike $k$, expiry $T$ call option can be expressed in the form
$$C_0(x) = \frac{1}{x}\left[\left(x - \tfrac{1}{2}ke^{-rT}\right)^+\right]^2$$
where $r$ is the risk free rate. Verify the Breeden–Litzenberger Equation of the previous question for this example.

9. Prove formally, i.e., by direct substitution, that
$$V(x, t) = e^{-r\tau}\, \mathbb{E}_Q\{F(X_T)\}$$
where $X_T = x \exp\{(r - \tfrac{1}{2}\sigma^2)\tau + \sigma\sqrt{\tau}Z\}$, $\tau = (T - t)$ and $Z \overset{d}{=} N(0, 1)$ satisfies the BS-pde
$$V_t = r V - rx\, V_x - \tfrac{1}{2}\sigma^2 x^2\, V_{xx}; \qquad V(x, T) = F(x)$$
*Hint:* First prove the Gaussian result $\mathbb{E}\{ZG(Z)\} = \mathbb{E}\{G'(Z)\}$ for any differentiable function $G(Z)$.

10. Derive the BS-pde (1.50) for a derivative security written on an asset which pays a continuous dividend of constant yield $q$. Follow the dynamic hedging argument of Section 1.7, suitably modified to include the effects of dividends.

11. (a) Prove that if $V_0(x, t)$ satisfies the BS-pde $V_t = rV - rxV_x - \tfrac{1}{2}\sigma^2 x^2 V_{xx}$ with TV, $V(x, T) = F(x)$, then $V_q(x, t) = V_0(xe^{-q(T-t)}, t)$ satisfies the pde $V_t = rV - (r - q)xV_x - \tfrac{1}{2}\sigma^2 x^2 V_{xx}$ with the same TV.

 (b) Prove also the result, that if we write $V_0 = V_0(x, t; r, \sigma)$, then
$$V_q(x, t; r, \sigma) = e^{-q(T-t)}\, V_0(x, t; r - q, \sigma).$$

12. A perpetual American put of strike price $k$ has no expiry date and so can be exercised at any time. The price $P_k(x)$ satisfies the time-independent BS-pde. Show that there is a critical asset price $c < k$, below which the put should be exercised and that the price of the perpetual put can be expressed in the form

$$P_k(x) = \begin{cases} (k - x) & \text{if} \quad x \le c \\ (k - c)\left(\frac{c}{x}\right)^\alpha & \text{if} \quad x > c \end{cases}$$

where $\alpha = 2r/\sigma^2$ and $c = \frac{\alpha k}{1+\alpha}$.

*Hint:* The price $P_k(x)$ and its derivative $P_k'(x)$ wrt $x$, are assumed to be bounded and continuous for all $x > 0$.

—ooOoo—

# Chapter 2

## *Mathematical Preliminaries*

This chapter is a review of the mathematical details that underpin many of the topics of the previous chapter. These include elements of probability theory, stochastic calculus and other issues that play an important role in the development of financial mathematics in general, and the Black–Scholes and EMM methods in particular. For a more detailed account on stochastic calculus and sde's, the reader is referred to the texts of Lamberton and Lapeyre [51], Karatzas and Shreve [41] and Øksendal [60]. A recent very readable text that is not too technical, and hence in the same vein as this book, is that by Wiersema [76].

Another useful reference which covers almost everything in this chapter, and the previous one, is the book by Jeanblanc et al. [39]. This text is much more technical and as such is perhaps more suitable for those with sufficiently advanced mathematical expertise. However, any serious student of Quantitative Finance ought to be familiar with this important contribution to the field.

We shall not present a fully integrated theory, which would take us far from our goal of pricing exotic options, but prefer to step lightly through various topics, identifying important concepts and earmarking important equations. Thus the reader should treat this chapter as a list of mathematical tools, rather than a coherent set of propositions and theorems. We offer only a few formal proofs in this chapter. Others, in most cases, can be found in the cited texts. It will take some skill to utilize the tools presented in this chapter, to price exotic options. Hopefully, therefore, readers of this book will not fall into the trap epitomized by the adage: *a bad craftsman blames his tools.*

---

## 2.1 Probability Spaces

We assume a market in which future prices up to a finite time horizon $T$, are random variables associated with a probability space $(\Omega, \mathbb{F}, P)$. $\Omega$ is the set of all possible price outcomes; $\mathbb{F}$ is a $\sigma$−algebra (a family of subsets of $\Omega$) containing all sets pertaining to future prices, and $P$ is a probability measure (called the *real-world measure*) which determines the probability of any event in $\mathbb{F}$. We also equip this probability space with a filtration $\mathcal{F}_t$. This is a non-decreasing family of sub-$\sigma$-algebras of $\mathbb{F}$, such that

$$\mathcal{F}_s \subset \mathcal{F}_t \subset \mathcal{F}_T \subset \mathbb{F} \quad \text{for all } 0 < s < t < T.$$

It is convenient to think of $\mathcal{F}_t$ as the price information available for all times up to and including $t$. We assume that the resulting filtered probability space $(\Omega, \mathbb{F}, P, \mathcal{F}_t)$ satisfies the so-called *usual conditions*. Technically, this means: $\mathbb{F}$ is $P$-complete, $\mathcal{F}_0$ contains all $P$-null sets of $\Omega$ and $\mathcal{F}_t$ is right-continuous.

A *stochastic price process* $X_t$ is then a family of random variables defined on $(\Omega, \mathbb{F}, P, \mathcal{F}_t)$. $X_t$ is said to be *adapted* to $\mathcal{F}_t$ if $X_t$ is $\mathcal{F}_t-$ measurable. Basically, this simply means that $X_t$ is known with certainty at time $t$.

An important operator on stochastic processes is the *conditional expectation* denoted by $\mathbb{E}\{X_T | \mathcal{F}_t\}$. By this we mean the expected value of $X_T$ given all information up to and including $t$, for $t \leq T$. The expectation operator is linear, so that if $(\alpha_t, \beta_t)$ are processes adapted to $\mathcal{F}_t$, then for any processes $(X_T, Y_T)$ with $t < T$,

$$\mathbb{E}\{\alpha_t X_T + \beta_t Y_T | \mathcal{F}_t\} = \alpha_t \mathbb{E}\{X_T | \mathcal{F}_t\} + \beta_t \mathbb{E}\{Y_T | \mathcal{F}_t\} \tag{2.1}$$

**The Tower Law**
Let $X_t$ be a stochastic process; then for all $s \leq t \leq T$,

$$\mathbb{E}\{X_T | \mathcal{F}_s\} = \mathbb{E}\{\mathbb{E}\{X_T | \mathcal{F}_t\} | \mathcal{F}_s\}. \tag{2.2}$$

This is also known as the *Law of Iterated Expectations* and demonstrates how information can be nested through a non-decreasing filtration sequence. The tower law has important ramifications for option pricing, as we saw in Section 1.9 and as we shall also demonstrate in the applications.

---

## 2.2   Brownian Motion

A standard $\mathcal{P}$ Brownian motion (or standard Wiener process) $B_t$ for $t > 0$ is the stochastic process satisfying

| |
|---|
| 1.   $B_t$ is continuous and $B_0 = 0$ |
| 2.   $B_t$ has stationary and independent increments |
| 3.   For fixed $t > 0$, $B_t \overset{d}{=} N(0, t)$ under probability measure $\mathcal{P}$ |

where $\overset{d}{=}$ means "equal in distribution." This characterization of Brownian motion is due to Kolmogorov, and is known to define it uniquely.

**REMARK 2.1**   We list here a number of properties of Brownian motion. We do not necessarily use all these properties, but it is nevertheless helpful to be familiar them.

1. Although the sample paths of $B_t$ are continuous, they are (with probability one) nowhere differentiable. The sample paths are therefore fractal and have similarity dimension $H = \frac{1}{2}$. That is, for all $c > 0$,

$$B_{ct} \overset{d}{=} c^H B_t = \sqrt{c}\, B_t.$$

2. The sample paths of $B_t$ are of unbounded variation, but have bounded quadratic variation. This means that if $\Delta_n = \{t_i^{(n)} : i = 1, 2, \ldots, n\}$ be a sequence of partitions of the interval $[0, t]$ such that

$$\lim_{n \to \infty} \max_i |t_{i+1}^{(n)} - t_i^{(n)}| = 0$$

with $t_0^{(n)} = 0$ and $t_{n+1}^{(n)} = t$, then it can be shown that

$$\lim_{n \to \infty} \sum_{i=1}^{n} |B_{t_{i+1}^{(n)}} - B_{t_i^{(n)}}| \to \infty$$

while

$$\lim_{n \to \infty} \sum_{i=1}^{n} |B_{t_{i+1}^{(n)}} - B_{t_i^{(n)}}|^2 = t.$$

The quadratic variation of $B_t$ is therefore equal to $t$.

3. It can also be shown that:

$$\lim_{t \to \infty} \frac{B_t}{t} = 0 \quad \text{a.s.}$$

$$\limsup_{t \to \infty} \frac{B_t}{\sqrt{2t \log \log t}} = 1 \quad \text{a.s.}$$

$$\liminf_{t \to \infty} \frac{B_t}{\sqrt{2t \log \log t}} = -1 \quad \text{a.s.}$$

The last two equations are collectively called the *Law of the Iterated Logarithm*. The symbol a.s. stands for "almost surely," which means the equations are probabilistic statements, with probability one.

4. The property of independent increments means that for any non-overlapping intervals $(s,t)$ and $(u,v)$, the increment $(B_t - B_s)$ is independent of the increment $(B_v - B_u)$. In particular, since $B_0 = 0$, the increment $(B_t - B_s)$ is independent of $B_s$ for all $s < t$.

The stationarity of these increments means that for any $h + s \geq 0$, we have $B_t - B_s \overset{d}{=} B_{t+h} - B_{s+h}$. In particular, taking $h + s = 0$ leads to the important conclusion

$$B_t - B_s \overset{d}{=} B_{t-s} \qquad (2.3)$$

5. For any fixed $t > 0$, $B_t$ is Gaussian with zero mean and variance equal to $t$. Hence it is possible to write

$$B_t \overset{d}{=} \sqrt{t}\, Z \qquad \text{where } Z \sim N(0,1). \qquad (2.4)$$

6. $B_t$ is a Markov process. That is, for all $s \leq t$,

$$\mathbb{P}\{B_t | \mathcal{F}_s\} = \mathbb{P}\{B_t | B_s\}$$

where $\mathcal{F}_s$ denotes the filtration of the probability space induced by $B_s$. Hence the Markov property implies that all the information at time $s \leq t$ is contained in the value of $B_s$ alone, independent of the history prior to time $s$. In this sense, Brownian motion is said to be a *zero-memory* process.

7. The covariance of Brownian motion is given by

$$\text{cov}\{B_s, B_t\} = \min(s,t) \qquad (2.5)$$

which leads to the correlation structure

$$\text{corr}\{B_s, B_t\} = \sqrt{s/t} \quad \text{for all } s < t \qquad (2.6)$$

Thus while non-overlapping Brownian increments are independent, the above implies that overlapping Brownian increments are dependent, with covariance equal to the duration of the overlap. The correlation coefficient (2.6), for Brownian motion at two distinct instants of time, plays an important role in dual and multi-period exotic options considered in later chapters. □

---

## 2.3 Stochastic DE's

A very readable account of sde's, and their numerical solution can be found in Kloeden and Platen [44] or Wiersema [76]. A more technical approach is given in the book by Øksendal [60]. We present here a simplified account.

The stochastic process $X_t$ is said to be an Itô process with instantaneous drift $\mu_t = \mu(X_t, t)$ and instantaneous variance $\sigma_t^2 = \sigma^2(X_t, t)$ if it satisfies the stochastic differential equation (sde)

$$dX_t = \mu_t dt + \sigma_t \, dB_t \qquad (2.7)$$

where $B_t$ is a $\mathcal{P}-$Brownian motion. Such Itô processes are also Markov processes.

Since $\mathbb{E}\{dB_t\} = 0$ and $\mathbb{V}\{dB_t\} = dt$, one way of interpreting this sde is to say that it is equivalent to the pair of statements:

$$\mu(x,t) = \lim_{dt \to 0} \frac{\mathbb{E}_P\{X_{t+dt} - X_t | X_t = x\}}{dt} \qquad (2.8)$$

$$\sigma^2(x,t) = \lim_{dt \to 0} \frac{\mathbb{V}_P\{X_{t+dt} - X_t | X_t = x\}}{dt}. \qquad (2.9)$$

Both limits are assumed to exist for well-defined Itô processes.

Associated with every sde of the form of Equation (2.7) there exists a transition probability density function (pdf) $f(x_0, t_0; x, t)$ that gives the probability that $X_t = x$ given $X_{t_0} = x_0$ for all $t_0 < t$. It can be shown that $f(x_0, t_0; x, t)$ satisfies a pair of pde's called the forward and backward Kolmogorov equations. The forward equation is

$$\frac{\partial f}{\partial t} = -\frac{\partial}{\partial x}[\mu(x,t)f] + \frac{\partial^2}{\partial x^2}[\tfrac{1}{2}\sigma^2(x,t)f] \qquad (2.10)$$

with initial condition $f(x_0, t_0; x, t) \to \delta(x - x_0)$ as $t \to t_0$; and the backward equation is

$$-\frac{\partial f}{\partial t_0} = \mu(x_0, t_0)\frac{\partial f}{\partial x_0} + \tfrac{1}{2}\sigma^2(x_0, t_0)\frac{\partial^2 f}{\partial x_0^2} \qquad (2.11)$$

with initial condition $f(x_0, t_0 | x, t) \to \delta(x_0 - x)$ as $t_0 \to t$.

Since these pde's are of the diffusion type (i.e., parabolic), Itô processes are also called *diffusion processes*.

## 2.3.1 Arithmetic Brownian Motion

Arithmetic Brownian Motion or aBm is defined to be the process satisfying a sde for $t > t_0$ of the form

$$dX_t = \mu dt + \sigma dB_t; \qquad X_{t_0} = x_0 \qquad (2.12)$$

where $\mu$ and $\sigma$ are constants. In this case the solution of the sde, using Equation (2.3), can be written as

$$X_t \overset{d}{=} x_0 + \mu(t - t_0) + \sigma B_{t-t_0} \tag{2.13}$$

and the transition pdf of $X_t$ is obviously the Gaussian density with mean (drift) $x_0 + \mu(t - t_0)$ and variance $\sigma^2(t - t_0)$:

$$f(x_0, t_0; x, t) = \frac{1}{\sigma\sqrt{t - t_0}} \phi \left[ \frac{x - x_0 - \mu(t - t_0)}{\sigma\sqrt{t - t_0}} \right], \tag{2.14}$$

where $\phi(z) = \frac{1}{\sqrt{2\pi}} e^{-\frac{1}{2}z^2}$ for $z \in \mathbb{R}$ is the density of a $N(0, 1)$ variate.

It is a straight forward matter to show that this $f(x_0, t_0; x, t)$ satisfies both the forward and backward Kolmogorov equations with constant $\mu$ and $\sigma$ and the given initial conditions. It should also be clear that aBm is simply a re-scaled Brownian motion with non-zero deterministic drift.

---

## 2.4 Stochastic Integrals

The sde expressed by Equation (2.7) has an alternative meaning in terms of stochastic integrals. This is the representation

$$X_t = X_0 + \int_0^t \mu_s \, ds + \int_0^t \sigma_s \, dB_s. \tag{2.15}$$

The last integral is an example of a stochastic integral — an integral with respect to a Brownian motion. This integral is formally defined by the limit

$$\int_0^t \sigma_s \, dB_s = \lim_{n \to \infty} \sum_{i=0}^n \sigma(t_i^{(n)}) [B(t_{i+1}^{(n)}) - B(t_i^{(n)})],$$

where $t_i^{(n)}$ determines the partition $\Delta_n$ of $[0, t]$ defined earlier. It is essential in this definition that $\sigma(t_i^{(n)})$ is evaluated at each left-hand point of the partition $[t_i^{(n)}, t_{i+1}^{(n)})$, because in this case, it is $\mathcal{F}_{t_i}$−measurable (i.e., adapted to the Brownian increment). It is this "non-anticipatory" feature that distinguishes the Itô integral from other stochastic integrals such as the Stratonovich integral. In any case, the Itô integral above is found to be the ideal one in financial applications. The next result, often referred to as Itô's Isometry, gives the mean and variance of a stochastic integral.

**Itô's Isometry**

If $f(t)$ is a deterministic function of time (in fact it can also be a stochastic process that is adapted to $B_t$), then $Z_t = \int_0^t f(s)\,dB_s$ for each fixed $t$, is a Gaussian random variable with zero mean and variance equal to

$$\mathbb{V}\{Z_t\} = \int_0^t |f(s)|^2\,ds. \tag{2.16}$$

---

## 2.5 Itô's Lemma

This is the stochastic extension of the chain rule for ordinary (deterministic) calculus. Let $X_t$ satisfy the sde $dX_t = \alpha dt + \beta dB_t$ for arbitrary (predictable processes) $\alpha = \alpha(X_t, t)$ and $\beta = \beta(X_t, t)$. If $F(x, t)$ is any $\mathbb{C}_{2,1}$ function (i.e., twice differentiable in $x$ and differentiable in $t$), then the random process $F(X_t, t)$ is an Itô process satisfying the sde

$$dF = (F_t + \alpha F_x + \tfrac{1}{2}\beta^2 F_{xx})dt + \beta F_x\,dB_t \tag{2.17}$$

where subscripts on the function $F(X_t, t)$ denote partial derivatives.

A useful alternate form of Itô's lemma is:

$$dF(X_t, t) = (F_t + \tfrac{1}{2}\beta^2 F_{xx})dt + F_x\,dX_t. \tag{2.18}$$

### 2.5.1 Geometrical Brownian Motion

A process $X_t$ is said to follow geometrical Brownian motion (gBm) with constant drift rate $\mu$ and volatility $\sigma$ if it satisfies the sde

$$dX_t = X_t(\mu dt + \sigma\,dB_t); \qquad X_{t_0} = x_0. \tag{2.19}$$

This sde can be solved explicitly by transforming it to simple aBm through Itô's Lemma. Let $F(X_t) = \log X_t$ independent of $t$. Then substituting

$$\alpha = \mu X_t; \;\; \beta = \sigma X_t; \;\; F_t = 0; \;\; F_x = 1/X_t; \;\; F_{xx} = -1/X_t^2$$

into Itô's formula Equation (2.17), there results the sde

$$dF = (\mu - \tfrac{1}{2}\sigma^2)dt + \sigma\,dB_t; \qquad F(X_{t_0}) = \log x_0.$$

This is aBm with solution

$$F(X_t) = \log X_t \overset{d}{=} \log x_0 + (\mu - \tfrac{1}{2}\sigma^2)(t - t_0) + \sigma B_{t-t_0}$$

The solution for $X_t$ is now obtained by exponentiation to yield the result

$$X_t \stackrel{d}{=} x_0 \exp\{(\mu - \tfrac{1}{2}\sigma^2)(t - t_0) + \sigma B_{t-t_0}\}. \tag{2.20}$$

The result is important because gBm is the basic asset price model in the Black–Scholes framework. Since, under gBm $X_t/x_0$ for fixed $t$, is the exponential of a Gaussian rv with mean $m(t_0, t) = (\mu - \tfrac{1}{2}\sigma^2)(t - t_0)$ and variance $v^2(t_0, t) = \sigma^2(t - t_0)$, its pdf is log-normal (see Section 3.6) with parameters $(m, v)$. The corresponding transition pdf of $X_t$ is therefore

$$f(x_0, t_0; x, t) = \frac{1}{xv(t, t_0)} \phi \left[ \frac{\log(x/x_0) - m(t_0, t)}{v(t_0, t)} \right]. \tag{2.21}$$

This transition pdf can also be shown to satisfy the forward and backward Kolmogorov equations with instantaneous drift $\mu x$ and instantaneous variance $(\sigma x)^2$.

### 2.5.2 Itô's Product and Quotient Rules

Let $X_t$ satisfy the sde $dX_t = \alpha dt + \beta dB_t$ and let $F(X_t, t)$ and $G(X_t, t)$ be two given $\mathbb{C}_{2,1}$ functions. Then Itô's product and quotient rules are given respectively by

$$d(FG) = (FdG + GdF) + \beta^2 F_x G_x \, dt \tag{2.22}$$

and

$$d(F/G) = \frac{GdF - FdG}{G^2} + \frac{\beta^2 G_x}{G^3}(FG_x - GF_x) \, dt. \tag{2.23}$$

These formulae (see Q6 in Exercise Problems) are interesting in that they show that stochastic calculus includes the extra $dt$ terms. Observe that these terms vanish when $\beta = 0$, and we recover the standard product and quotient rules for ordinary Newtonian calculus.

---

## 2.6 Martingales

A $(P, \mathcal{F}_t)$–*martingale* $M_t$ is a stochastic process satisfying

$$\mathbb{E}_P\{M_t | \mathcal{F}_s\} = M_s \quad \text{for all } s \leq t. \tag{2.24}$$

Martingales are important in financial mathematics because they are the stochastic entities that capture the notion of a "fair game." In other words, they are intimately associated with the no-arbitrage assumption of idealized markets. For example, if $M_t$ denotes the time $t$ random wealth of a gambler playing a fair game, then Equation (2.24) above tells us that expected future wealth is equal to current wealth. It is in this sense, that the game is said to be fair.

### Example 2.1

All the following

$$B_t, \quad B_t^2 - t \quad \text{and} \quad e^{\sigma B_t - \frac{1}{2}\sigma^2 t}$$

are well-known martingales (see Q3(c) in Exercise Problems). ⬚

### 2.6.1 Martingale Representation Theorem

This essentially states that if $X_t$ satisfies the zero drift sde $dX_t = \sigma(X_t, t)dB_t$ then $X_t$ is an $\mathcal{F}_t-$ martingale[1].

This is easily seen as follows. Assume $0 < s < t$, then in terms of stochastic integrals,

$$X_t = X_0 + \int_0^t \sigma(X_u, u)dB_u = X_s + \int_s^t \sigma(X_u, u)dB_u$$

Hence $\mathbb{E}\{X_t | \mathcal{F}_s\} = X_s$ since the last stochastic integral has zero mean.

The converse is also true. Thus if $X_t$ and $Y_t$ are $\mathcal{F}_t-$local martingales, there exists an $\mathcal{F}_t-$adapted process $c_t$ such that $dY_t = c_t dX_t$.

Itô's Lemma now provides the following corollary. If $f(x, t)$ is a given $\mathbb{C}_{2,1}$ function, then $f(X_t, t)$, where $dX_t = \alpha(x, t)dt + \beta(x, t)dB_t$, is an $\mathcal{F}_t-$martingale if and only if $f(x, t)$ satisfies the pde

$$f_t + \alpha(x, t)f_x + \tfrac{1}{2}\beta^2(x, t)f_{xx} = 0. \tag{2.25}$$

This is just the condition that the drift term vanishes in the sde for $f(X_t, t)$.

The three processes in example 2.1 are now readily seen to be martingales because

$$f(x, t) = x, \quad f(x, t) = x^2 - t \quad \text{and} \quad f(x, t) = e^{\sigma x - \frac{1}{2}\sigma^2 t}$$

all satisfy the pde (2.25) with $(\alpha = 0, \beta = 1)$.

---

[1]Strictly speaking, only a *local* martingale, a distinction we shall not elucidate further.

## 2.7 Feynman–Kac Formula

Let $\alpha(x,t), \beta(x,t)$ and $r(x,t)$ be given functions of $(x,t)$ for $t < T$ and $x \in D$, where $D$ is some subset of $\mathbb{R}^+$, and let $\mathcal{A}$ be the differential operator defined by $\mathcal{A}u(x,t) = \alpha\dfrac{\partial u}{\partial x} + \frac{1}{2}\beta^2\dfrac{\partial^2 u}{\partial x^2}$. Then, subject to technical conditions, the unique solution of the pde

$$\frac{\partial u}{\partial t} + \mathcal{A}u - ru = 0; \qquad x \in D, \quad 0 \le t \le T$$

with terminal value $u(x,T) = g(x)$, is given by

$$u(x,t) = \mathbb{E}\left\{e^{-\int_t^T r(X_s,s)ds}\, g(X_T)|X_t = x\right\}, \tag{2.26}$$

where the expectation is taken with respect to the transition density induced by the sde $dX_t = \alpha(X_t,t)dt + \beta(X_t,t)dB_t$.

### REMARK 2.2

1. The variables $\alpha, \beta^2$ are the instantaneous mean and variance of an underlying Itô diffusion. The variable $r$ plays the role of a time and state dependent interest rate.

2. The operator $\mathcal{A}$ is the infinitesimal generator of the diffusion $X_t$ and is often called the *Dynkin operator.*

3. The technical conditions alluded to are required to make the proof, outlined below, rigorous. In particular, the conditions allow interchange of integration and differentiation and also allow us to bring the limit $t \to T$ inside the integration.

4. The FK-formula provides an important link between the two principal methods used to price options and derivatives. These are the PDE-method and the EMM-method alluded to in Chapter 1.

☐

**Proof of FK-Formula**

Let $f(x,t;X,T)$ denote the transition pdf of the process $X_t$. Then Equation (2.26) is equivalent to

$$u(x,t) = \int_D e^{-\int_t^T r_s ds}\, g(X)f(x,t;X,T)\, dX.$$

Formally differentiating inside the integral, we get

$$\frac{\partial u}{\partial t} + \mathcal{A}u - ru = \int_D e^{-\int_t^T r_s ds} g(X) \left[ rf + \frac{\partial f}{\partial t} + \mathcal{A}f - rf \right] dX$$

$$= \int_D e^{-\int_t^T r_s ds} g(X) \left[ \frac{\partial f}{\partial t} + \mathcal{A}f \right] dX$$

$$= 0.$$

The last line follows from the observation that $-\frac{\partial f}{\partial t} = \mathcal{A}f$ is the backward Kolmogorov equation (2.11). It remains to demonstrate that the terminal condition is satisfied.

$$u(x, T) = \lim_{t \to T} \int_D e^{-\int_t^T r_s ds} g(X) f(x, t; X, T) \, dX$$

$$= \int_D g(X) \lim_{t \to T} f(x, t; X, T) \, dX$$

$$= \int_D g(X) \, \delta(X - x) \, dX$$

$$= g(x) \quad \text{for } x \in D.$$

This completes the proof. □

It should now be clear that the FTAP described by equation (1.46) is just a special case of the FK-formula.

## 2.8 Girsanov's Theorem

If $B_t$ is a standard $\mathcal{P}$–Brownian motion and $\lambda(t)$ is an adapted (i.e., $\mathcal{F}_t$–measurable) process satisfying the Novikov condition,

$$\mathbb{E}_P\{\exp(\tfrac{1}{2} \int_0^T \lambda^2(t)dt)\} < \infty,$$

then there exists a measure $\mathcal{P}^*$ such that

1. $\mathcal{P}^*$ is equivalent to $\mathcal{P}$
2. $\dfrac{d\mathcal{P}^*}{d\mathcal{P}} = \exp\left( -\int\limits_0^T \lambda(t)dW_t - \tfrac{1}{2}\int\limits_0^T \lambda^2(t)dt \right)$
3. $B_t^* = B_t + \int_0^t \lambda(s)ds$ is a $\mathcal{P}^*$–Brownian motion

The converse states that if $B_t$ is a $\mathcal{P}$−Brownian motion and $\mathcal{P}^*$ is a measure equivalent to $\mathcal{P}$, then there exists an adapted process $\lambda(t)$ such that

$$B_t^* = B_t + \int_0^t \lambda(s)ds$$

is a $\mathcal{P}^*$−Brownian motion.

**REMARK 2.3**

1. Two probability measures are said to be equivalent if they have the same null sets. That is $\mathcal{P}(A) = 0 \Leftrightarrow \mathcal{P}^*(A) = 0$.

2. The term $\mathcal{R}_t = \dfrac{d\mathcal{P}^*}{d\mathcal{P}}$ is called the Radon-Nikodym derivative and gives the factor needed in computing the change of measure formula: for any random process $X_t$

$$\mathbb{E}_{P^*}\{X_t|\mathcal{F}_s\} = \mathcal{R}_s^{-1}\mathbb{E}_P\{\mathcal{R}_t X_t|\mathcal{F}_s\} \quad \text{for all } s \le t. \tag{2.27}$$

3. For the sde $dX_t = \mu_t dt + \sigma_t dB_t$, let $B_t^* = B_t + \int_0^t (\mu_s/\sigma_s)\, ds$. Then $dX_t = \sigma_t dB_t^*$ and $X_t$, by the martingale representation theorem, is therefore a $\mathcal{P}^*$−martingale. □

Suppose we have a stochastic process satisfying an arbitrary sde with respect to a measure $P$. This process will not in general be a martingale under $P$. Girsanov's theorem shows, by changing the drift as above, how to find a new measure $\mathcal{P}^*$, equivalent to $P$, under which the process is a martingale.

---

## 2.9 Time Varying Parameters

Under the EMM, the sde for the asset price $X_t$ when the risk-free rate $r_t$, dividend yield $q_t$ and volatility $\sigma_t$ are deterministic functions of $t$, is given by

$$dX_s = X_s\left[(r_s - q_s)\, ds + \sigma_s\, dB_s\right]; \quad (s > t, \ X_t = x). \tag{2.28}$$

We assume that $r_s$, $q_s$ and $\sigma_s$ are piecewise continuous. The solution of this sde is readily obtained as (e.g., see Problem 8(b) in the Exercise Problems)

$$X_T = x\, \exp\left[\int_t^T (r_s - q_s - \tfrac{1}{2}\sigma_s^2)ds + \int_t^T \sigma_s\, dB_s\right] \tag{2.29}$$

Now let us define

$$\bar{r} = \frac{1}{T-t} \int_t^T r_s \, ds; \qquad \bar{q} = \frac{1}{T-t} \int_t^T q_s \, ds \qquad (2.30)$$

$$\hat{\sigma} = \left[ \frac{1}{T-t} \int_t^T \sigma_s^2 \, ds \right]^{\frac{1}{2}}. \qquad (2.31)$$

Then, $(\bar{r}, \bar{q})$ are the mean risk-free rate and mean dividend yield over $[t, T]$, and $\hat{\sigma}$ is the root mean square (rms) volatility over $[t, T]$. Furthermore, by Itô's Isometry, (2.16), we have

$$\int_t^T \sigma_s \, dB_s \overset{d}{=} N\left[ 0, \int_t^T \sigma_s^2 \, ds \right]$$

$$= N(0, \hat{\sigma}^2 \tau); \quad \tau = (T - t)$$

$$\overset{d}{=} \hat{\sigma}\sqrt{\tau} \, Z; \quad Z \sim N(0, 1).$$

Hence, under the EMM, the asset price at time $T$ has the representation

$$X_T = x \, e^{(\bar{r} - \bar{q} - \frac{1}{2}\hat{\sigma}^2)\tau + \hat{\sigma}\sqrt{\tau} \, Z}. \qquad (2.32)$$

Comparing this expression with the stock price formula (1.51) for conditions of constant parameters, we see that the case of time varying deterministic parameters has exactly the same mathematical structure. It follows, that if $V(x, t; r, q, \sigma)$ is the price of a European derivative with fixed expiry date $T$, under constant parameters $(r, q, \sigma)$, then $V(x, t; \bar{r}, \bar{q}, \hat{\sigma})$ will be the corresponding price when the parameters are deterministic functions of time.

## 2.10   The Black–Scholes PDE

We mentioned in Section 1.7 that the BS-pde for $V(x, t)$

$$V_t = rV - rx \, V_x - \tfrac{1}{2}\sigma^2 x^2 \, V_{xx}; \qquad V(x, T) = f(x)$$

can be transformed by two distinct variable changes into the the standard heat equation. We begin by converting the pde from a backward to a forward pde by through $\tau = (T - t)$, under which the time partial derivative becomes $V_t = -V_\tau$. Thus, $V(x, \tau)$ satisfies the *forward*, initial value problem (IVP)

$$V_\tau = -rV + rx \, V_x + \tfrac{1}{2}\sigma^2 x^2 \, V_{xx}; \qquad V(x, 0) = f(x)$$

**Scheme 1**
Let

$$y = \log x; \qquad V(x, \tau) = U(y, \tau) \, e^{-\frac{1}{2}\alpha y - \beta \tau} \qquad (2.33)$$

This is the transformation scheme of Wilmott et al. [77]. The factor $\frac{1}{2}$ in the exponent is introduced for later identification with the image prices defined in Section 7.2 on barrier options. To see how the transformation works, observe that when $y = \log x$, we have

$$V_x = V_y \frac{\partial y}{\partial x} = \frac{1}{x} V_y \qquad \Rightarrow \qquad x V_x = V_y \tag{2.34}$$

$$V_{xx} = -\frac{1}{x^2} V_y + \frac{1}{x^2} V_{yy} \qquad \Rightarrow \qquad x^2 V_{xx} = V_{yy} - V_y \tag{2.35}$$

In terms of the variable $y$, the new pde for $V(y, \tau)$ becomes

$$V_\tau = -rV + \left(r - \tfrac{1}{2}\sigma^2\right) V_y + \tfrac{1}{2}\sigma^2 \, V_{yy}; \qquad V(y, 0) = f(e^y).$$

It is clear then, that the transformation $y = \log x$ converts the pde with non-constant coefficients into one with constant coefficients. Next, observe that in terms of $U(y, \tau)$,

$$V_\tau = (U_\tau - \beta U)\, e^{-\frac{1}{2}\alpha y - \beta \tau}$$
$$V_y = (U_y - \tfrac{1}{2}\alpha U)\, e^{-\frac{1}{2}\alpha y - \beta \tau}$$
$$V_{yy} = (U_{yy} - \alpha U_y + \tfrac{1}{4}\alpha^2 U)\, e^{-\frac{1}{2}\alpha y - \beta \tau}.$$

So substituting into the new pde for $U(y, \tau)$, we are free to choose the constant parameters $(\alpha, \beta)$ and do so by making the coefficients of $U$ and $U_y$ both equal to zero. This yields, after a little algebra,

$$\alpha = \frac{2r}{\sigma^2} - 1 \qquad \text{and} \qquad \beta = \frac{(r + \tfrac{1}{2}\sigma^2)^2}{2\sigma^2}. \tag{2.36}$$

The pde for $U(y, \tau)$ then reduces to

$$U_\tau = \tfrac{1}{2}\sigma^2 \, U_{yy}; \qquad U(y, 0) = e^{\frac{1}{2}\alpha y} f(e^y). \tag{2.37}$$

This is the standard heat equation with "thermal conductivity" equal to $\frac{1}{2}\sigma^2$ and initial "temperature" equal to $e^{\frac{1}{2}\alpha y} f(e^y)$. Note that appropriate domain for this pde is $[\tau > 0; \; y \in \mathbb{R}]$, so even though $x > 0$ in the BS pde, the variable $y$ may be any real number, positive or negative.

### Scheme 2

Perhaps less well-known is this second scheme, which transforms the forward BS-pde via the new variables $y$ and $U(y, \tau)$ through

$$y = \log x + (r - \tfrac{1}{2}\sigma^2)\tau; \qquad V(y, \tau) = e^{-r\tau} U(y, \tau). \tag{2.38}$$

We omit the details, which follow similar calculations as in Scheme 1. The new pde that results from this transformation is

$$U_\tau = \tfrac{1}{2}\sigma^2 \, U_{yy}; \qquad U(y, 0) = f(e^y) \tag{2.39}$$

This is the same heat equation, as for Scheme 1, but with a different initial value.

**REMARK 2.4**   Some texts price exotic options by transforming the BS-pde to the heat equation, as above, and then solve the simpler looking pde that results. However, that is really an illusion. With the right tools, as we shall demonstrate throughout this book, it is just as easy to solve the BS-pde as it stands. This avoids the need to both forward and back transform the associated variables. Nevertheless, there are some situations where it is useful to consider the heat equation transformation.

Once we have transformed the BS-pde into the heat equation, we have at our disposal the vast theory that has been devoted to it. For example, the text by Cannon [11], gives many important results about the heat equation. While we don't often use these results explicitly in this book, we may do so implicitly. For example, we shall assume that the existence and uniqueness of solutions of the heat equation are inherited by the corresponding BS-pde. This means, that if we are able to construct a solution of the BS-pde by some clever tricks, then we can be assured that this is the only solution. We actually use this idea in several places in later chapters.

Another use of the heat equation is to get a better understanding of the image option, defined in Chapter 7 on barrier options. The image solution for the BS-pde looks rather complicated, but is easy to describe and understand for the heat equation. ⬚

---

## 2.11   The BS Green's Function

Here is a specific example of where the heat equation transformation can be used to good effect. The *fundamental solution* or *Green's Function* for the heat equation $U_\tau = \frac{1}{2}\sigma^2 U_{yy}$ is well-known and is given by the formula

$$G(y, \tau; y_0) = \frac{1}{\sigma\sqrt{2\pi\tau}} e^{-\frac{(y-y_0)^2}{2\sigma^2\tau}} = \frac{1}{\sigma\sqrt{\tau}} \phi\left(\frac{y - y_0}{\sigma\sqrt{\tau}}\right). \qquad (2.40)$$

This is the unique solution expressed in terms of the Gaussian density $\phi(z)$, with initial value $U(y, 0) = \delta(y - y_0)$, where $\delta(y - y_0)$ is again the Dirac delta function.

If we now invert this expression for either of the above transformation schemes, we arrive at the Green's Function for the BS-pde. This leads to the important formula

$$G(x,t;\xi,T) = \frac{e^{-r\tau}}{\sigma\sqrt{\tau}}\,\phi\left(\frac{\log(x/\xi) + (r - \frac{1}{2}\sigma^2)\tau}{\sigma\sqrt{\tau}}\right) \tag{2.41}$$

where we have written $\xi$ for $\log y_0$. The terminal condition satisfied by this Green's Function is $G(x,T;\xi,T) = \delta(x/\xi - 1)$.

But as mentioned previously, given the right tools, we could also derive this result directly. We proceed to develop these tools in what follows.

---

## 2.12  Log-Volutions

We show in the present section how to obtain solutions of the BS-pde in terms of an operator we call the logarithmic convolution or *log-volution* for short.

Let $f(x), g(x)$ and $h(x)$ denote arbitrary functions defined on $x \in \mathbb{R}^+$. We allow these to be generalized functions such as Dirac delta functions and their derivatives.

**DEFINITION 2.1**   *The log-volution of $f(x)$ and $g(x)$, written as $f(x) \star g(x)$, is defined by*

$$f(x) \star g(x) = \int_0^\infty f(y)\, g\left(\frac{x}{y}\right) \frac{dy}{y} \tag{2.42}$$

The log-volution, as demonstrated next, is indeed a "logarithmic convolution." Let $x' = \log x$; $y' = \log y$ and write $F(x') = f(e^{x'})$; $G(x') = g(e^{x'})$. Then

$$f(x) \star g(x) = \int_{-\infty}^\infty F(y')G(x' - y')\, dy' = F(x') * G(x').$$

The expression $F(x') * G(x')$ is the usual way of writing a standard or linear convolution. Hence, the log-volution in the variable $x$ is seen to be equal to a convolution in the variable $x' = \log x$, and hence the name.

The log-volution has many interesting properties, some of which are listed in the table on the next page. In this table, $(\alpha, \beta)$ are arbitrary scalars; $k > 0$ is a positive scalar and $D$ denotes the differential operator

$$D = x\frac{d}{dx}. \tag{2.43}$$

Further, $L(x) = f(x) \star g(x)$ denotes the log-volution of $f(x)$ and $g(x)$.

| Log-Volution Properties | |
|---|---|
| L1. | $f \star (\alpha g + \beta h) = \alpha(f \star g) + \beta(f \star h)$ |
| L2. | $f \star g = g \star f$ |
| L3. | $f \star (g \star h) = (f \star g) \star h$ |
| L4. | $L(kx) = f(kx) \star g(x) = f(x) \star g(kx)$ |
| L5. | $[x^\alpha f(x)] \star [x^\alpha g(x)] = x^\alpha [f(x) \star g(x)]$ |
| L6. | $L(x^{-1}) = f(x^{-1}) \star g(x^{-1})$ |
| L7. | $D[f \star g] = (Df) \star g = f \star (Dg)$ |
| L8. | $f(x) \star \delta(x - 1) = f(x)$ |

Thus log-volution is linear, commutative, and associative, and the identity element under log-volution is the Dirac delta function $\delta(x-1)$. The log-volution also has useful scaling, power, inversion, and derivative rules as given by properties L4 to L7 inclusively. The proofs of these properties are relatively straightforward using standard integration rules alone, but a simpler method utilizes the Mellin Transform, which we introduce later in this section.

Consider now the BS-pde given by Equation (1.50) and repeated here in terms of relative time $\tau = (T - t)$,

$$V_\tau = -rV + (r - q)x\, V_x + \tfrac{1}{2}\sigma^2 x^2\, V_{xx}.$$

Then if $D$ is the operator defined by (2.43), it is a simple matter to show that

$$V_\tau = \tfrac{1}{2}\sigma^2 D^2 V + (r - q - \tfrac{1}{2}\sigma^2)DV - rV = Q(-D)V, \qquad (2.44)$$

where $Q(s)$ is the quadratic function $Q(s) = \tfrac{1}{2}\sigma^2 s^2 - (r - q - \tfrac{1}{2}\sigma^2)s - r$. The minus sign in $Q(-D)$ is included for later convenience.

Thus, while the BS-pde for $V(x, \tau)$ in terms of the differential operator $d/dx$ has non-constant coefficients, under the operator $D = xd/dx$, it now has constant coefficients.

The next theorem shows how to build solutions of the BS-pde from more basic ones using log-volutions.

### THEOREM 2.1
*Let $F(x)$ be any function independent of time $\tau$ defined on $x > 0$, such that the log-volution $F(x) \star U(x, \tau)$ exists. Then if $U(x, \tau)$ is a solution of the*

*BS-pde, so is*

$$V(x, \tau) = F(x) \star U(x, \tau). \tag{2.45}$$

**PROOF**   Consider formally the operator $\partial_\tau - Q(-D)$ applied to $V(x, \tau)$,

$$V_\tau - Q(-D)V = F(x) \star [U_\tau - Q(-D)U]$$

by the linearity and derivative properties of log-volution. However, this expression is equal to zero, since by assumption, $U(x, \tau)$ satisfies the BS-pde $U_\tau = Q(-D)U$. □

The next theorem shows how to price European derivatives with an arbitrary payoff $f(x)$, in terms of log-volutions.

### COROLLARY 2.1

*Let $G(x, \tau)$ denote the Green's Function of the BS-pde (2.44) with the initial value $G(x, 0) = \delta(x - 1)$. Then the solution of the BS-pde with initial value (i.e., payoff) $V(x, 0) = f(x)$, is given by*

$$V(x, \tau) = f(x) \star G(x, \tau). \tag{2.46}$$

**PROOF**   The proof of (2.46) follows directly from theorem 2.1 and log-volution property L8. which yields, $V(x, 0) = f(x) \star \delta(x - 1) = f(x)$. □

## 2.12.1   The Mellin Transform

For any function $f(x)$ on $x > 0$, for which the integral below exists, the Mellin Transform of $f(x)$ denoted by $F(s)$, is defined by

$$F(s) = \mathcal{M}_s[f(x)] = \int_0^\infty f(x) x^{s-1} \, dx. \tag{2.47}$$

The parameter $s$ is generally taken as a complex variable. The Mellin Transform is closely related to the Laplace Transforms, and just like the latter, has an inverse transform, which is given (with $c$ real) by

$$f(x) = \mathcal{M}_x^{-1}[F(s)] = \frac{1}{2\pi i} \int_{c-i\infty}^{c+i\infty} F(s) x^{-s} \, ds. \tag{2.48}$$

Properties and tables of Mellin Transforms can be found in Erdélyi et al. [22]. We present some of the more important ones in the following. With the same notation we used in the log-volution table, we have the general properties shown in the next table. Of particular importance is property M6, that the transform of a log-volution of two functions is the product of their transforms. In property M7, the function $\phi$ denotes the Gaussian pdf of Equation (3.1).

| | **Mellin Transform Properties** | |
|---|---|---|
| | *x-domain function* | *Mellin Transform* |
| M1. | $\alpha f(x) + \beta g(x)$ | $\alpha F(s) + \beta G(s)$ |
| M2. | $f(kx)$ | $k^{-s} F(s)$ |
| M3. | $f(x^\alpha)$ | $|\alpha|^{-1} F(s/\alpha)$ |
| M4. | $x^\alpha f(x)$ | $F(s+\alpha)$ |
| M5. | $(-D)^n f(x)$ | $s^n F(s)$ |
| M6. | $f(x) \star g(x)$ | $F(s) G(s)$ |
| M7. | $\phi(\log x)$ | $e^{\frac{1}{2}s^2}$ |
| M8. | $x^\gamma \mathbb{I}(x<k)$ | $k^{s+\gamma}(s+\gamma)^{-1}$ |

The Mellin Transform provides a quick and efficient way of deriving an explicit expression for the Green's Function $G(x,\tau)$, defined in corollary 2.1. The function $G(x,\tau)$ satisfies the IVP

$$G_\tau = Q(-D)G; \qquad G(x,0) = \delta(x-1).$$

Taking the Mellin Transform of this pde leads to the ode, for $\hat{G}(s,\tau) = \mathcal{M}_s[G(x,\tau)]$,

$$\hat{G}_\tau = Q(s)\,\hat{G}(s,\tau); \qquad \hat{G}(s,0) = 1$$

which follows directly from the derivative property. This has unique solution $\hat{G}(s,\tau) = e^{Q(s)\tau}$, which can be written in the form,

$$\hat{G}(s,\tau) = e^{-r\tau} \cdot k^{-s} \cdot e^{\frac{1}{2}(\sigma\sqrt{\tau}s)^2}$$

with $k = e^{(r-q-\frac{1}{2}\sigma^2)\tau}$. Several of the Mellin Transform properties then lead to the result

$$G(x,\tau) = \frac{e^{-r\tau}}{\sigma\sqrt{\tau}}\, \phi\left(\frac{\log x + (r-q-\frac{1}{2}\sigma^2)\tau}{\sigma\sqrt{\tau}}\right). \qquad (2.49)$$

This expression (which now includes the dividend yield $q$) agrees with the result (2.41) which we derived by first transforming the BS-pde to the heat equation.

### PROPOSITION 2.1
*Let $G(x,\tau)$ defined by (2.49) be the BS Green's Function. Then for any $k > 0$,*

$$x\mathbb{I}(x>k) \star G(x,t) = xe^{-q\tau} \mathcal{N}(d_1)$$
$$\mathbb{I}(x>k) \star G(x,t) = e^{-r\tau} \mathcal{N}(d_2)$$

*where*

$$d_{1,2}(x,\tau) = \frac{\log(x/k) + (r - q \pm \frac{1}{2}\sigma^2)\tau}{\sigma\sqrt{\tau}}.$$

We leave the proof as Q14 in the Exercise Problems to this chapter.

Sometimes we are interested in solving the inhomogeneous BS-pde (see Equation (2.50)), where the added term represents cash instalment rates per unit time to be paid if $h < 0$, or to be received if $h > 0$. The next theorem shows how to price derivatives with such continuous instalments.

**THEOREM 2.2**

*Let $V(x,\tau)$ satisfy the inhomogeneous BS-pde*

$$V_\tau = Q(-D)\,V + h(x,\tau); \qquad V(x,0) = f(x). \qquad (2.50)$$

*Then, subject to the existence of the log-volutions, the solution of this pde is given by*

$$V(x,\tau) = f(x) \star G(x,\tau) + \int_0^\tau h(x,\tau') \star G(x, \tau - \tau')\,d\tau'. \qquad (2.51)$$

**PROOF**   Take the Mellin Transform of the pde to arrive at

$$\hat{V}_\tau = Q(s)\hat{V}(s,\tau) + \hat{h}(s,\tau); \qquad \hat{V}(s,0) = \hat{f}(s).$$

This ode has solution

$$\hat{V}(s,\tau) = \hat{f}(s)e^{Q(s)\tau} + \int_0^\tau e^{Q(s)(\tau-\tau')}\hat{h}(s,\tau')\,d\tau'.$$

Inverting the Mellin Transforms using, $\mathcal{M}_s[G(x,\tau)] = e^{Q(s)\tau}$, and several entries stated in table of Mellin Transform Properties, leads to the given result. $\qquad\Box$

Curiously, the second term in Equation (2.51) is simultaneously both a convolution (in time) and a log-volution (in price).

*Example 2.2* **(American Call)**

Consider a standard American call option on a dividend paying asset, with constant dividend yield $q$. Then the price $V(x,\tau)$ with IV, $V(x,0) = f(x) = (x-k)^+$, satisfies

$$\begin{aligned}
\mathcal{L}V = V_\tau - Q(-D)V = 0 & \quad \text{in } x < b(\tau) \\
V = x - k & \quad \text{in } x > b(\tau)
\end{aligned}$$

where $\mathcal{L}$ denotes the BS-pde operator and $x = b(\tau) \geq k$ is the early exercise boundary. The domain $x < b(\tau)$ is called the *continuation region*, in which we continue to hold the option; the complementary domain $x > b(\tau)$ is the *stopping region* in which we exercise the option before expiry. Both equations are encompassed by the single inhomogeneous BS-pde

$$\mathcal{L}V(x, \tau) = -Q(-D)[x - k] \cdot \mathbb{I}(x > b(\tau)) = (qx - rk)\mathbb{I}(x > b(\tau)).$$

Using theorem 2.2 and proposition 2.1, we therefore have the solution

$$V(x, \tau) = (x - k)^+ \star G(x, \tau) + \int_0^\tau [(qx - rk)\mathbb{I}(x > b(\tau - s)) \star G(x, s)] \, ds$$

$$= C_k(x, \tau) + \int_0^\tau [qxe^{-qs}\mathcal{N}(z_1) - rke^{-rs}\mathcal{N}(z_2)] \, ds,$$

where $C_k(x, \tau)$ denotes the European call option price, and

$$z_{1,2}(x, \tau, s) = \frac{\log(x/b(\tau - s)) + (r - q \pm \frac{1}{2}\sigma^2)s}{\sigma\sqrt{s}}.$$

The last expression for $V(x, \tau)$ determines the American call option price in terms of the (yet) unknown early exercise boundary $b(\tau)$. The continuity condition, $V(x, \tau)$ is continuous across the boundary $x = b(\tau)$, then yields the following integral equation for $b(\tau)$

$$b(\tau) - k = C_k(b(\tau), \tau) + \int_0^\tau [qb(\tau)e^{-qs}\mathcal{N}(z_1) - rke^{-rs}\mathcal{N}(z_2)] \, ds$$

where now, $z_1$ and $z_2$ are evaluated at $x = b(\tau)$.

This representation of the price of a vanilla American call option was first given by Kim [43] and Jamshidian [37]. A survey of similar results can be found in Chiarella et al. [15]. It is obviously not a closed form solution because first the integral equation for $b(\tau)$ needs to be solved, and second the price also involves an unknown integral in terms of $b(\tau)$. $\quad\Box$

---

## 2.13 Summary

This chapter dealt mainly with the mathematical and statistical tools needed to price options in the Black–Scholes framework. Most of these tools were introduced in a fairly non-technical way, consistent with the general approach taken in this book. Proofs of the well-known results were generally ignored, as these may be looked up in the quoted texts.

The most important topics we covered included:

| Tower Law | Feynman–Kac Formula |
|---|---|
| Brownian motion | Girsanov's Theorem |
| SDE's | Time-Varying Parameters |
| Itô's Lemma | BS Green's Function |
| Martingales | Log-Volutions |
| | Mellin Transforms |

The section on Time-Varying Parameters is important, as it has a direct consequence on the pricing of Asian options considered in Chapter 9.

The applications of the Mellin Transform and log-volutions to the BS-pde contain significant unpublished material, which also will have an impact on exotic option pricing in later chapters, particularly for reflecting barrier options.

We have seen that the FTAP applied to the BS model can be expressed entirely in terms of Gaussian rv's as the only stochastic variable required. It is therefore of important to have a good understanding of Gaussian rv's and the next chapter is devoted exclusively to their study. This chapter includes properties of univariate, bivariate and multi-variate Gaussian random variables, with corresponding extensions for the BS-pde and FTAP.

---

## Exercise Problems

1. Let $V(x, t)$ denote the price of any European derivative on a single underlying asset $X$, for $t < T$, that pays $F(x)$ at expiry $T$. Now suppose you are offered a contract at time $t$ which pays at time $s$, the amount $V(x, s)$ for $t < s < T$. Use the Tower Law to prove that the price of the derivative at time $t$ should equal $V(x, t)$.

2. Let $\mathbb{E}_s\{X_t\}$ denote the conditional expectation $\mathbb{E}\{X_t | \mathcal{F}_s\}$ for $s < t$. If $B_t$ is a standard Brownian motion, show that

$$\mathbb{E}_s\{B_t^3\} = B_s^3 + 3(t - s)B_t.$$

Verify the Tower Law, for the example $\mathbb{E}_u\{\mathbb{E}_s\{B_t^3\}\} = \mathbb{E}_u\{B_t^3\}$ for $u < s < t$.

3. Prove by direct methods the following results for standard Brownian motions $B_t$:

   (a) $\text{cov}\{B_s, B_t\} = \min(s, t)$ and for $s < t$, $\text{corr}\{B_s, B_t\} = \sqrt{s/t}$

   (b) $B_t$ is self-similar with similarity dimension $H = \frac{1}{2}$;

   i.e., $B_{ct} \stackrel{d}{=} c^H B_t$ for all $c > 0$.

(c) $B_t$, $B_t^2 - t$ and $e^{\sigma B_t - \frac{1}{2}\sigma^2 t}$ are $\mathcal{F}_t$−martingales.

4. For the general linear sde

$$dX_t = (a_t X_t + \alpha_t)dt + (b_t X_t + \beta_t)dB_t; \qquad X_0 = x$$

show that

$$\frac{dE_t}{dt} = a_t E_t + \alpha_t; \qquad E_0 = x$$

$$\frac{dV_t}{dt} = (2a_t + b_t^2)V_t + (b_t E_t + \beta_t)^2; \qquad V_0 = 0$$

where $E_t = \mathbb{E}\{X_t\}$ and $V_t = \mathbb{V}\{X_t\}$.

5. Use Itô's Lemma to show that the sde

$$dX_t = \tfrac{1}{2}h(X_t)h'(X_t)dt + h(X_t)dB_t$$

is reducible to simple aBm by the transformation $Y_t = f(X_t)$ where $f(x) = \int^x [1/h(u)]du$. Hence solve the sde's

(a) $dX_t = \frac{1}{3}X_t^{1/3}dt + X_t^{2/3}dB_t$, and

(b) $dX_t = a^2\,dt + 2a\sqrt{X_t}\,dB_t$ with $a > 0$ and $X_0 = 1$ . Hence find the probability that $X_t < 1$ at any time $t > 0$.

6. Let $X_t$ satisfy the sde $dX_t = \alpha_t dt + \beta_t dB_t$ and suppose, $F(X_t, t)$ and $G(X_t, t)$ are two $\mathbb{C}_{2,1}$ functions. Derive Itô's Product and Quotient Rules

$$d(FG) = FdG + GdF + \beta_t^2(F_x G_x)dt$$

$$d(F/G) = \frac{GdF - FdG}{G^2} + \frac{\beta_t^2 G_x}{G^3}(FG_x - GF_x)dt.$$

7. Prove that $f(B_t, t)$, where $B_t$ is a standard Brownian motion, is a (local) $\mathcal{F}_t$ martingale if $f(x,t)$ satisfies the pde $f_t + \frac{1}{2}f_{xx} = 0$.

(a) Hence show that $X_t = t^{n/2}\,H_n\left(\frac{B_t}{\sqrt{2t}}\right)$, where $H_n(x)$ are Hermite polynomials, is an $\mathcal{F}_t$ martingale.

(b) Obtain polynomials of orders one through four, in $B_t$ that are $\mathcal{F}_t$ martingales.

(c) Use the Hermite polynomial generating function

$$e^{2xs - s^2} = \sum_{n=0}^{\infty} \frac{1}{n!}H_n(x)s^n$$

to show that $e^{\sigma B_t - \frac{1}{2}\sigma^2 t}$ is an $\mathcal{F}_t$ martingale.

8. (a) Use the 2D Itô's Lemma to prove that if $X = X_t$ and $Y = Y_t$ are two Itô processes relative to the same Brownian motion, with instantaneous variances $\sigma_x^2$ and $\sigma_y^2$, then

$$d(X/Y) = \frac{Y\,dX - X\,dY}{Y^2} + \frac{\sigma_y(\sigma_y X - \sigma_x Y)}{Y^3}\,dt.$$

(b) Let $X_t$ satisfy the *general linear sde*

$$dX_t = (a_t X_t + \alpha_t)dt + (b_t X_t + \beta_t)\,dB_t$$

where $a_t, \alpha_t, b_t, \beta_t$ are deterministic functions of $t$. Show that the solution of this sde is given by

$$X_t = \Phi_0^t X_0 + \int_0^t (\alpha_s - b_s\beta_s)\Phi_s^t\,ds + \int_0^t \beta_s\Phi_s^t\,dB_s$$

where

$$\Phi_s^t = \exp\left(\int_s^t (a_u - \tfrac{1}{2}b_u^2)du + \int_s^t b_u\,dB_u\right).$$

*Hint:* Solve the sde $dY_t = Y_t(a_t dt + b_t dB_t)$; $Y_0 = 1$ and use Itô's Quotient Rule above, to solve for $Z_t = X_t/Y_t$.

9. The mean-reverting OU (Ornstein-Uhlenbeck) process is the solution of the sde

$$dX_t = a(\gamma - X_t)dt + \sigma dB_t; \qquad X_0 = x$$

where $a, \gamma, \sigma$ are positive constants. Solve this sde and hence show that $X_t$ is Gaussian with mean and variance

$$\mathbb{E}\{X_t\} = \gamma + (x - \gamma)e^{-at}$$
$$\mathbb{V}\{X_t\} = \frac{\sigma^2}{2a}(1 - e^{-2at}).$$

10. Suppose the underlying asset $X_t = x$ and an associated derivative $V_t = V(X_t, t)$ satisfy the sde's, in the real-world measure,

$$dX_t = \mu_X dt + \sigma_X dB_t; \qquad dV_t = \mu_V dt + \sigma_V dB_t$$

Show that, in order to avoid arbitrage,

$$\frac{\mu_X - rX_t}{\sigma_X} = \frac{\mu_V - rV_t}{\sigma_V}.$$

Hence derive the BS-pde when $X_t$ follows gBm.

11. Show that the transformation (Scheme-2)

$$y = \log x + (r - \tfrac{1}{2}\sigma^2)\tau; \quad \tau = T - t; \quad V(x,t) = e^{-r\tau}U(y,\tau)$$

reduces the BS-pde for $V(x,t)$, with TV, $V(x,T) = f(x)$ to the IV heat equation

$$U_\tau = \tfrac{1}{2}\sigma^2 U_{yy}; \qquad U(y,0) = f(e^y).$$

12. Derive the log-volution properties L1 to L8 by

   (a) direct integration, and

   (b) using Mellin Transforms.

13. Derive the BS solutions as given in Section 2.9, for time-varying parameters, by applying the Mellin Transform directly to the the BS-pde.

14. Show that if $G(x,t)$ denotes the Green's Function for the BS-pde, then

$$\mathbb{I}(x>k) \star G(x,t) = e^{-r\tau}\mathcal{N}(d_2)$$
$$x\mathbb{I}(x>k) \star G(x,t) = xe^{-q\tau}\mathcal{N}(d_1)$$

where

$$d_{1,2} = \frac{\log(x/k) + (r - q \pm \tfrac{1}{2}\sigma^2)\tau}{\sigma\sqrt{\tau}}$$

and $q$ is the constant dividend yield.

—ooOoo—

# Chapter 3

## Gaussian Random Variables

We have seen that one powerful method of pricing options in the BS-framework, on a dividend paying asset, is via the pair of formulae (1.51), repeated below for convenience:

$$V(x,t) = e^{-r\tau}\,\mathbb{E}_Q\{F(X_T)\,|\,\mathcal{F}_t\}; \qquad X_T = x\,e^{((r-q)-\frac{1}{2}\sigma^2)\tau+\sigma\sqrt{\tau}Z}.$$

Essentially, this is the mathematical statement of the FTAP applied to the BS economy. The function $F(X_T)$ is the payoff of the derivative at expiry $T$, and the expression for $X_T$ is the random asset price at time $T$, given its current value $x$ at time $t < T$, as seen under the EMM, $Q$. This random price depends only on a single Gaussian random variable $Z$, although for each $t$ there will be a different such $Z$.

This chapter summarizes many of the important statistical properties of Gaussian random variables. Our summary will include specifically the univariate and bivariate cases in anticipation of pricing both first-order and second-order exotic options. We shall also include some important results for multi-variate Gaussian vectors to facilitate our analysis of the multi-period, multi-asset binaries considered later in this book.

The reader is particularly directed to the corresponding Gaussian Shift Theorems for these different scenarios. The GST is the device that makes it possible to price exotic options without the necessity of doing a formal integration, or the requirement to solve a pde.

---

## 3.1 Univariate Gaussian Random Variables

A standard Gaussian rv, denoted by $Z \overset{d}{=} N(0,1)$ is a normal variate with zero mean and unit variance. The probability density function (pdf) is given by

$$\phi(z) = \frac{1}{\sqrt{2\pi}}\,e^{-\frac{1}{2}z^2}; \qquad z \in \mathbb{R}. \tag{3.1}$$

This is the density often referred to as the *bell-shaped curve*. The corresponding cumulative density function (or cdf) is far more important for BS option pricing and is defined by

$$\mathcal{N}(d) = \int_{-\infty}^{d} \phi(z)\, dz. \tag{3.2}$$

This higher transcendental function cannot be expressed in terms of algebraic or other elementary functions. However, it is closely related to the *Error Function* erf$(z)$, through

$$\mathcal{N}(d) = \tfrac{1}{2} \left[ 1 + \mathrm{erf}\left( \frac{d}{\sqrt{2}} \right) \right]. \tag{3.3}$$

The function $\mathcal{N}(d)$ represents the probability $\mathbb{P}\{Z < d\}$ and so can also be usefully written as the following expectation

$$\boxed{\mathcal{N}(d) = \mathbb{E}\{\mathbb{I}(Z < d)\}} \tag{3.4}$$

We shall often have recourse to this formula in pricing options. The function $\mathcal{N}(d)$ is what mathematicians call a *nice* function, since it is infinitely differentiable. In fact, the first derivative recovers the Gaussian density,

$$\mathcal{N}'(d) = \phi(d) \quad \text{for all } d \in \mathbb{R}. \tag{3.5}$$

Figure 3.1 displays graphs of $\phi(z)$ and $\mathcal{N}(z)$.

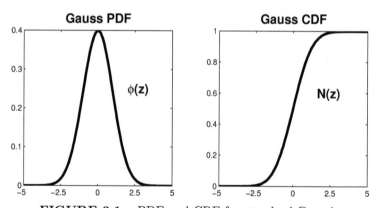

**FIGURE 3.1:** PDF and CDF for standard Gaussian

If $Z$ is standard Gaussian, then $F(Z)$, where $F(Z)$ is a measurable function, represents another rv with expectation given by the integral

$$\mathbb{E}\{F(z)\} = \int_{-\infty}^{\infty} F(z)\phi(z)\, dz. \tag{3.6}$$

Since $\phi(z)$ is a symmetric function, i.e., $\phi(-z) = \phi(z)$, it is clear that

$$\mathbb{E}\{F(-Z)\} = \mathbb{E}\{F(Z)\} \tag{3.7}$$

for any measurable function $F(Z)$ with finite expectation. Thus in any such expectation of a standard Gaussian, we can always replace $Z$ by $-Z$ and vice versa. The corresponding symmetry property for the Gaussian cdf is

$$\mathcal{N}(d) + \mathcal{N}(-d) = 1. \tag{3.8}$$

---

## 3.2  Gaussian Shift Theorem

Option pricing in the BS economy is closely connected to integrals such as (3.6), through the FTAP (1.46). However, as we shall presently see, we are largely able to bypass the direct calculation of such integrals through a clever device which we call the *Gaussian Shift Theorem* or simply GST for short.

**THEOREM 3.1**
*Let $Z \stackrel{d}{=} N(0,1)$ be a standard Gaussian rv, $c$ any real constant and $F(Z)$ a measurable function of $Z$ with finite expectation. Then*

$$\boxed{\mathbb{E}\{e^{cZ} F(Z)\} = e^{\frac{1}{2}c^2} \mathbb{E}\{F(Z+c)\}} \tag{3.9}$$

**PROOF**   The proof is rather elementary, and relies on the identity

$$e^{cy}\phi(y) \equiv e^{\frac{1}{2}c^2} \phi(y - c)$$

which follows from (3.1) by direct algebra. We then obtain

$$\mathbb{E}\{e^{cZ} F(Z)\} = \int_{-\infty}^{\infty} e^{cy} F(y)\phi(y)\, dy = e^{\frac{1}{2}c^2} \int_{-\infty}^{\infty} F(y)\phi(y - c)\, dy$$

$$= e^{\frac{1}{2}c^2} \int_{-\infty}^{\infty} F(z + c)\phi(z)\, dz = e^{\frac{1}{2}c^2} \mathbb{E}\{F(Z+c)\}.$$

In the second line we have changed the variable of integration through $z = (y - c)$. This completes the proof.  □

**REMARK 3.1**   The *Gaussian Shift Theorem*, despite its simplicity, has some very significant consequences for what follows. First, as mentioned above, the formula allows us to bypass many Gaussian integrations when pricing not only simple European options, but also many exotic options as well. Second, other techniques including change of measure formulae, are often found in the literature to price derivatives. The GST allows us to eschew these approaches too. The name incidentally derives from the observation that $(Z + c)$, which appears on the rhs of the theorem, is a *shifted* Gaussian with mean $c$ and unit variance.  ☐

### Example 3.1

We demonstrate that the BS asset price in the EMM, i.e.,

$$X_T = xe^{(r-\frac{1}{2}\sigma^2)\tau + \sigma\sqrt{\tau}Z}; \qquad \tau = (T - t),$$

indeed satisfies the Martingale Restriction. Taking expectations, we find using the GST with $c = \sigma\sqrt{\tau}$ and $F(Z) = 1$,

$$\mathbb{E}_Q\{X_T\} = xe^{(r-\frac{1}{2}\sigma^2)\tau}\, \mathbb{E}_Q\{e^{\sigma\sqrt{\tau}Z}\}$$
$$= xe^{(r-\frac{1}{2}\sigma^2)\tau} \cdot e^{\frac{1}{2}\sigma^2\tau}\mathbb{E}_Q\{1\}$$
$$= xe^{r\tau}.$$

This is the required martingale restriction.  ☐

## 3.3  Rescaled Gaussians

Let $Y = \sigma Z + \mu$ where $\sigma > 0$ and $Z$ is a standard Gaussian. Then $Y$ is also Gaussian but with mean $\mu$ and variance $\sigma^2$. That is, $Y \overset{d}{=} N(\mu, \sigma^2)$. On the other hand, if $Y \sim N(\mu, \sigma^2)$ then the variate $(Y - \mu)/\sigma \overset{d}{=} N(0, 1)$. The corresponding pdf and cdf of the rescaled Gaussian $Y$, are respectively

$$\phi(y; \mu, \sigma) = \frac{1}{\sigma}\phi\left(\frac{y - \mu}{\sigma}\right) = \frac{1}{\sqrt{2\pi}\sigma}e^{-(y-\mu)^2/2\sigma^2}, \qquad (3.10)$$

$$\Phi(y; \mu, \sigma) = \mathcal{N}\left(\frac{y - \mu}{\sigma}\right). \qquad (3.11)$$

## 3.4 Gaussian Moments

The quantity $\mu_k = \mathbb{E}\{Z^k\}$ where $k$ is a positive integer, is called the $k$th moment of $Z$ about the origin and is given explicitly by

$$\mu_k = \int_{-\infty}^{\infty} x^k \phi(x)\, dx = \begin{cases} 0 & k = \text{odd} \\ (k-1)(k-3)\cdots 1 & k = \text{even} \end{cases} \tag{3.12}$$

Clearly the mean is $\mu_1 = 0$ and the variance is $\mu_2 = 1$. For the standard Gaussian rv with zero mean and unit variance, the third and fourth moments are respectively called the *skewness* and *kurtosis*. The moment formula above yields $\mu_3 = 0$ and $\mu_4 = 3$. Distributions with zero skewness are called symmetric distributions. Distributions with $\mu_4 > 3$ are called *fat-tailed* or *leptokurtic*; those with $\mu_4 < 3$ are called *thin-tailed* or *platykurtic*. Real stock price returns typically show strong leptokurtic behavior.

The moment-generating function (mgf) of a rv $Y$ is defined to be $M_Y(\theta) = \mathbb{E}\{e^{\theta Y}\}$. For the rescaled Gaussian $Y \overset{d}{=} N(\mu, \sigma^2) = \mu + \sigma Z$, this is given by the GST as

$$M_Y(\theta) = e^{\mu\theta}\mathbb{E}\{e^{\theta\sigma Z}\} = e^{\mu\theta + \frac{1}{2}\sigma^2\theta^2}. \tag{3.13}$$

If $X_i$ $(i = 1, 2, \ldots, n)$ are a set of independent random variables and $Y = \sum_i X_i$ then

$$M_Y(\theta) = \prod_i M_{X_i}(\theta). \tag{3.14}$$

That is, the mgf of a sum of independent rv's is the product of their individual mgf's.

### 3.4.1 Sums of Independent Gaussians

Let $X_i$ $(i = 1, 2, \ldots n)$ be a set of independent Gaussian rv's with mean $\mu_i$ and variance $\sigma_i^2$. Then using (3.14), one can show

$$Y = \sum_{i=1}^{n} c_i X_i \sim N\left(\sum_i c_i \mu_i, \sum_i c_i^2 \sigma_i^2\right). \tag{3.15}$$

Thus a sum of Gaussians is itself Gaussian. Distributions with this invariance property are said to be *stable* distributions.

Note that if the $X_i$ are not independent (considered later), then $Y$ is still Gaussian but the expression for the variance is more complicated.

## 3.5   Central Limit Theorem

This celebrated theorem explains the ubiquity of the normal distribution in many practical situations. We state and prove it (as usual) in non-technical terms.

**THEOREM 3.2**
*If $X_i$ $(i = 1, \ldots, n)$ are iid random variables with the same mean $\mu$ and variance $\sigma^2$ then, no matter what the common distribution, $X = \sum_i X_i$ is asymptotically Gaussian $N(n\mu, n\sigma^2)$.*

Our non-technical proof is based on mgf's.

**PROOF**   The variate $Y_i = X_i - \mu$ has zero mean and variance $\sigma^2$ and therefore has mgf of the form

$$M_i(\theta) = 1 + \tfrac{1}{2}\sigma^2\theta^2 + \quad \ldots \quad + \text{hot}.$$

Since $M_{aX}(\theta) = M_X(a\theta)$ and the mgf of a sum of independent rv's is the product of their mgf's, the variate $Y = \frac{\sum Y_i}{\sigma\sqrt{n}} = \frac{X - \mu n}{\sigma\sqrt{n}}$ has mgf

$$M_Y(\theta) = \left[M_i\left(\frac{\theta}{\sigma\sqrt{n}}\right)\right]^n = \left(1 + \frac{1}{2n}\theta^2 + \quad \ldots \quad + \text{hot}\right)^n.$$

Recall the well-known limit $\lim_{n\to\infty}(1 + x/n)^n = e^x$. Thus $M_Y(\theta) \to e^{\frac{1}{2}\theta^2}$ as $n \to \infty$ and this is the mgf of a $N(0,1)-$variate. Hence

$$\frac{X - \mu n}{\sigma\sqrt{n}} \sim N(0,1) \quad \text{as} \quad n \to \infty$$

and the result follows.  ⬚

## 3.6   Log-Normal Distribution

One of the main defects of using Gaussians to describe physical models is their domain $x \in \mathbb{R}$. Many naturally occurring phenomena require $x$ to be non-negative (e.g., number of species, stock prices etc.). To circumvent this problem, many statistical modelers prefer to employ the *log-normal distribution*. This was historically the principal motivation for using the log-normal to model stock price processes. A comprehensive treatise on the log-normal distribution is Aitchison and Brown [2].

If $Y$ is Gaussian $N(\mu, \sigma^2)$, then the rv $X = \exp(Y)$ is said to be a log-normal variate with parameters $(\mu, \sigma^2)$. The pdf of $X$ is found to be

$$f(x) = \frac{1}{x\sigma\sqrt{2\pi}} \exp\left\{-\frac{(\ln x - \mu)^2}{2\sigma^2}\right\}; \qquad x > 0 \qquad (3.16)$$

and the corresponding cdf is give by

$$F_X(x) = \mathcal{L}(x|\mu, \sigma^2) = \mathcal{N}[(\log x - \mu)/\sigma]. \qquad (3.17)$$

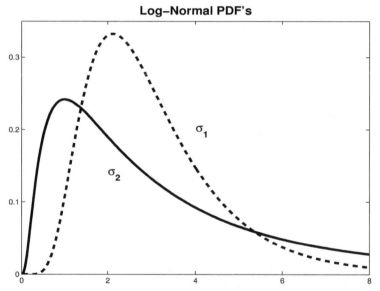

**FIGURE 3.2:** The log-normal density functions with $\mu = 1$ and $(\sigma_1, \sigma_2) = (0.5, 1.0)$

| Properties of the Log-Normal |
| :---: |
| Mean $= e^{\mu + \frac{1}{2}\sigma^2}$ |
| Variance $= e^{2\mu + \sigma^2}(e^{\sigma^2} - 1)$ |
| Skewness $= (e^{\sigma^2} + 2)\sqrt{e^{\sigma^2} - 1}$ |
| Moments $\mu_k = e^{k\mu + \frac{1}{2}k^2\sigma^2}$ |

Note that the log-normal distribution has positive skewness which depends only on the $\sigma$ parameter.

## 3.7 Bivariate Normal

Many exotic options, including dual expiry options and two-asset rainbow options have BS prices that can be expressed in terms of the bivariate normal. We therefore include here quite a bit of useful information about this statistical distribution.

Let $X$ and $Y$ be any two rv's. Their *covariance* is defined to be

$$\text{cov}\{X,Y\} = \mathbb{E}\{XY\} - \mathbb{E}\{X\}\mathbb{E}\{Y\}. \tag{3.18}$$

Their *correlation coefficient* (or ccf) is a normalized covariance defined by

$$\rho = \text{corr}\{X,Y\} = \frac{\text{cov}\{X,Y\}}{\sqrt{\mathbb{V}\{X\}\mathbb{V}\{Y\}}}. \tag{3.19}$$

The normalization ensures that $-1 \le \rho \le 1$. If $\rho = \pm 1$, then $X$ and $Y$ are said to be perfectly correlated and there will exist a linear relationship between them of the form $Y = \alpha X + \beta$ with $\alpha > 0$ if $\rho = 1$ and $\alpha < 0$ if $\rho = -1$. If $X$ and $Y$ are independent then $\rho = 0$; however, the converse is not necessarily true. But if $X, Y$ are Gaussian rv's then their independence implies they are uncorrelated and vice versa.

Let $X, Y \sim N(0, 1; \rho)$ be joint Gaussian with ccf $\rho$ and assume $|\rho| < 1$. Then their joint pdf is the function

$$\phi(x, y; \rho) = \frac{1}{2\pi\beta} e^{-(x^2 - 2\rho xy + y^2)/2\beta^2}; \qquad (x, y) \in \mathbb{R}^2. \tag{3.20}$$

where $\beta = \sqrt{1 - \rho^2}$. The corresponding joint cdf i.e., $\mathbb{P}\{X < x, Y < y\}$ is called the *bivariate normal* and is written as:

$$\mathcal{N}(x, y; \rho) = \mathbb{E}\{\mathbb{I}(X < x)\mathbb{I}(Y < y)\} = \int_{-\infty}^{x} \int_{-\infty}^{y} \phi(u, v; \rho) \, du dv. \tag{3.21}$$

**Special values:**

$$\mathcal{N}(x, -\infty; \rho) = 0$$
$$\mathcal{N}(x, \infty; \rho) = \mathcal{N}(x)$$
$$\mathcal{N}(x, y; 0) = \mathcal{N}(x)\mathcal{N}(y)$$
$$\mathcal{N}(x, y; 1) = \mathcal{N}[\min(x, y)]$$
$$\mathcal{N}(x, y; -1) = \begin{cases} [\mathcal{N}(x) - \mathcal{N}(-y)] & \text{if } x + y > 0 \\ 0 & \text{otherwise.} \end{cases}$$

**Symmetries:**

$$\mathcal{N}(x,y;\rho) - \mathcal{N}(y,x;\rho) = 0$$
$$\mathcal{N}(x,y;\rho) + \mathcal{N}(x,-y;-\rho) = \mathcal{N}(x)$$
$$\mathcal{N}(x,y;\rho) + \mathcal{N}(-x,y;-\rho) = \mathcal{N}(y)$$
$$\mathcal{N}(x,y;\rho) - \mathcal{N}(-x,-y;\rho) = \mathcal{N}(x) + \mathcal{N}(y) - 1$$

### 3.7.1 Gaussian Shift Theorem (Bivariate Case)

The following theorem is an extension of the univariate GST considered in Section 3.2.

### THEOREM 3.3

*Let $X, Y \sim N(0,1;\rho)$ be joint Gaussian with ccf $\rho$; $a, b$ any constants and $F(X,Y)$ any measurable function of $X, Y$. Then*

$$\boxed{\mathbb{E}\{e^{aX+bY} F(X,Y)\} = e^{\frac{1}{2}c^2} \mathbb{E}\{F(X+a', Y+b')\}} \qquad (3.22)$$

*where $c^2 = a^2 + b^2 + 2\rho ab$, $a' = a + \rho b$; $b' = b + \rho a$.*

Since this theorem is a special case of the multi-variate GST considered later, we will not offer a proof here, but wait till we meet the more general case.

The next proposition plays an important role in pricing exchange options and other two-asset and dual expiry options.

### PROPOSITION 3.1

*Let $X, Y \sim N(0,1;\rho)$. Then if $a, b, \sigma_1, \sigma_2 > 0$*

$$\mathbb{E}\left\{ \left( ae^{\sigma_1 X - \frac{1}{2}\sigma_1^2} - be^{\sigma_2 Y - \frac{1}{2}\sigma_2^2} \right)^+ \right\} = a\mathcal{N}(d_1) - b\mathcal{N}(d_2) \qquad (3.23)$$

*with*

$$d_{1,2} = \frac{1}{\sigma}\log(a/b) \pm \tfrac{1}{2}\sigma \qquad and \qquad \sigma = \sqrt{\sigma_1^2 + \sigma_2^2 - 2\rho\sigma_1\sigma_2}. \qquad (3.24)$$

We leave the proof of this interesting result to the Exercise Problems (see Q8) at the end of this chapter.

## 3.8 Multi-Variate Gaussian Statistics

Let $\boldsymbol{Z}$ denote an $m$-dimensional Gaussian random vector such that for each $1 \leq i \leq m$, $\mathbb{E}\{Z_i\} = 0$, $\mathbb{V}\{X_i\} = 1$. Further, let the correlation matrix $\boldsymbol{R}$ have components $\rho_{ij} = \mathrm{corr}\{Z_i, Z_j\}$ for $i \neq j$; $\rho_{ii} = 1$ when $i = j$. Matrix $\boldsymbol{R}$ is symmetric since $\rho_{ij} = \rho_{ji}$. We write this Gaussian vector in the suggestive notation $\boldsymbol{Z} \sim N(0, 1; \boldsymbol{R})$. The joint pdf of $\boldsymbol{Z}$ is the $m$ dimensional scalar function

$$\phi_m(\boldsymbol{z}; \boldsymbol{R}) = \frac{e^{-\frac{1}{2}\boldsymbol{z}'\boldsymbol{R}^{-1}\boldsymbol{z}}}{(2\pi)^{m/2}\sqrt{\det(\boldsymbol{R})}} \; ; \qquad \boldsymbol{z} \in \mathbb{R}^m . \tag{3.25}$$

Naturally we assume here that the correlation matrix $\boldsymbol{R}$ is positive definite, so that $\boldsymbol{R}^{-1}$ exists and $\det(\boldsymbol{R}) > 0$. Indeed, the pdf will not exist if the matrix $\boldsymbol{R}$ is singular.

The associated cdf is written as

$$\mathcal{N}_m(\boldsymbol{z}; \boldsymbol{R}) = \int_{-\infty}^{\boldsymbol{z}} \phi_m(\boldsymbol{u}; \boldsymbol{R})\, d\boldsymbol{u} \tag{3.26}$$

where the following short-hand notation has been used:

$$\int_{-\infty}^{\boldsymbol{z}} d\boldsymbol{u} = \int_{-\infty}^{z_1} du_1 \int_{-\infty}^{z_2} du_2 \cdots \int_{-\infty}^{z_m} du_m .$$

When $m = 1$: $\boldsymbol{R} = 1$ so that $\phi_1(\boldsymbol{z}, \boldsymbol{R}) = \phi(z)$ and $\mathcal{N}_1(\boldsymbol{z}, \boldsymbol{R}) = \mathcal{N}(z)$.

When $m = 2$, $\boldsymbol{R} = \begin{bmatrix} 1 & \rho \\ \rho & 1 \end{bmatrix}$. It is therefore simpler to write $\phi_2(\boldsymbol{z}, \boldsymbol{R})$ as $\phi(x, y; \rho)$ and $\mathcal{N}_2(\boldsymbol{z}, \boldsymbol{R})$ as $\mathcal{N}(x, y; \rho)$. Note also that when $m = 2$,

$$\det \boldsymbol{R} = \beta^2; \quad \boldsymbol{R}^{-1} = \frac{1}{\beta^2}\begin{bmatrix} 1 & -\rho \\ -\rho & 1 \end{bmatrix}; \quad \boldsymbol{z}'\boldsymbol{R}^{-1}\boldsymbol{z} = (x^2 + y^2 - 2\rho xy)/\beta^2$$

leading to the expression in (3.20) for the bivariate pdf.

**REMARK 3.2** Computation of the univariate and multi-variate normal cdf's requires specialized routines which are readily available in the literature. Most mathematical software packages will have a routine to compute the univariate normal cdf, or perhaps the Error Function, from which the cdf can be obtained through Equation (3.3). A good numerical routine to compute the bivariate normal cdf, due to Drezner [21], has been published in several financial works. We mention Appendix A2 of Haug [29] as one example.

Computation of multi-variate normal cdf's of dimension greater than two, is a much more difficult proposition. The reader is referred to Genz [24] as an appropriate place to learn how such computations may be performed. ☐

**Sums of dependent Gaussian rv's**

Let $X_i$ be a set of $m$ Gaussian rv's with mean $\mu_i$, variance $\sigma_i^2$ and covariance matrix $\Gamma_{ij} = \rho_{ij}\sigma_i\sigma_j$. Then the rv $Y = \sum_i c_i X_i$ is Gaussian $N(\mu, \sigma^2)$ with

$$\mu = \sum_{i=1}^{m} c_i \mu_i = c'\mu \qquad \text{and} \qquad \sigma^2 = \sum_{i=1}^{m}\sum_{j=1}^{m} c_i c_j \Gamma_{ij} = c'\Gamma c.$$

Readers will recognize these last results as being the cornerstone of Markowitz portfolio theory, which assumes correlated, multi-variate Gaussian returns for the component assets of the portfolio.

## 3.9   Multi-Variate Gaussian Shift Theorem

The next result is the multi-variate version of the Gaussian Shift Theorem.

### THEOREM 3.4

Let $Z \sim N(0, 1; R)$. Then if $c$ is an $m-$dimensional constant vector and $F(Z)$ is a measurable scalar function of $Z$,

$$\boxed{\mathbb{E}\{e^{c'Z} F(Z)\} = e^{\frac{1}{2}c'Rc} \, \mathbb{E}\{F(Z + Rc)\}} \tag{3.27}$$

**PROOF**    The proof depends on the easily proved matrix identity

$$\tfrac{1}{2}(z - Rc)' R^{-1}(z - Rc) \equiv \tfrac{1}{2}z'R^{-1}z + \tfrac{1}{2}c'Rc - c'z.$$

We have also used the fact that $R$ is a symmetric matrix. It then follows that

$$
\mathbb{E}\{e^{c'Z} F(Z)\} = \int_{-\infty}^{\infty} e^{c'z} F(z) \frac{e^{-\frac{1}{2}z'R^{-1}z}}{(2\pi)^{m/2}\sqrt{\det(R)}}\,dz
$$

$$
= e^{\frac{1}{2}c'Rc} \int_{-\infty}^{\infty} \frac{e^{-\frac{1}{2}(z-Rc)'R^{-1}(z-Rc)}}{(2\pi)^{m/2}\sqrt{\det(R)}}\,F(z)\,dz
$$

$$
= e^{\frac{1}{2}c'Rc} \int_{-\infty}^{\infty} \frac{e^{-\frac{1}{2}u'R^{-1}u}}{(2\pi)^{m/2}\sqrt{\det(R)}}\,F(u+Rc)\,du
$$

$$
= e^{\frac{1}{2}c'Rc}\,\mathbb{E}\{F(Z+Rc)\},
$$

where in the third line above we have changed the variable of integration through the transformation $u = z - Rc$. This completes the proof.  □

### Example 3.2

For the case $m = 2$ we obtain the bivariate GST stated in theorem 3.3. With

$$
Z = \begin{bmatrix} X \\ Y \end{bmatrix}; \qquad c = \begin{bmatrix} a \\ b \end{bmatrix}; \qquad R = \begin{bmatrix} 1 & \rho \\ \rho & 1 \end{bmatrix}
$$

we find

$$
\tfrac{1}{2}c'Rc = \tfrac{1}{2}(a^2 + b^2 + 2\rho ab) = \tfrac{1}{2}c^2
$$

and

$$
Rc = \begin{bmatrix} 1 & \rho \\ \rho & 1 \end{bmatrix}\begin{bmatrix} a \\ b \end{bmatrix} = \begin{bmatrix} a + \rho b \\ b + \rho a \end{bmatrix} = \begin{bmatrix} a' \\ b' \end{bmatrix}
$$

and the result follows.  □

---

## 3.10   Multi-Variate Itô's Lemma and BS-PDE

This result has important implications for rainbow (multi-variate) exotic options considered later in this book. Suppose there are $n$ assets $A_i$, ($i = 1, 2, \ldots, n$) satisfying the $n$ sde's (each asset price follows gBm)

$$
dX_i = x_i(\mu_i\,dt + \sigma_i\,dB_i^P), \tag{3.28}
$$

where $x_i = X_i(t)$, ($\mu_i$, $\sigma_i$) are the growth rates and volatilities of the $n$ assets and $B_i^P$ are $n$ correlated Brownian motions under the real-world measure $P$. We assume

$$
\mathbb{E}_P\{dB_i^P\,dB_j^P\} = \rho_{ij}\,dt \tag{3.29}
$$

so that $\rho_{ij}$ is the correlation coefficient of the logarithmic returns of assets $A_i$ and $A_j$. Obviously $\rho_{ij} = 1$ for $i = j$.

The multi-variate Itô's Lemma then states:

**LEMMA 3.1**
*If $V(x_i, t)$ is a scalar function, twice differentiable in the $n$ variables $x_i$ and differentiable in time $t$, then $V(X_i, t)$ satisfies the sde*

$$dV = \left( V_t + \sum_{i=1}^{n} \mu_i x_i V_i + \tfrac{1}{2} \sum_{i=1}^{n} \sum_{j=1}^{n} \rho_{ij} \sigma_i \sigma_j x_i x_j V_{ij} \right) dt + \sum_{i=1}^{n} \sigma_i x_i V_i \, dB_i^P$$

(3.30)

*where*

$$V_i = \frac{\partial V}{\partial x_i} \qquad and \qquad V_{ij} = \frac{\partial^2 V}{\partial x_i \partial x_j}.$$

**Multi-Variate Black–Scholes PDE**

**THEOREM 3.5**
*Suppose we have a rainbow option on the $n$ assets with expiry payoff $V(x_i, T) = F(x_i)$. Then $V(x_i, t)$ satisfies the multi-variate BS-pde*

$$\boxed{V_t = rV - r \sum_{i=1}^{n} x_i V_i - \tfrac{1}{2} \sum_{i=1}^{n} \sum_{j=1}^{n} \rho_{ij} \sigma_i \sigma_j x_i x_j V_{ij}}$$

(3.31)

The proof is left as an exercise problem at the end of this chapter. The multi-variate BS-pde can also be written in vector-matrix form, using the gradient operator

$$\nabla_i = \mathrm{grad}_i = \frac{\partial}{\partial x_i}.$$

Then (3.31) has the equivalent representation

$$V_t = rV - r\boldsymbol{x}' \nabla V - \tfrac{1}{2} \mathrm{tr}\left[ (X\Gamma X)(\nabla \nabla' V) \right]$$

(3.32)

where $\Gamma_{ij} = (\Sigma' R \Sigma)_{ij} = \rho_{ij} \sigma_i \sigma_j$ is the covariance matrix of the asset logarithmic returns. Also $X = \mathrm{diag}(\boldsymbol{x})$ and $\Sigma = \mathrm{diag}(\boldsymbol{\sigma})$ are the diagonal matrices

$$X = \begin{bmatrix} x_1 & & & \\ & x_2 & & \\ & & \ddots & \\ & & & x_n \end{bmatrix} \qquad and \qquad \Sigma = \begin{bmatrix} \sigma_1 & & & \\ & \sigma_2 & & \\ & & \ddots & \\ & & & \sigma_n \end{bmatrix}$$

and $\mathrm{tr}(A) = \sum_i A_{ii}$ denotes the trace of the matrix $A = A_{ij}$. Note that the dash on a vector or matrix represents its transpose.

## 3.11    Linear Transformations of Gaussian RVs

The next Lemma, concerning transformations of multi-variate Gaussian random variables, is required to price the general multi-period, multi-asset binaries studies in Chapter 10.

**LEMMA 3.2**

*Let $Z$ be a standard $n-$dimensional Gaussian random vector $N(0, 1; R)$ with correlation matrix $R$ and let $B$ be an $(m \times n)$ constant matrix with rank $m \leq n$. Then $Y = BZ$ is an $m-$dimensional Gaussian random vector with zero mean and covariance matrix $\Gamma = BRB'$.*

*Let $D = \sqrt{diag(\Gamma)}$ denote the $(m \times m)$ diagonal matrix obtained from the square-root of principal diagonal of $\Gamma$. Then, these diagonal elements contain the standard deviation of each component of $Y$ and the corresponding correlation matrix will be*

$$C = corr\{Y\} = D^{-1}\Gamma D^{-1}; \qquad \Gamma = BRB'. \tag{3.33}$$

## 3.12    Summary

This chapter on Gaussian random variables completes Part I, the Technical Background, for the book. Again we have been mainly content with stating well-known results rather than offering formal proofs. Nevertheless, the reader should be aware, even from this short review, that Gaussian rv's have a very deep mathematical structure. Indeed, it is this very structure that ultimately allows closed form solutions to many exotic option pricing formulae. Further information on multi-variate Gaussian rv's can be found in the text by Long [54].

The importance of the Gaussian Shift Theorems (univariate, bivariate and multi-variate cases) cannot be under-estimated. Ultimately, the existence of these theorems is the reason we can price exotic options in the BS framework, without recourse to any formal integrations implied by the FTAP, or formal pde solving, required by the BS-pde.

Part II of the book is concerned with specific applications to pricing exotic options and derivatives. A good grasp of the Technical Background is essential to understanding how to price such securities within the framework of a

Black–Scholes economy. Many of the results presented in Part I will have a direct bearing on the specialized pricing techniques we introduce in subsequent chapters.

---

## Exercise Problems

1. (a) Show that if $Z \sim N(0,1)$ then

$$\mathbb{E}\{e^{aZ}\mathbb{I}(Z>b)\} = e^{\frac{1}{2}a^2}N[(a-b)].$$

(b) Hence show that if $a, b$ and $\sigma$ are positive constants, then

$$\mathbb{E}\{\left(ae^{\sigma Z-\frac{1}{2}\sigma^2} - b\right)^+\} = aN(d_1) - bN(d_2)$$

where $d_{1,2} = \frac{1}{\sigma}\log(a/b) \pm \frac{1}{2}\sigma$.

2. (a) Let $Y \sim N(\mu, \sigma^2)$. Derive the mgf $M_Y(\theta) = \mathbb{E}\{e^{\theta Y}\} = e^{\mu\theta+\frac{1}{2}\sigma^2\theta^2}$.

(b) Prove that mgf of the sum of two independent rv's is equal to the product of their mgf's. Use this result to show that if $Y_i \sim N(\mu_i, \sigma_i)$ are independent Gaussians, then

$$Y = \sum_i c_i Y_i \sim N\left(\sum_i c_i\mu_i, \sum_i c_i^2\sigma_i^2\right).$$

3. Let $\mathcal{L}(y|\mu, \sigma^2)$ denote the cdf of the log-normal distribution. Derive the log-normal results:

$$\mathbb{E}\{Y^k\} = \mu_k = e^{k\mu+\frac{1}{2}k^2\sigma^2}$$
$$\mathbb{E}\{Y^k\mathbb{I}(Y<x)\} = \mu_k\,\mathcal{L}(x|\mu + k\sigma^2, \sigma^2)$$
$$\mathbb{E}\{Y^k\mathbb{I}(Y>x)\} = \mu_k\,\mathcal{L}(x^{-1}|-(\mu + k\sigma^2), \sigma^2).$$

Deduce that $\mathcal{L}(x^{-1}|\mu, \sigma^2) = 1 - \mathcal{L}(x|-\mu, \sigma^2)$.

4. (a) Let $\mathcal{N}(x, y; \rho)$ denote the bivariate normal cdf. Show that

$$\int_{-\infty}^{b} N\left(\frac{a - \rho x}{\sqrt{1-\rho^2}}\right) \phi(x)\, dx = \mathcal{N}(a, b; \rho).$$

(b) Use part(a) to prove the following special cases:

$$\mathcal{N}(x, y; 0) = \mathcal{N}(x)\mathcal{N}(y)$$
$$\mathcal{N}(x, y; 1) = \mathcal{N}[\min(x, y)]$$
$$\mathcal{N}(x, y; -1) = \begin{cases} [\mathcal{N}(x) - \mathcal{N}(-y)] & \text{if } x + y > 0 \\ 0 & \text{otherwise.} \end{cases}$$

5. Prove that if $X_i$ are correlated Gaussian rv's with mean $\mu_i$, variances $\sigma_i$ and covariance matrix $\Gamma_{ij} = \sigma_i\sigma_j\rho_{ij}$ ($\rho_{ij}$ are the correlation coefficients), then $Y = \sum_i c_i X_i$ is univariate Gaussian with mean and variance given respectively by

$$\mu = \sum_{i=1}^{d} c_i\mu_i = \boldsymbol{c}'\boldsymbol{\mu} \qquad \text{and} \qquad \sigma^2 = \sum_{i=1}^{d}\sum_{j=1}^{d} c_i c_j \Gamma_{ij} = \boldsymbol{c}'\boldsymbol{\Gamma}\boldsymbol{c}.$$

*Hint*: Find the mgf of $Y$ using the multi-variate Gaussian Shift Theorem.

6. If $X, Y$ are joint $N(0, 1; \rho)$ rv's with correlation coefficient $\rho$, prove that

$$\mathbb{E}\{Y f(X)\} = \rho\,\mathbb{E}\{f'(X)\}$$

where $f(X)$ is any differentiable, measurable function of $X$.

7. Let $X_t, Y_t$ satisfy the sde's

$$dX_t = X_t(\mu_x dt + \sigma_x dB_t); \qquad dY_t = Y_t(\mu_y dt + \sigma_y dB_t).$$

Use the 2D Itô's Lemma, to derive the alternate form of the Product and Quotient Rules,

$$d(X_t Y_t) = X_t Y_t[(\mu_x + \mu_y + \sigma_x\sigma_y)dy + (\sigma_x + \sigma_y)dB_t]$$
$$d(X_t/Y_t) = (X_t/Y_t)[(\mu_x - \mu_y - \sigma_y(\sigma_x - \sigma_y))dt + (\sigma_x - \sigma_y)dB_t]$$

8. Prove the result of proposition 3.1 for two correlated Gaussian rv's $(X, Y)$,

$$\mathbb{E}\left\{\left(ae^{\sigma_1 X - \frac{1}{2}\sigma_1^2} - be^{\sigma_2 Y - \frac{1}{2}\sigma_2^2}\right)^+\right\} = a\mathcal{N}(d_1) - b\mathcal{N}(d_2)$$

with

$$d_{1,2} = \frac{1}{\sigma}\log(a/b) \pm \tfrac{1}{2}\sigma \qquad \text{and} \qquad \sigma = \sqrt{\sigma_1^2 + \sigma_2^2 - 2\rho\sigma_1\sigma_2}.$$

9. If $Z \sim N(0, 1)$, prove that

$$\mathbb{E}\{\mathcal{N}(a_1 Z + b_1)\mathcal{N}(a_2 Z + b_2)\} = \mathcal{N}(c_1, c_2; \rho)$$

where

$$c_i = \frac{b_i}{\sqrt{1+a_i^2}}; \qquad \rho = \frac{a_1 a_2}{\sqrt{1+a_1^2}\sqrt{1+a_2^2}}.$$

Hence determine $\mathbb{E}\{\mathcal{N}(aZ+b)\}$.

10. (a) Prove the identity

$$\mathcal{N}(a,b;\rho) = \mathcal{N}(a,c;\rho_1) + \mathcal{N}(b,-c;\rho_2)$$

where

$$\rho_1 = \frac{-\sigma_1 + \rho\sigma_2}{\sigma}; \qquad \rho_2 = \frac{-\sigma_2 + \rho\sigma_1}{\sigma}$$

$$c = \frac{\sigma_2 b - \sigma_1 a}{\sigma}; \qquad \sigma^2 = \sigma_1^2 + \sigma_2^2 - 2\rho\sigma_1\sigma_2.$$

*Hint:* Let $Y = \alpha X + \beta$ be a line passing through $(a,b)$, so the rectangular region $X < a$, $Y < b$ is divided into two triangular regions.

*Note:* This identity was discovered when comparing different but equivalent representations of an option price (see Section 6.5 for details).

(b) Hence, prove

$$\mathcal{N}(a,b,;\rho) = \mathcal{N}(a,c;-\beta) + \mathcal{N}(b)\mathcal{N}(-c)$$

where

$$c = \frac{\rho b - a}{\beta}; \qquad \beta = \sqrt{1-\rho^2}.$$

11. Derive the multi-variate version (3.31) and its vector counterpart (3.32), of the BS-pde for the price of a derivative security written on $n$ assets, whose dynamics follow $n$ correlated geometric Brownian motions.

—ooOoo—

# Part II

# Applications to Exotic Option Pricing

# Chapter 4

## Simple Exotic Options

We refer to *simple* exotic options as those that are path-independent and involve only a single underlying asset $X$ and a single expiration date $T$. Generally, such simple exotic options will have a payoff function $V(x, T) = f(x)$ which is different from a vanilla European call or put option. Often the function $f(x)$ can be decomposed into a sum of simpler contracts that have already been priced. In that case, the price of the derivative is then, by the Principle of Static Replication, equal to the price of the replicating portfolio. We shall meet many examples of this basic idea, not only in this chapter, but also in later chapters on more complex derivatives.

Two useful collections of papers and technical articles on exotic options can be found in the *Risk* publications: Jarrow (ed.) [38] and Lipton (ed.) [53].

This chapter begins with a definition of a first-order binary option and shows how it can be priced in a standard BS economy. Binary options figure prominently in this book, and for a very good reason. They are indeed the building blocks of many, even very complex, exotic options. The chapter also includes several examples of single asset, single period derivatives, which are by far the easiest to price. One of these applications is to pricing corporate (or defaultable) bonds under Merton's firm value model. The topic of credit derivatives is a vast one, so we are content to analyze only the simplest cases of default. We also include in this chapter the pricing of derivatives on a binomial tree. Results of this presentation are used elsewhere in the book.

The final example investigates a fascinating instrument belonging to the class of options on a traded account. Only a simple example of this option is considered here. The more complex examples, which include passport and vacation options, involve stochastic optimization methods and are therefore outside the scope of this book. Details can be found in Hyer et al. [35] and Shreve and Vecer [69].

## 4.1 First-Order Binaries

We mentioned in the first chapter that binary options play an important role in pricing many derivative securities, since they are often the basic building blocks of the derivatives in question. This idea is of course not new. It was perhaps first used in the above sense by Rubinstein and Reiner [65] and later elaborated on by Ingersoll [36]. Later in the book we shall introduce second-order binaries and the also the general multi-period, multi-asset binary, which we refer to as the M-binary.

There are two generic first-order binary options, which can be represented as follows.

**DEFINITION 4.1**   *We say that a derivative on a single underlying asset is an up/down-type first-order binary option with payoff $f(x)$ and exercise price $\xi > 0$, if at time $T$ in the future it pays*

$$V_\xi^s(x, T) = f(x)\,\mathbb{I}(sx > s\xi) \tag{4.1}$$

*where $s = $ "$+$" for the up-type binary and $s = $ "$-$" for the down-type binary.*

Thus an up-type binary option pays the amount $f(x)$ at expiry $T$ iff the asset price at expiry is *above* the exercise price $\xi$. The down-type binary option pays the amount $f(x)$ at expiry $T$ iff the asset price at expiry is *below* the exercise price $\xi$. Note that when $s$ is a plus-sign, the indicator function reads: $\mathbb{I}(x > \xi)$ and when $s$ is a minus-sign, it reads $\mathbb{I}(x < \xi)$.

**REMARK 4.1**   Up-type and down-type binaries have also been referred in the literature as *call-type* and *put-type* binaries respectively. We prefer to use the terms up and down, since for more complex derivatives, the connection between calls and puts can become very tenuous. □

In equations (1.4) and (1.5) of Chapter 1, we introduced the two best known first-order binaries, namely the asset binary with payoff $f(x) = x$, i.e., one unit of the underlying asset, and the bond binary with payoff $f(x) = 1$, i.e., one unit of the relevant currency (which wlog we take as the dollar). We write the pv of the asset and bond binaries as $A_\xi^\pm(x, t)$ and $B_\xi^\pm(x, t)$ respectively and note that, at expiry $T$

$$A_\xi^+(x, T) = x\,\mathbb{I}(x > \xi); \qquad A_\xi^-(x, T) = x\,\mathbb{I}(x < \xi)$$
$$B_\xi^+(x, T) = \mathbb{I}(x > \xi); \qquad B_\xi^-(x, T) = \mathbb{I}(x < \xi).$$

The next section shows how to price these binaries in a BS economy, without performing any integrations or transforming a pde.

## 4.2 BS-Prices for First-Order Asset and Bond Binaries

Since the price calculation is somewhat simpler for bond binaries, we start with these and do the asset binaries afterwards. The following Lemma will help in the calculation.

**LEMMA 4.1**
If $X_T = x\, e^{(r-\frac{1}{2}\sigma^2)\tau + \sigma\sqrt{\tau}Z}$, where $\tau = (T-t)$ and $Z \sim N(0,1)$, then $X_T \gtrless k$ according as $Z \gtrless -d'_k(x,\tau)$ where

$$d'_k(x,\tau) = \frac{\log(x/k) + (r - \frac{1}{2}\sigma^2)\tau}{\sigma\sqrt{\tau}}. \tag{4.2}$$

The proof of this result is just simple algebra. $\qquad\qquad\square$

**The bond binary**
Using the FTAP in the BS representation (1.46), we price the up-type bond binary as follows.

$$\begin{aligned}
B^+_\xi(x,t) &= e^{-r\tau}\,\mathbb{E}_Q\{\mathbb{I}(X>\xi)\,|\,\mathcal{F}_t\} \\
&= e^{-r\tau}\,\mathbb{E}_Q\{\mathbb{I}(Z>-d'_\xi(x,\tau))\} \\
&= e^{-r\tau}\,\mathbb{E}_Q\{\mathbb{I}(Z<d'_\xi)\} \\
&= e^{-r\tau}\,\mathcal{N}(d'_\xi).
\end{aligned}$$

The second, third and fourth lines above follow from lemma (4.1) and equations (3.7) and (3.4) respectively. The corresponding down-type bond binary follows similarly and we find $B^-_\xi(x,t) = e^{-r\tau}\,\mathcal{N}(-d'_\xi)$. We can combine these two expressions into a single one, and then obtain

$$\boxed{B^s_\xi(x,t) = e^{-r\tau}\,\mathcal{N}(sd'_\xi)} \tag{4.3}$$

This is the unique, arbitrage-free, BS price of the up $(s = +)$ and down $(s = -)$ type, 1st-order bond binary of exercise price $\xi$.

**The asset binary**
The corresponding calculation for the asset binary is a little more complicated,

in that it uses the GST. Starting from (1.46) again, we obtain:

$$A_\xi^+(x,t) = e^{-r\tau} \, \mathbb{E}_Q\{X\mathbb{I}(X>\xi) \,|\, \mathcal{F}_t\}$$
$$= x \, e^{-\frac{1}{2}\sigma^2\tau} \, \mathbb{E}_Q\{e^{\sigma\sqrt{\tau}Z}\mathbb{I}(Z>-d'_\xi)\}$$
$$= x \, \mathbb{E}_Q\{\mathbb{I}(Z + \sigma\sqrt{\tau} > -d'_\xi)\}$$
$$= x \, \mathbb{E}_Q\{\mathbb{I}(Z < d'_\xi + \sigma\sqrt{\tau})\}$$
$$= x \, \mathcal{N}(d_\xi).$$

The third line has used the GST given by equation (3.9), while the remaining calculations follow the same pattern as for the bond binary. In addition we have defined the new parameter

$$d_\xi(x,\tau) = d'_\xi(x,\tau) + \sigma\sqrt{\tau} = \frac{\log(x/\xi) + (r + \frac{1}{2}\sigma^2)\tau}{\sigma\sqrt{\tau}}. \tag{4.4}$$

The corresponding down-type asset binary follows similarly and we find $A_\xi^-(x,t) = x\mathcal{N}(-d_\xi)$. As for the bond binaries, we can combine these two expressions to obtain

$$\boxed{A_\xi^s(x,t) = x\,\mathcal{N}(sd_\xi)} \tag{4.5}$$

This is the unique, arbitrage-free, BS price of the up ($s = +$) and down ($s = -$) type, 1st-order asset binary of exercise price $\xi$.

---

## 4.3   Parity Relation

Let $V_0(x,t)$ denote the derivative price when the payoff at time $T$ is $V_0(x,T) = f(x)$. This is the same payoff as for the binary options stated in (4.1), except that no exercise condition is specified. We might call this derivative the *standard derivative*, with the same payoff as the binary options which also pay $f(x)$, but only when the asset price satisfies the exercise condition.

It is clear from the PSR and the identity

$$f(x)\mathbb{I}(x>\xi) + f(x)\mathbb{I}(x<\xi) \equiv f(x),$$

that for all $t < T$, we must have

$$\boxed{V_\xi^+(x,t) + V_\xi^-(x,t) = V_0(x,t)} \tag{4.6}$$

This is the parity relation for first-order binary options.

In the special cases of asset and bond binaries, it reduces to the results we derived in example 1.2. These results also follow immediately from the (Gaussian) symmetry property (3.8), since

$$A_k^+(x,t) + A_k^-(x,t) = x[\mathcal{N}(d_k) + \mathcal{N}(-d_k)] = x$$

and

$$B_k^+(x,t) + B_k^-(x,t) = e^{-r\tau}[\mathcal{N}(d_k') + \mathcal{N}(-d_k')] = e^{-r\tau}.$$

**REMARK 4.2**  There is one technical point associated with the parity relation (4.6) that needs to be clarified. More correctly, the identity which gives rise to the parity relation, is:

$$\mathbb{I}(x>\xi) + \mathbb{I}(x\leq\xi) \equiv 1 \quad \text{or} \quad \mathbb{I}(x\geq\xi) + \mathbb{I}(x<\xi) \equiv 1.$$

However, since in the BS framework, the stock price $x$ is a *continuous* variable, there is no loss of generality by taking $\mathbb{I}(x>\xi) + \mathbb{I}(x<\xi) \equiv 1$. ⬚

---

## 4.4  European Calls and Puts

Now that we have determined the pv's of the asset and bond binaries in the BS economy, we are in a position to state the celebrated results referred to as the BS option pricing formulae for European calls and puts.

Using the PSR formulae (1.6) and (1.7) and equations (4.3) and (4.5) for the asset and bond binaries, we immediately arrive at

$$\boxed{C_k(x,t) = x\,\mathcal{N}(d_k) - ke^{-r\tau}\,\mathcal{N}(d_k')} \tag{4.7}$$

for the call option, and

$$\boxed{P_k(x,t) = -x\,\mathcal{N}(-d_k) + ke^{-r\tau}\,\mathcal{N}(-d_k')} \tag{4.8}$$

for the put option.

**REMARK 4.3**  As a matter of notation, we sometimes find it more convenient to write the prices of calls and puts (and other derivatives as well) using the *relative* time $\tau = (T-t)$, rather than *absolute* calendar time $t$. Thus we often write $V(x,\tau)$ in place of $V(x,t)$. The context will generally determine which notation we are using, and so ensure that no ambiguity exists. ⬚

## 4.5   Gap and Q-Options

Simple extensions of standard European calls and puts are the contracts known as *gap calls* and *gap puts*. These have also been referred to as *threshold* calls and puts in the literature. They are essentially calls and puts in which the strike price is different from their exercise price. Thus if $k$ denotes the strike price and $\xi$ the exercise price, then the gap call and gap put payoff functions are respectively

$$C(x,T) = (x-k)\mathbb{I}(x>\xi); \qquad P(x,T) = (k-x)\mathbb{I}(x<\xi). \qquad (4.9)$$

It follows that the corresponding prices at time $t < T$ will be

$$C(x,t) = A_\xi^+(x,\tau) - kB_\xi^+(x,\tau) \qquad (4.10)$$

$$P(x,t) = -A_\xi^-(x,\tau) + kB_\xi^-(x,\tau). \qquad (4.11)$$

Obviously these simple extensions reduce to the usual expressions for for European calls and puts when $\xi = k$, that is, when the strike and exercise prices coincide.

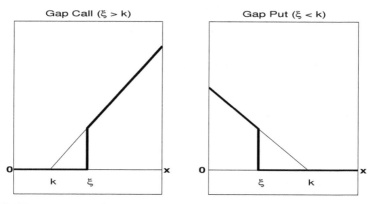

**FIGURE 4.1:**   Left panel shows payoff of a gap call with $\xi > k$. Right panel shows payoff of a gap put with $\xi < k$.

Generally, we will have $\xi \geq k$ for a gap call option and $\xi \leq k$ for a gap put option, as shown in Figure 4.1. These conditions, which are not an essential requirement, will ensure however, that the payoffs at expiry will then always be non-negative. The quantity $|\xi - k|$ is called the gap, and when the gap is non-zero, the expiry payoff will actually be a discontinuous function.

We shall see in later chapters, that many exotic options are built from underlying calls and puts. These include compound options, barrier options, lookback options and many others. Hence, to simplify the notation, which can quickly get out of hand, it helps to introduce a new symbol to represent either a call-like option or a put-like option. We shall refer to this representation as a $Q$-option. Such $Q$-0ptions will appear as 1st-order $Q$-options, 2nd-order $Q$-options, and even higher order or multi $Q$-options can be similarly defined.

**DEFINITION 4.2**    *A first order $Q$-option of exercise price $\xi$ and strike price $k$ is the derivative with expiry $T$ payoff*

$$Q_\xi^s(x,T;k) = s(x-k)\,\mathbb{I}(sx>s\xi); \qquad (s=\pm). \tag{4.12}$$

Thus $Q_\xi^+(x,t;k)$ denotes the pv of a strike $k$, exercise $\xi$ gap call option, while $Q_\xi^-(x,t;k)$ denotes the pv of a similar gap put option. Being a portfolio of asset and bond binaries, the pv of a $Q$-option can be written down by inspection as,

$$\boxed{\begin{aligned} Q_\xi^s(x,t;k) &= sA_\xi^s(x,t) - skB_\xi^s(x,t) \\ &= sx\mathcal{N}(sd_\xi) - ske^{-r\tau}\mathcal{N}(sd_\xi') \end{aligned}} \tag{4.13}$$

Second order $Q$-options are defined in the next chapter on dual-expiry options.

---

## 4.6   Capped Calls and Puts

European calls have potentially an unlimited payoff at expiry, if the underlying asset price finishes well above the strike price. This feature accounts for part of the premium of a call option. One way of reducing the premium is to cap the call option payoff at expiry. The payoff will then be given by

$$C(x,T) = \max[(x-k)^+,\ c]$$

where $c>0$ is the cap price of the option. This means that if the asset price at expiry exceeds $(k+c)$, then the payoff is capped at the value $c$. It is not difficult to show the capped payoff can also be written in the equivalent form

$$C(x,T) = (x-k)^+ - (x-k-c)^+.$$

This payoff is a simple portfolio of two standard call options: a long call with strike price $k$, and a short call with strike strike price $(k+c)$. The pv of the capped call is therefore given by the expression

$$\boxed{C(x,t) = C_k(x,\tau) - C_{k+c}(x,\tau)} \tag{4.14}$$

Here $C_k(x, \tau)$ denotes the pv of a standard European call option of strike price $k$ and time $\tau = (T - t)$ remaining to expiry.

The corresponding case of a put option capped at the level $p < k$, has payoff and pv given by the expressions below.

$$P(x, T) = \max[(k - x)^+, \; p] = (k - x)^+ - (k - p - x)^+$$

and

$$\boxed{P(x, t) = P_k(x, \tau) - P_{k-p}(x, \tau)} \tag{4.15}$$

**REMARK 4.4** Capped calls and puts are traded in the market place, but more usually as American style options. Thus, the capped calls and puts are typically exercised as soon as the cap value is reached. The capped calls and puts considered above are examples of European style capped options. ▯

---

## 4.7 Range Forward Contracts

These are simple examples of portfolios of calls and puts, which are popular in the FX markets. A long, range forward contract is the portfolio:

$$\mathcal{R} = \begin{cases} \text{long} \;\; 1 \; \text{put of strike} \;\; k_1 \\ \text{short} \; 1 \; \text{call of strike} \;\; k_2 \end{cases}$$

In practice, both options expire at $T$, and $k_1 < k_2$. Further, the strikes are chosen at initiation (at $t = 0$ say) so that the two options have the same premium. Hence, like other forward contracts, the net cost is initially zero.

The payoff at expiry is given by the expression

$$R(x, T) = (k_1 - x)^+ - (x - k_2)^+$$

which is seen to have the profile: $(k_1 - x)$ for $x < k_1$, zero for $k_1 < x < k_2$ and $(k_2 - x)$ for $x > k_2$. The pv of the range forward is obviously given by the expression

$$\boxed{R(x, t) = P_{k_1}(x, \tau) - C_{k_2}(x, \tau)} \tag{4.16}$$

The range forward value at time $0 < t \leq T$ may thus be either positive or negative. A short, range forward contract simply reverses the two trades above, so that the low strike put is held short and the high strike call is held long. The pv will then simply be the negative of the long range forward value.

Clearly, there are many other simple exotics which are just portfolios of European calls and puts. Examples include straddles, strangles, strips, straps and spreads. This class of exotic options is sometimes referred to as a *packaged option* or even more simply as a *package*. The pv of such a package is clearly just the pv of the replicating portfolio.

## 4.8 Turbo Binary

A generalization of the asset and bond binaries is the *turbo* or *power* binary, with payoff function $f(x) = x^p$ (if exercised) for some real value of $p$, which may be positive or negative. That is,

$$P_\xi^s(x, T; p) = x^p \, \mathbb{I}(sx > s\xi).$$

We leave it as an exercise problem at the end of this chapter, for the reader to show that the up ($s = +$) and down ($s = -$) type turbo binaries have pv's:

$$\boxed{P_\xi^s(x, t; p) = x^p \, e^{\lambda(p)\tau} \, \mathcal{N}[s(d_\xi' + p\sigma\sqrt{\tau})]} \tag{4.17}$$

where $\lambda(p)$ is the quadratic function of $p$, defined by

$$\lambda(p) = \tfrac{1}{2}\sigma^2 p^2 + (r - \tfrac{1}{2}\sigma^2)p - r = (\tfrac{1}{2}\sigma^2 p + r)(p - 1). \tag{4.18}$$

Observe that the special cases $p = 1$ and $p = 0$, reduce to the asset and bond binaries respectively. That is,

$$P_\xi^s(x, t; 1) = A_\xi^s(x, t) \qquad \text{and} \qquad P_\xi^s(x, t; 0) = B_\xi^s(x, t)$$

as indeed they must. The corresponding parity relation for up-down turbo binaries is

$$P_\xi^+(x, t; p) + P_\xi^-(x, t; p) = x^p \, e^{\lambda(p)\tau}. \tag{4.19}$$

Clearly, $P_\xi(x, t; p) = x^p \, e^{\lambda(p)\tau}$ is the pv of a contract that pays $x^p$ at expiry, independent of any exercise condition (see Q1(a) in Exercise Problems).

## 4.9 The Log-Contract

As an example of an option which is not a package (at least not a finite one), we consider next the case of the log-contract. This is a European derivative with expiry $T$ payoff function $V(x, T) = \log(x/k)$, where $k$ is a positive constant. Thus the payoff is positive if at expiry, $x > k$ and will be negative (i.e., the holder must pay this amount) if $x < k$. This contract was analyzed in detail by Neuberger [59] who showed that the contract permits investors to trade views on future volatility. This idea was further expanded with regard to pricing volatility and variance swaps in Demeterfi et al. [19].

To price the log-contract in the BS-framework is an easy task. Using the FTAP (1.46) we obtain

$$
\begin{aligned}
V(x,t) &= e^{-r\tau} \mathbb{E}_Q\{\log(X/k)\}; \quad X = xe^{(r-\frac{1}{2}\sigma^2)\tau + \sigma\sqrt{\tau}Z} \\
&= e^{-r\tau} \mathbb{E}_Q\{\log(x/k) + (r - \tfrac{1}{2}\sigma^2)\tau + \sigma\sqrt{\tau}Z\} \\
&= e^{-r\tau} \left[\log(x/k) + (r - \tfrac{1}{2}\sigma^2)\tau\right]
\end{aligned}
\tag{4.20}
$$

since $\mathbb{E}_Q\{Z\} = 0$. The question is often asked: is it possible to statically hedge an exotic option with traded contracts? It would seem impossible to do so for the log-contract, but if we permit continuously weighted portfolios, we may be able to achieve this outcome. Of course such portfolios are infinite dimensional, so there would be great impracticalities to implement such a hedge.

Consider a continuously weighted portfolio $\mathcal{P}$ of European calls and puts in which the weights depend only on the strike price $y$. The puts have weights $w_1(y) = y^{-2}\mathbb{I}(y < k)$, while the calls have weights $w_2(y) = y^{-2}\mathbb{I}(y > k)$, for some fixed positive constant $k$. In both cases, the weights are proportional to the inverse square of the strike. The payoff at expiry of such a portfolio is then given by

$$
\mathcal{P}(x,T) = \int_0^k (y-x)^+ \frac{dy}{y^2} + \int_k^\infty (x-y)^+ \frac{dy}{y^2}.
$$

The two integrals are elementary, and result in

$$
\begin{aligned}
I_1 &= \left[\log(x/k) + k^{-1}(x-k)\right]\mathbb{I}(x < k) \\
I_2 &= \left[\log(x/k) + k^{-1}(x-k)\right]\mathbb{I}(x > k)
\end{aligned}
$$

respectively. So when added together, we obtain

$$
\mathcal{P}(x,T) = I_1 + I_2 = \log(x/k) + k^{-1}(x-k).
$$

It follows that $\log(x/k) = \mathcal{P}(x,T) - k^{-1}(x-k)$ and therefore we have indeed achieved our aim of replicating the payoff of a log-contract with traded securities. The log-contract is statically hedged, by going long the continuous portfolio $\mathcal{P}$ and short $k^{-1}$ units of a forward contract with delivery price $k$. This observation is the basis for pricing variance swaps, but of course in practice, some approximation has to be made to the continuously weighted portfolio $\mathcal{P}$.

## 4.10 Pay-at-Expiry and Money-Back Options

**Pay-at-expiry option**
There exist in the market certain options which cost nothing at initiation time, but a premium is paid at the expiry date $T$ of the option. This premium is paid, however, only on the condition that the option finishes in-the-money (itm) at time $T$. If the option finishes out-of-the-money (otm) at time $T$, then it expires worthless. Such options are called *pay-at-expiry* or pax options. They are also known by the names *contingent premium* options and *cash-on-delivery* options. The best known example of a pax option is a forward contract on an underlying asset. In most forwards, the delivery or settlement price is set to ensure that the initial cost is zero.

To price pax options, let $p$ denote the premium paid at time $T$ and let $t_0$ denote the initiation date of the option. Let $x_0$ be the asset price at time $t_0$. Then the payoff of a general pax option, which gives its holder an amount $F(x)$ at time $T$, can be written as

$$\boxed{V(x,T) = [F(x) - p]\,\mathbb{I}(F(x) > 0)} \qquad (4.21)$$

This payoff captures the conditions stated above. The holder receives $F(x)$, pays the premium $p$, but only if $F(x)$ at expiry is itm, i.e., positive. The premium $p$ is determined by the condition of zero cost at initiation, i.e., from the condition $V(x_0, t_0) = 0$.

The total pax payoff is therefore a binary option, possibly either up or down type depending on the exercise condition $F(x) > 0$. The actual payoff, however, may be positive or negative depending on whether $F(x) \gtrless p$ at expiry.

**Example 4.1**
Consider the case of a pax call option of exercise price $k$ and expiry date $T$. According to (4.21), the payoff at $T$ will be

$$V(x,T) = [(x-k)^+ - p]\,\mathbb{I}(x > k),$$

since the call finishes itm when $x > k$. This payoff simplifies to

$$V(x,T) = (x-k)^+ - p\,\mathbb{I}(x > k).$$

Hence the pv, for $t_0 < t < T$, of this payoff can be written in the form

$$V(x,t) = C_k(x,t) - p\,B_k^+(x,t).$$

The premium $p$ is determined from the condition $V(x_0, t_0) = 0$ and reduces to

$$p = \frac{C_k(x_0, t_0)}{B_k^+(x_0, t_0)} = \frac{A_k^+(x_0, t_0)}{B_k^+(x_0, t_0)} - k.$$

The second expression for $p$ derives from the decomposition (1.6) for a European call option. ▯

It is also possible to price a *reverse* pax option, which requires a payment at expiry only if the option finishes out-of-the-money. Here the payoff function will be

$$V(x, T) = F(x)\mathbb{I}(F(x) > 0) - p\,\mathbb{I}(F(x) < 0).$$

**Money-Back option**

The *money-back* option is a related exotic option, which demands a premium at initiation, but repays the premium at expiry if the option finishes itm. Hence a money-back call option, for example, would have an expiry payoff

$$V(x, T) = (x - k)^+ + p_0\mathbb{I}(x > k),$$

where $p_0$ is the initial premium. The pv of this option, including now both $t$ and $T$ for clarity, is just

$$V(x, t) = C_k(x, t; T) + p_0 B_k^+(x, t; T) \tag{4.22}$$

which demonstrates, as expected, that the money-back call is worth more than a standard call option. To determine the initial ($t_0$ say) premium, we simply evaluate (4.22) at the initial time, to obtain $p_0 = C_k(x_0, t_0; T) + p_0 B_k^+(x_0, t_0; T)$. It follows, that

$$p_0 = \frac{C_k(x_0, t_0; T)}{1 - B_k^+(x_0, t_0; T)}. \tag{4.23}$$

Of course $p_0 > C_k(x_0, t_0; T)$ since $0 < B_k^+(x_0, t_0; T) < 1$ for all values of its argument with $t_0 < T$.

See Zhang [78] for more details on pax and money-back options.

---

## 4.11   Corporate Bonds

In this section we present what has come to be known as Merton's Firm Model for assessing the credit risk of a company. This model assumes that the company issues debt in the form of a zero-coupon bond $D_k(x, t)$ of face value $k$, which becomes due at sometime $T$ in the future. The company defaults at time $T$ if the value $x$ of its assets is less than the promised amount $k$ of the debt. Such a bond is obviously a defaultable bond and hence the notation $D_k(x,t)$ to denote its value at any time $t \leq T$. We can price $D_k(x,t)$ in the BS-framework assuming that the firm's asset process $X$ satisfies the usual assumptions of gBm. In this case, the arbitrage free price of the debt $D_k(x,t)$ at any time $t < T$ will satisfy the BS-pde with expiry condition

$$D_k(x,T) = \min(x,\ k). \tag{4.24}$$

This payoff captures the notion of default, since if $x < k$ at time $T$, the company's assets will be liquidated to pay out the bondholders, who have first claim ahead of the equity or shareholders. If, on the other hand $x > k$ at time $T$, the firm survives and the bondholders can be paid in full.

The payoff (4.24) can be expressed in the equivalent form

$$D_k(x,T) = \min(x,k) = k - (k-x)^+$$

which shows it can be replicated by a default-free bond of face-value $k$ and a short put option of strike price $k$. Hence, by the Principle of Static Replication, the value of the debt at time $t < T$ is given by

$$D_k(x,t) = B_k(t,T) - P_k(x,\tau). \tag{4.25}$$

Here $B_k(t,T)$ denotes the current price of the default-free bond, which obviously depends on the term structure of interest rates. For the simplest case of constant continuous compounding at the risk-free rate $r$, we would take $B_k(t,T) = ke^{-r\tau}$, with $\tau = (T-t)$. This is an interesting result, because it shows that the corporate (or defaultable) bond is cheaper than a corresponding treasury (or default-free) bond with the same face-value. The difference is equal to the price of a put option on the assets of the issuing company.

It is now possible to price the company's equity. According to the Modigliani–Miller Theorem, a company's assets are invariant to its debt-equity structure. This means that if $E(x,t)$ denotes the equity at time $t$, then

$$E(x,t) + D_k(x,t) = x.$$

Hence, issuing more or less debt means reducing or increasing the firm's equity without changing the total asset value. It follows that at the maturity $T$ of the debt, we have

$$E(x,T) = x - D_k(x,T) = x - \min(x,k) = (x-k)^+ .$$

The conclusion is that the equity is equivalent to a call option of strike price $k$ (the face-value of the debt) on the firm's assets.

Merton's Firm Model above has been criticized on several grounds. First, that a single debt structure and recovery process is too simplistic; second, that the credit spreads predicted by the model are too low; and third, that default is basically an unpredictable event. The Merton model, being a diffusion model, will in general contain few surprises. If the firm is close to default, the probability of default increases. To handle this latter problem, the more modern approach to pricing defaultable bonds utilizes intensity based models which permit pure jumps in the underlying asset process.

### Subordinated debt

Despite the above criticisms, it is not too difficult to include more complex debt structures, such as subordinated debt. Consider the case of two issues of debt with face value $k_1$ and $k_2$, and to keep thing simple, assume both mature at the same date $T$. Let $k_1$ be the senior debt and $k_2$ the junior debt. If the company defaults at time $T$ (i.e., if $x < k$; $k = k_1 + k_2$), the senior debt must be paid out in full (if possible) before any junior debt is paid. The value of the two debts at maturity will then be given by,

$$D_1(x, T) = \min(x,\ k_1)$$
$$D_2(x, T) = \min(x - k_1,\ k_2)\,\mathbb{I}(x > k_1).$$

Since $D_1(x, T) = x - (x - k_1)^+$ and $D_2(x, T) = (x - k_1)^+ - (x - k)^+$ where $k$ is the total debt, we can price the debts at any time $t < T$ as:

$$D_1(x, t) = x - C_{k_1}(x, \tau) \tag{4.26}$$
$$D_2(x, t) = C_{k_1}(x, \tau) - C_k(x, \tau). \tag{4.27}$$

Here we have written the debts in terms of call options, but because of put-call parity we could just as well have written them in terms of put options. Note that the total debt is

$$D(x, t) = D_1 + D_2 = x - C_k(x, \tau) = x - E(x, t),$$

where $E(x, t)$ is the firm's equity and again we have consistency with the Modigliani–Miller Theorem. The inclusion of more than two debt issues is readily obtained by extending the above analysis.

Attempts to make the Merton model's default process more realistic lead to more complex structures. For example, suppose $x = b$ denotes a default barrier which represents the minimum asset value below which the firm is technically insolvent. If the firm's total asset value falls below this barrier at *any* time before the maturity date of the debt, then the firm defaults at this time and the bondholders are immediately compensated by the remaining asset value after liquidation. This model of default, in the presence of a default barrier, is obviously more complex than the simple default-at-maturity models considered above. Indeed, the price of the debt now becomes a barrier option pricing problem, which is consider in Chapter 7 of this book.

This brief look at pricing defaultable corporate bonds belongs to the field of pricing credit derivatives in general. This is a vast topic on which several books have been written (for example see Schönbucher [66]) and is mostly beyond the scope of this particular book.

---

## 4.12   Binomial Trees

A simple model for pricing derivative securities is the binomial model, which we briefly looked at in example 1.4. Although these models are basically computational models, they are easy to construct and can often price derivatives for which there are no closed-form solutions in the Black–Scholes framework. Examples include American and Bermudan options, which are readily computed on a binomial tree. Binomial models were first introduced by Cox, Ross and Rubinstein in [17] as an *ersatz* for understanding the Black–Scholes model. However, the CRR model, as it is sometimes called, turned out to be a very useful one in its own right, to the extent that most financial institutions still maintain a binomial pricing engine for risk analysis purposes.

We shall consider in this section some rudimentary aspects of the binomial model and how the model relates to the Black–Scholes framework for pricing derivatives. Indeed, the discrete binomial model is usually constructed so that in the limit as the step size decreases and the number of steps increases, it asymptotically approaches the continuous BS model.

Let $x$ denote the current asset price and suppose that after a small time interval $\Delta$, the asset prices becomes either $ux$, in the up-state, or $dx$ in the down-state. We assume that the up-factor $u > 1$, so that in the up-state, the asset price actually increases. Similarly, assume $0 < d < 1$ so the stock price decreases in the down-state. Such a model is referred to as a multiplicative binomial model. It is also possible to consider additive binomial models in which the asset prices in the two states become $x + u$ and $x - d$ for some $(u, d)$ assumed both positive. However this model runs into difficulties when after several down jumps, the asset price may become negative. The multiplicative binomial model never experiences this problem. After $n$ consecutive down jumps, the asset price will be $d^n x$ which, although possibly very small, is never negative.

Another very important feature of the multiplicative binomial model is that after $n$ time steps, the asset price will be only one of $(n + 1)$ possible values, namely (in increasing order):

$$\left[ d^n x, \; d^{n-1} u x, \; d^{n-2} u^2 x, \; \cdots, \; du^{n-1} x, \; u^n x \right].$$

This follows from the observation that the corresponding binomial tree is re-combining: the asset price after an up and a down jump $(ud\,x)$ is exactly the same as after a down and an up jump $(du\,x)$. This situation is displayed in Figure 4.2. Recombining binomial trees are sometimes referred to as a bino-mial lattice.

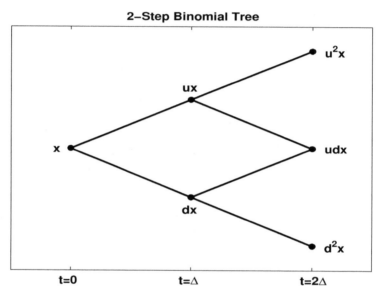

**2–Step Binomial Tree**

**FIGURE 4.2**: The recombining 2-step binomial lattice for the asset price.

If the tree was not recombining, there would be $2^n$ possible values of the asset price after $n$ steps. Since $n$, in applications, may be several hundreds or even thousands, this would limit the binomial model considerably as a prac-tical pricing algorithm.

It is natural to ask, what probability structure should we impose on our binomial lattice? We know that as far as pricing derivatives in an arbitrage free way is concerned, real-world probabilities are irrelevant. We need the risk-neutral or equivalent martingale measure for each branch of the tree. Hence, apply the Martingale Restriction to any one step on the lattice, and assume successive jumps are statistically independent. This leads to

$$\mathbb{E}_Q\{X_\Delta | X_0 = x\} = p(ux) + q(dx) = xe^{r\Delta}$$

where $(p, q)$ are the up and down probabilities under the EMM, $Q$. Together with the condition $p + q = 1$, we then obtain the important result on which all binomial models depend:

$$p = \frac{\rho - d}{u - d}; \qquad q = \frac{u - \rho}{u - d} \tag{4.28}$$

In these expressions, we have replaced $e^{r\Delta}$ by the parameter $\rho$, which represents the 1-step future value factor. Note that $\rho = e^{r\Delta}$ only if $r$ is interpreted as the continuously compounded risk free rate. However, it is also quite common to let $r$ be the discrete rate over one time step. In that case, we simply take $\rho = 1 + r\Delta$. Naturally, the two expressions are nearly the same for small interest rates $r$ and small time steps $\Delta$. Observe further, that for this simple binomial model, every up branch has a RN probability $p$, and every down branch has a RN probability $q = 1 - p$. It is in fact possible to construct generalized binomial trees in which the probabilities vary from branch to branch.

Note further, that under the independence assumption, the (risk-neutral) expected asset price after $n$ steps will be

$$\mathbb{E}_Q\{X_{n\Delta}|X_0 = x\} = \sum_{j=0}^{n} \binom{n}{j} p^j q^{n-j} \cdot u^j d^{n-j} x$$

$$= x \sum_{j=0}^{n} \binom{n}{j} (pu)^j (qd)^{n-j}$$

$$= x(pu + qd)^n = x\rho^n$$

since after $j$ up-jumps, the asset price will be $u^j d^{n-j} x$ with probability given by the binomial distribution, $B_j(n; p) = \binom{n}{j} p^j q^{n-j}$ for $j = 0, 1,, \ldots, n$.

To price any derivative on the binomial tree, we need only specify the payoff at expiry (after $n$ time steps, say), and apply the FTAP appropriately. Let $F(x)$ denote the expiry payoff function for some European derivative, then the FTAP yields the result,

$$V_0(x) = \rho^{-n} \mathbb{E}_Q\{F(X_n)|X_0 = x\}$$

$$= \rho^{-n} \sum_{j=0}^{n} \binom{n}{j} p^j q^{n-j} F(u^j d^{n-j} x).$$

While this sum may be computed directly, in practice it is simpler to obtain the derivative price $V_0(x)$ by backward recursion on the tree. One starts from expiry where the $(n + 1)$ possible payoff values are known, then works backward (in time) through the lattice, node by node, using the recursive formula

$$V_i = \rho^{-1} \left[ pV_{i+1}^u + qV_{i+1}^d \right] \tag{4.29}$$

where $V_{i+1}^{u,d}$ are derivative values at the up and down nodes at time step $(i+1)\Delta$ emanating from the node at time step $i\Delta$. The recursion stops at the single node corresponding to $i = 0$, and the derivative value $V_0$ at this node is the required price.

This basic recursion is readily modified to incorporate special features, such as early exercise in American options. At each node of the recursion, one only needs to determine if the immediate exercise value exceeds the continuation value of the derivative. If it does, attach the exercise value at this node, otherwise continue the recursion. For Bermudan options, where early exercise takes place at certain fixed times of the lattice, the comparison of continuation and stopping values should operate only at the nodes occurring at these times.

**Relating Binomial to BS Parameters**

Using the Central Limit Theorem (see Section 3.5), it is not too difficult to show that the multiplicative binomial asset model described above has a log-normal limit, when $n \to \infty$, $\Delta \to 0$ such that $n\Delta \to T$. In other words, the binomial model for small $\Delta$ and large $n$ is an approximation to the Black–Scholes model. However, the two models have very different parameters. The Black–Scholes model uses volatility $\sigma$, while the binomial model uses up and down factors $(u, d)$. The risk free rate $r$ (notwithstanding possibly different interpretations mentioned above) and expiry date $T = n\Delta$ are common to both models.

The relationship between $(u, d)$ and $\sigma$ is usually determined by finding the 1-step binomial standard deviation of the asset price return and equating it to $\sigma\sqrt{\Delta}$. This leads to

$$\mathbb{V}_Q\{X_\Delta/x\} = \sigma^2\Delta$$
$$(pu^2 + qd^2) - (pu + qd)^2 = \sigma^2\Delta$$
$$pq(u^2 + d^2 - 2ud) = \sigma^2\Delta$$
$$\text{or} \quad (u - d) = \sigma\sqrt{\Delta/pq}.$$

Together with the Martingale Restriction $pu + qd = \rho$ and probability condition $p + q = 1$, we have three equations in the four unknowns $(u, d, p, q)$. Hence these parameters will not be uniquely determined in terms of $\sigma$ and $\Delta$. However, they imply the exact relationships

$$\boxed{\begin{aligned} u &= \rho + \sigma\sqrt{q\Delta/p} \\ d &= \rho - \sigma\sqrt{p\Delta/q} \end{aligned}} \tag{4.30}$$

Here the binomial model literature deviates along different directions, depending on how one wishes to uniquely tie down the four binomial parameters. Some authors prefer to fix $(u, d)$ and use (4.30) to find $(p, q)$, at least to order

$O(\Delta)$.

But, it should be clear that the most elegant solution is simply to fix $(p,q)$, subject of course to $p+q=1$, and use (4.30) to determine $(u,d)$. And what could be simpler than the symmetric case $p=q=\frac{1}{2}$, leading to the result

$$\boxed{\begin{aligned}u &= \rho + \sigma\sqrt{\Delta} \\ d &= \rho - \sigma\sqrt{\Delta}\end{aligned}}\qquad(4.31)$$

One need only be careful that the binomial time step $\Delta$ be chosen small enough to ensure that $d>0$. That is, we must select $\Delta$ such that

$$\sigma\sqrt{\Delta} < 1+r\Delta \quad(\text{or}\quad e^{r\Delta})$$

depending on how the risk free rate $r$ is interpreted.

**Relating Binomial Recursion to BS-pde**
If $V(x,t)$ denotes the derivative price at any nodal time $t$ and nodal asset price $x$, then the binomial recursion (4.29) is seen to be equivalent to the functional recursion

$$V(x,t) = \rho^{-1}\left[pV(ux,t+\Delta) + qV(dx,t+\Delta)\right].\qquad(4.32)$$

We now demonstrate that in the limit $\Delta \to 0$, for the parameter set (4.30) and $\rho = 1+r\Delta$, how this recursion reduces to the standard Black–Scholes pde. To begin, let

$$u' = u-1;\qquad d' = d-1$$

be small parameters, so the recursion becomes,

$$(1+r\Delta)V(x,t) = pV(x+u'x,t+\Delta) + qV(x+d'x,t+\Delta).$$

Now perform a Taylor expansion of the rhs to order[1] $O(\Delta)$, to obtain

$$(1+r\Delta)V(x,t) = p\left[V + \Delta V_t + (u'x)V_x + \tfrac{1}{2}(u'x)^2 V_{xx} + \cdots\right]$$
$$+q\left[V + \Delta V_t + (d'x)V_x + \tfrac{1}{2}(d'x)^2 V_{xx} + \cdots\right]$$

where $V$ and its partial derivatives are evaluated at $(x,t)$ and the dots represent higher order terms. Note that to order $O(\Delta)$, we need to expand the $x$ terms to second order. The following relationships

$$p+q=1;\quad pu'+qd'=r\Delta;\quad p(u')^2+q(d')^2 = \sigma^2\Delta + \cdots$$

[1]The terms $O(\Delta)$ and $o(\Delta)$ have different meanings. We say $X$ is $O(\Delta)$ if $\lim_{\Delta\to0} X/\Delta$ is bounded and non-zero; while $X$ is said to be $o(\Delta)$ if $\lim_{\Delta\to0} X/\Delta = 0$.

are easily derived from (4.30) and when applied to the above expansion, yield (after cancelling the common term $V(x,t)$ from both sides and dividing by $\Delta$)

$$rV(x,t) = V_t(x,t) + rxV_x(x,t) + \tfrac{1}{2}\sigma^2 x^2 V_{xx}(x,t) + \cdots o(\Delta).$$

Since terms of order $o(\Delta)$ vanish in the limit as $\Delta \to 0$, we finally obtain

$$V_t = rV - rxV_x - \tfrac{1}{2}\sigma^2 x^2 V_{xx}$$

which, as predicted, is the BS-pde. Thus, one step on the binomial tree for small time step $\Delta$, should be an approximation to the BS value, for any European style derivative security. While errors may build up on a tree of many time steps, we should still get a good approximation to the BS derivative price, if we choose $\Delta$ small enough.

The graph in Figure 4.3 shows the percentage error of a binomial call option price relative to the BS price, as a function of the number of binomial steps to expiry. The figure shows the slow convergence to the BS value (where the error equals zero). The maximum error is less than 0.2% for all time steps in the range 100 to 500; the mean error is $-0.01\%$ and after 500 steps the error is less than 0.001% At each step, the error tends to oscillate, often over-shooting the BS value, then under-shooting it.

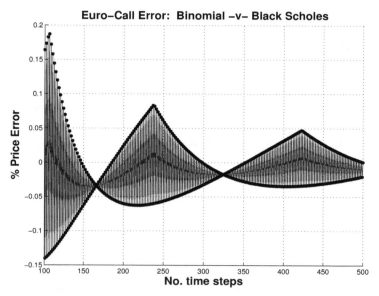

**FIGURE 4.3:** Percentage call price error of the binomial model relative to the BS model for parameters: $[x = k = 10,\ T = 0.5,\ r = 0.10,\ \sigma = 0.25]$. Note how the error oscillates between even and odd numbers of steps.

## 4.13 Options on a Traded Account

Consider an asset $X$ following gBm in the usual BS model, so that under the EMM, $Q$,

$$dX_t = X_t(rdt + \sigma \, dB_t). \tag{4.33}$$

Now suppose an investor holds a trading account $Y$ and has purchased a fixed number $\theta$, of units of asset $X$. Assume that the traded account pays and receives cash flows generated by changes in the asset price $X_t$, and that this account earns interest at the risk free rate $r$. Then the sde satisfied by the amount $Y_t$ in the account is given by

$$dY_t = \theta dX_t + r(Y_t - \theta X_t)dt = rY_t dt + \theta \sigma X_t \, dB_t. \tag{4.34}$$

The term $\theta dX_t$ represents the change in the traded account due to changes in the asset price $dX_t$, while $r(Y_t - \theta X_t)dt$ is the interest earned on the net position. Thus, even though we appear to have two state variables, $X_t$ and $Y_t$, they are in fact driven by a single source of randomness, namely the Brownian motion $B_t$.

While it is possible to consider a variety of payoffs depending on both $X_T$ and $Y_T$ at the expiry date $T$, we shall investigate the simple case where $V(X_T, Y_T, T) = [Y_T]^+$. Thus, at time $T$, the trader is forgiven all the losses in the traded account, and gets to keep all the profits. Note that while $X_T$ must be positive, there is no such restriction on $Y_T$. The problem is to value the price $V(x, y, t)$ of this deal at any time $t < T$, $x > 0$ and $y \in \mathbb{R}$.

The general problem is actually to find an optimal trading strategy $\theta = \theta_t$, which maximizes the payoff at time $T$. In particular, when the trading strategy $\theta$ is restricted to the range $[-1, 1]$ the contract is a called a passport option; to the range $[0, 1]$ a vacation call; and to the range $[-1, 0]$ a vacation put. These are problems in stochastic optimal control, which are somewhat beyond the scope of this book. Surprisingly, however, the optimal problems do have closed form solutions in the BS framework. The interested reader is directed to the papers by Hyer et al. [35] and Shreve and Vecer [69] for details.

As mentioned above, we shall be content to solve the pricing problem for the *fixed* trading strategy $\theta \in \mathbb{R}$, where $\theta$ is constant. To solve the pricing problem, note first from Equation (4.33), that we can write in the usual way,

$$X_T = xe^{(r - \frac{1}{2}\sigma^2)\tau + \sigma\sqrt{\tau}Z} \tag{4.35}$$

where $\tau = (T - t)$ and $Z \sim N(0, 1)$.

Now let us first define a new stochastic variable $R_t = Y_t/X_t$, the ratio of the traded account value and the underlying asset price. We use Itô's Quotient Rule (2.23) to find the sde satisfied by $R_t$. Take

$$F = Y_t; \quad G = X_t; \quad G_x = 1; \quad \beta = \sigma X_t$$

and observe (from Itô's Lemma for $dF$ and $dG$) that the ratio of the coefficients of the common Brownian increment $dB_t$ is

$$\frac{F_x}{G_x} = \frac{\theta \sigma X_t}{\sigma X_t} = \theta \quad \Rightarrow \quad F_x = \theta.$$

Further, using (4.33) and (4.34),

$$\frac{G dF - F dG}{G^2} = \frac{X_t dY_t - Y_t dX_t}{X_t^2} = -\sigma(R_t - \theta)dt$$

and

$$\frac{\beta G_x}{G^3}[F G_x - G F_x] = \frac{\sigma^2 X_t^2 \cdot 1}{X_t^3}[Y_t \cdot 1 - X_t \cdot \theta] = \sigma^2(R_t - \theta)dt.$$

Hence, $R_t$ satisfies the sde

$$dR_t = (R_t - \theta)[\sigma^2 \, dt - \sigma dB_t]. \tag{4.36}$$

The solution of this sde at time $T$, given $X_t = x$ and $Y_t = y$ is,

$$R_T = \theta + (R_t - \theta)e^{\frac{1}{2}\sigma^2 \tau - \sigma\sqrt{\tau} Z} \tag{4.37}$$

where $R_t = x/y$. Since we have both $X_T$ and $R_T$, we also get $Y_T$. Thus

$$Y_T = X_T R_T = \theta X_T + (y - \theta x)e^{r\tau}. \tag{4.38}$$

The FTAP, (1.35) now gives the price of the derivative as

$$V(x, y, t) = e^{-r\tau}\mathbb{E}_Q\{Y_T^+\} = e^{-r\tau}\mathbb{E}_Q\{[\theta X_T + (y - \theta x)e^{r\tau}]^+\}.$$

There are four distinct cases to consider.

1. $\theta > 0; \; y > \theta x$
Here both terms in the expectation are positive, so we may simply remove the plus-function, to get, using the MR, $\mathbb{E}_Q(X_T) = xe^{r\tau}$,

$$V(x, y, t) = \theta x + (y - \theta x) = y.$$

2. $\theta > 0; \; y < \theta x$
Then the price is

$$V(x, y, t) = \theta e^{-r\tau}\mathbb{E}_Q\{(X_T - k)^+\}; \qquad k = (x - y/\theta)e^{r\tau}.$$

In this case the option is equal to $\theta$ units of a strike $k = k(x, y, t)$ call option. Hence, $V(x, y, t) = \theta\, C_k(x, \tau)$.

### 3.  $\theta < 0;\ y < \theta x$

Here both terms are negative, so $V(x, y, t) = 0$.

### 4.  $\theta < 0;\ y > \theta x$

Let $\phi = -\theta = |\theta| > 0$. Then, the price can be written as

$$V(x, y, t) = \phi e^{-r\tau}\mathbb{E}_Q\{(k - X_T)^+\}; \qquad k = (x + y/\phi)e^{r\tau}.$$

This represents $\phi$ units of a strike $k = k(x, y, t)$ put option. That is, $V(x, y, t) = \phi\, P_k(x, \tau)$.

Note that when $\theta = 1$, we get

$$V(x, y, t) = \begin{cases} y & \text{if } y > x; \\ C_k(x, \tau) & \text{if } y < x; \end{cases} \qquad k = (x - y)e^{r\tau} \tag{4.39}$$

and when $\theta = -1$,

$$V(x, y, t) = \begin{cases} 0 & \text{if } y < -x; \\ P_k(x, \tau) & \text{if } y > -x; \end{cases} \qquad k = (x + y)e^{r\tau} \tag{4.40}$$

**REMARK 4.5**  For the passport option, the optimal trading strategy $\theta_t \in [-1, 1]$ is found to be the bang-bang solution $\theta_t = -\text{sgn}(Y_t)$. That is, go long when behind and short when ahead. Even given this strategy, pricing the passport option is no walk in the park. See [35] and [69] for some of the details or the PhD thesis [56] for a comprehensive treatment. $\Box$

## 4.14  Summary

This chapter considered the simplest types of exotic option, namely those which are path-independent and involve only a single asset and a single future expiry date. Many of these exotics can be expressed in terms of first-order asset and bond binary options. We showed how these could be readily priced in the BS framework, with the help of the univariate Gaussian Shift Theorem.

We introduced in the chapter the notion of a $Q$-option, which basically captures both a gap call and a gap put within a single formula given by (4.13). These $Q$-options will make appearances in several of our later chapters, so it is worthwhile having an understanding of what they represent. Indeed, as we progress, our exotic options tend to become more and more complex, with a concomitant increase in notational complexity. In order to assuage this complexity, we find ourselves continually introducing new notation. Thus at the most basic level, option prices are sums or portfolios of normal distribution functions; these are grouped into asset and bond binary classes; portfolios of asset and bond binaries define new classes such as the $Q$-options; simple $Q$-options are extended to second order and higher $Q$-options; and so on it goes.

We included in this chapter a brief discussion of the binomial option pricing model and how it is related to the Black–Scholes model. This may seem a little out of place in a book of this nature, but it is included for a couple of reasons. First, the binomial model is a discrete version of the BS model, and indeed has a limit in distribution that is identical to the BS model. Second, we shall revisit the binomial model in Chapter 7 on Barrier Options. There we will show how to use the binomial model to price barrier options, without the need for a continuity correction, as is usually argued must be done.

The final application on Options on a Traded Account may seem far from being a simple exotic. Indeed, they can be very complicated, but the examples we have chosen to price fall into the category of *fixed* strategy options and are therefore quite amenable to our methods of pricing. While these options appear to have two underlying assets or state variables, they are in fact both driven by the same Brownian motion and hence are instantaneously, perfectly correlated.

The next chapter looks at the class of dual-expiry options. These options have payoffs which depend on what the underlying asset does on two future dates. Typically the situation is something like this: on the first date the holder, instead of receiving cash or some units of the asset, may receive a financial contract. This contract has a finite life which expires at the later, second date. Examples include compound options (such as calls-on-calls), chooser options, and several others. Their prices will generally depend on the bivariate normal distribution function, but this is not guaranteed.

---

## Exercise Problems

1. Consider the following derivatives in the BS economy, with expiry $T$ payoff functions

   (a) $V(x, T) = x^p$;    $p$ real.

(b) $V(x,t) = [\log(x/k)]^+$;    $(k > 0)$.

Price (a) by solving the BS-pde, and (b) by using the FTAP.

2. Show that the derivative with expiry $T$ payoff $V(x,T) = |x-k|$ is a long straddle. Hence determine the pv of the derivative in terms of standard European call and put prices.

3. Let $(x, \tau, k, r, q, \sigma)$ denote the asset price, time remaining to expiry, strike price, risk-free rate, dividend yield and volatility, of a European option. Show that the BS formulae for the prices of European calls and puts satisfy the symmetry relation

$$C(x, \tau; k, r, q, \sigma) = P(k, \tau; x, q, r, \sigma).$$

4. If $x$ is the price of a non-dividend paying asset and $y$ is the pv of the strike price, derive the (European) put-call symmetries

   (a) $C(x,y) = P(y,x)$ for the BS model, and

   (b) $C_n(x,y) = P_n(y,x)$ for the $n$ step binomial model, provided the lattice probabilities satisfy $\sqrt{pq} = \rho$.

5. A *super-share* is a contract that pays at time $T$ in the future one unit of an asset, but only if the asset price at this time lies in the range $[a, b]$. Show that the pv of the super-share can be expressed as either a portfolio of up-type asset binaries, or as an equivalent portfolio of down-type asset binaries.

6. A security pays at expiry $T$ either a call option payoff $(x - k)^+$ or a put option payoff $(h - x)^+$, where $h > k$, whichever is the greater. Show that its pv is that of strike $k$ gap call and strike $h$ gap put, both with exercise price $\xi = \frac{1}{2}(k + h)$.

7. Consider pricing the derivative with time $T$ payoff $V(x,T) = \sqrt{x}$, in the BS framework, by solving the corresponding BS-pde. Derive the pv of this derivative, assuming a separable solution of the form $V(x,t) = A(\tau)\sqrt{x}$, where $\tau = (T - t)$. Verify this solution by also calculating the price using the FTAP method.

8. Repeat the previous question for the derivative with payoff $V(x,T) = x^p \log(x/k)$, assuming a separable solution of the form

$$V(x,t) = x^p [A(\tau) + B(\tau)\log(x/k)]; \qquad \tau = (T - t).$$

9. Price expiry $T$, pay-at-expiry options on

   (a) the underlying asset $X$; and

    (b) a European put option on $X$.

Identify the contract in part (a) above.

10. Find the pv of a reverse pay-at-expiry call option of strike price $k$ and expiry date $T$.

11. Determine the the time $t_0$ premium on a money-back put option of strike price $k$ and expiry date $T$. Hence determine the pv of the option for any time $t$ such that $t_0 < t < T$.

12. Verify the price of the turbo binary given in Equation (4.17). That is,

$$P_\xi^s(x, t; p) = x^p \, e^{\lambda(p)\tau} \, \mathcal{N}[s(d_\xi' + p\sigma\sqrt{\tau})]$$

where $\lambda(p) = \frac{1}{2}\sigma^2 p^2 + (r - \frac{1}{2}\sigma^2)p - r$.

—ooOoo—

# Chapter 5

# Dual Expiry Options

Second-order exotic options are perhaps the most direct extension of the simple first-order options considered in the previous chapter. There are two basic types: (a) single asset, two-period exotics which we call *dual expiry* options (considered in this chapter); and (b) single period, two-asset exotics, commonly referred to as two-asset *rainbow* options (considered in the next chapter).

As the name implies, the dual expiry options have a payoff structure that depends on the underlying asset price on two future dates $T_1$ and $T_2$ with $T_1 < T_2$. Examples, which we analyze in this chapter include: forward start options, compound options, chooser options and several others.

If $x_1, x_2$ are the asset prices at times $T_1$ and $T_2$, then at time $T_2$, the dual expiry option payoff will have the form $V(x_i; T_2) = f(x_1, x_2)$. What makes payoffs of this type somewhat intriguing is that the asset price $x_1$, having already been observed at $T_1$, is effectively constant over the whole time interval $T_1 < t < T_2$. Dual expiry options were analyzed in detail, using methods of this chapter, by the author [8].

Many, second-order exotics have prices in the BS-economy which depend on the bivariate normal distribution function. In this context, the bivariate Gaussian Shift Theorem (GST) will play an important role in pricing the corresponding binaries. Just as the first order $Q$-options help reduce notational burden, so will the second order $Q$-options defined in the chapter.

---

## 5.1  Forward Start Calls and Puts

A forward start call (resp. put) option gives its holder at time $T_1$ in the future an at-the-money (atm) European call (resp. put) option, which expires at the later time $T_2 > T_1$. Forward start call options were originally given to company executives as part of their remuneration packages. As such, they are what we call today, examples of an *executive stock option* or ESO.

The term atm means that the strike price of these options is set equal to the asset price at time $T_1$. Hence at time $t < T_1$, the strike price is unknown, so forward start options are examples of European calls and puts with stochastic strike prices.

It turns out that the prices of these options in the BS economy can be expressed as simple decompositions of 1st-order asset and bond binaries. Consider, for example, the forward start call option. The payoff at time $T_1$, using relative time notation (see Remark 4.3), is

$$F_c(x_1, T_1) = [C_k(x_1, \tau)]_{k=x_1} ; \qquad \tau = (T_2 - T_1)$$

This simply expresses mathematically that the payoff at time $T_1$ is a strike $k = x_1$ call option with time $\tau = (T_2 - T_1)$ remaining to expiry. Here $x_1$ obviously denotes the asset price at time $T_1$.

However, using (4.7) for the price $C_k(x, \tau)$ of a call option leads to the equivalent result $F_c(x_1, T_1) = g(\tau)x_1$, where the function $g(\tau)$ defined below, is constant, independent of the asset price $x_1$. Hence the payoff of the forward start call at $T_1$ is equivalent to receiving $g(\tau)$ (defined below) units of the underlying asset at time $T_1$. Thus its price for $t < T_1$ is given by

$$\boxed{\begin{aligned} F_c(x, t) &= g(\tau)\, x \\ g(\tau) &= \mathcal{N}(a\sqrt{\tau}) - e^{-r\tau} \mathcal{N}(a'\sqrt{\tau}) \end{aligned}} \tag{5.1}$$

and $[a, a'] = \frac{r}{\sigma} \pm \frac{1}{2}\sigma$.

The corresponding pv for $t < T_1$, of a forward start put option is found to be

$$\boxed{\begin{aligned} F_p(x, t) &= h(\tau)\, x \\ h(\tau) &= -\mathcal{N}(-a\sqrt{\tau}) + e^{-r\tau} \mathcal{N}(-a'\sqrt{\tau}) \end{aligned}} \tag{5.2}$$

If we let $x_2$ denote the asset price at time $T_2$, then it is interesting to note that the forward start call and put have time $T_2$ payoffs given respectively by

$$F_c(x_1, x_2, T_2) = (x_2 - x_1)^+ \qquad \text{and} \qquad F_p(x_1, x_2, T_2) = (x_1 - x_2)^+ . \tag{5.3}$$

As mentioned above, the asset price $x_1 = X(T_1)$ having been observed at time $T_1$, is effectively a fixed constant over the time interval $(T_1, T_2)$. We could use these $T_2$ payoffs to price the forward start options, employing the results of the next section. Often, as these forward start options demonstrate, it is actually easier to price dual expiry options by considering their $T_1$ payoffs, rather than their $T_2$ payoffs. Sometimes it is very instructive from an analytical point of view to do both and show how the two approaches lead to the same price.

## 5.2 Second-Order Binaries

First-order binaries (see definition 4.1) are derivatives that pay an amount $f(x)$ at expiry $T$, but only if the asset price at expiry $T$ is above or below a specified exercise price. There are just two basic types, which we called the up-type and the down-type, depending on whether the exercise condition is above or below this price.

Second-order binaries are binaries-on-binaries, or compound binaries, and are defined as follows.

**DEFINITION 5.1** *We say a derivative on a single underlying asset is a second-order binary, if its payoff at time $T_1$ is a binary option on a contract which itself is a 1st-order binary option, with expiry at $T_2 > T_1$. Thus at time $T_1$, the payoff is*

$$V_{\xi_1 \xi_2}^{s_1 s_2}(x_1, T_1) = V_{\xi_2}^{s_2}(x_1, \tau) \mathbb{I}(s_1 x_1 > s_1 \xi_1) \tag{5.4}$$

*where $x_1$ is the asset price at time $T_1$, $s_{1,2} = \pm$ denotes binary type, and $V_{\xi_2}^{s_2}(x_1, \tau)$ is the price of a 1st-order binary with exercise price $\xi_2$ and time $\tau = (T_2 - T_1)$ remaining to expiry.*

Thus 2nd-order binaries have two exercise prices $\xi_1$ and $\xi_2$, which come into effect at times $T_1$ and $T_2$ respectively. The four possible combinations of signs for $s_1$ and $s_2$ imply that there are four corresponding types of 2nd-order binaries. We might refer to these as respectively: up-up, up-down, down-up and down-down according to $(s_1 s_2) = [(++), (+-), (-+), (--)]$.

It is instructive to write down the payoff of the 2nd-order binary defined above, at the final expiry date $T_2$. If $f(x)$ denotes the associated payoff amount when both the exercise conditions at $(T_1, T_2)$ are met, then we have

$$V_{\xi_1 \xi_2}^{s_1 s_2}(x_1, x_2, T_2) = f(x_2) \mathbb{I}(s_1 x_1 > s_1 \xi_1) \, \mathbb{I}(s_2 x_2 > s_2 \xi_2) \tag{5.5}$$

where $x_2$ denotes the asset price at time $T_2$.

**REMARK 5.1** When pricing a 2nd-order binary option we shall generally value it at a time $t$ satisfying $t < T_1 < T_2$. But recall, that during the interval between $T_1$ and $T_2$, the asset price $x_1$ at time $T_1$ has already been observed, so remains a fixed constant in this interval.

Note further, according to definition 5.1, that if we can represent a derivative's payoff at time $T_1$ as a 1st-order binary, multiplied by an indictor function

$\mathbb{I}(s_1 x_1 > s_1 \xi_1)$, then this derivative must be a 2nd-order binary option. We shall use this simple observation to help price certain dual expiry options in this chapter. ⬚

---

## 5.3 Second-Order Asset and Bond Binaries

In this section we show how to price 2nd-order asset and bond binaries in the BS economy, when $f(x) = x$ and $f(x) = 1$ respectively, in equation (5.5). These calculations depend on the time $T_1$ and time $T_2$ asset price representations

$$X_i = x\, e^{(r - \frac{1}{2}\sigma^2)\tau_i + \sigma\sqrt{\tau_i}\, Z_i}; \qquad (i = 1, 2) \tag{5.6}$$

where $\tau_i = (T_i - t)$ and the $Z_i \sim N(0, 1; \rho)$ are joint zero mean, unit variance, Gaussian rv's with correlation coefficient (ccf)

$$\rho = \sqrt{\frac{\tau_1}{\tau_2}}. \tag{5.7}$$

The ccf $\rho$ is therefore time dependent, and follows from equation (3.19), which gives the ccf for a Brownian motion at two distinct instances of time. In the current circumstance, $Z_i$ are standard Gaussian rv's associated with the Brownian motions $B_{\tau_i} = \sqrt{\tau_i} Z_i$.

We shall also use the conditions (see lemma 4.1) that $X_i \gtrless \xi_i$ corresponds to $Z_i \gtrless -d'_i(x, \tau_i)$ where for $(i = 1, 2)$

$$[d_i,\, d'_i] = \frac{\log(x/\xi_i) + (r \pm \frac{1}{2}\sigma^2)\tau_i}{\sigma\sqrt{\tau_i}}. \tag{5.8}$$

**2nd-Order Bond Binary**
We price here the down-down type 2nd-order bond binary as an illustration. Using the FTAP (1.46) with the time $T_2$ payoff, we find

$$
\begin{aligned}
B^{--}_{\xi_1 \xi_2}(x, t) &= e^{-r\tau_2}\, \mathbb{E}_Q\{\mathbb{I}(X_1 < \xi_1)\, \mathbb{I}(X_2 < \xi_2) \mid \mathcal{F}_t\} \\
&= e^{-r\tau_2}\, \mathbb{E}_Q\{\mathbb{I}(Z_1 < -d'_1)\, \mathbb{I}(Z_2 < -d'_2)\} \\
&= e^{-r\tau_2}\, \mathcal{N}(-d'_1, -d'_2;\, \rho). \qquad \square
\end{aligned}
$$

The last line uses the representation (3.21) for the bivariate normal. When we include the other bond binary types, we obtain the general formula

$$\boxed{B^{s_1 s_2}_{\xi_1 \xi_2}(x, t) = e^{-r\tau_2}\, \mathcal{N}(s_1 d'_1,\, s_2 d'_2;\, s_1 s_2 \rho)} \tag{5.9}$$

Note that the ccf in this formula is $s_1 s_2 \rho$ where the product $s_1 s_2$ is interpreted as "+" when $s_1, s_2$ are the same sign; and as "−" when they are of opposite sign. Thus the mixed up-down and down-up binaries have ccf $= -\rho$. We shall see how that comes about in the next calculation.

**2nd-Order Asset Binary**
We price the down-up 2nd-order asset binary to again illustrate the method, and quote the prices of the other types. This calculation uses the bivariate GST of theorem 3.3.

$$
\begin{aligned}
A^{-+}_{\xi_1 \xi_2}(x, t) &= e^{-r\tau_2} \, \mathbb{E}_Q \{ X_2 \, \mathbb{I}(X_1 < \xi_1) \, \mathbb{I}(X_2 > \xi_2) \mid \mathcal{F}_t \} \\
&= x \, e^{-\frac{1}{2}\sigma^2 \tau_2} \, \mathbb{E}_Q \{ e^{\sigma\sqrt{\tau_2} \, Z_2} \mathbb{I}(Z_1 < -d'_1) \, \mathbb{I}(Z_2 > -d'_2) \} \\
&= x \, \mathbb{E}_Q \{ \mathbb{I}(Z_1 + \sigma\sqrt{\tau_1} < -d'_1) \, \mathbb{I}(Z_2 + \sigma\sqrt{\tau_2} > -d'_2) \} \\
&= x \, \mathbb{E}_Q \{ \mathbb{I}(Z_1 < -d_1) \, \mathbb{I}(Z_2 > -d_2) \} \\
&= x \, \mathbb{E}_Q \{ \mathbb{I}(Z_1 < -d_1) \, \mathbb{I}(\tilde{Z}_2 < d_2) \} \\
&= x \, \mathcal{N}(-d_1, d_2; -\rho). \qquad \square
\end{aligned}
$$

In the above calculation, we make the following observations.

- When employing the bivariate GST of theorem 3.3 in line 3 above we have used the parameters: $a = 0$, $b = \sigma\sqrt{\tau_2}$ and $\rho = \sqrt{\tau_1 / \tau_2}$. This leads to
$$
c = \sigma\sqrt{\tau_2}; \quad a' = \rho\sigma\sqrt{\tau_2} = \sigma\sqrt{\tau_1}; \quad b' = \sigma\sqrt{\tau_2}.
$$

- In line 4 we have used $d_i = d'_i + \sigma\sqrt{\tau_i}$ for $i = (1, 2)$.

- In line 5 we have defined $\tilde{Z}_2 = -Z_2$ and note $\tilde{Z} \overset{d}{=} N(0, 1)$.

- Finally, in line 6 we have used the fact that if $(Z_1, Z_2)$ are joint Gaussian with ccf $= \rho$, then $(Z_1, -Z_2)$ are joint Gaussian with ccf $= -\rho$.

The reader should also note that the above calculation has not performed a single integration. Essentially, we have only done some algebraic manipulations and carried out the implied integration through the use of the bivariate GST.

The complete family of four 2nd-order asset binaries is given by the expression

$$
\boxed{A^{s_1 s_2}_{\xi_1 \xi_2}(x, t) = x \, \mathcal{N}(s_1 d_1, s_2 d_2; \, s_1 s_2 \rho)} \tag{5.10}
$$

## 5.4 Second-Order Q-Options

Because many exotic options are built from underlying calls and puts, we introduced in section 4.5 the first-order $Q$-option. We present here its extension to second-order $Q$-options, which are similarly motivated to simplify notation.

**DEFINITION 5.2** *A second-order Q-option of exercise prices $(\xi_1, \xi_2)$ and strike price $k$ is the compound derivative with $T_1$ and $T_2$ payoffs*

$$Q^{s_1 s_2}_{\xi_1 \xi_2}(x_1, T_1; k) = Q^{s_2}_{\xi_2}(x_1, \tau; k)\mathbb{I}(s_1 x_1 > s_1 \xi_1) \tag{5.11}$$

$$Q^{s_1 s_2}_{\xi_1 \xi_2}(x_1, x_2, T_2; k) = s_2(x_2 - k)\mathbb{I}(s_1 x_1 > s_1 \xi_1)\mathbb{I}(s_2 x_2 > s_2 \xi_2) \tag{5.12}$$

*where $\tau = (T_2 - T_1)$ and $s_i = \pm$, determines the Q-option type.*

We saw that first-order $Q$-options are related to gap calls and puts, which are identical to standard European calls and puts, except that their strikes and exercise prices may be different. Second-order $Q$-options can be similarly interpreted as binary options on a gap call or put.

Since second-order $Q$-options are simple portfolios of second-order asset and bond binaries, their prices are readily determined using the PSR. Thus for all $t < T_1 < T_2$,

$$\boxed{Q^{s_1 s_2}_{\xi_1 \xi_2}(x, t; k) = s_2 A^{s_1 s_2}_{\xi_1 \xi_2}(x, t) - s_2 k B^{s_1 s_2}_{\xi_1 \xi_2}(x, t)} \tag{5.13}$$

We saw earlier in this chapter that although forward start options are technically dual expiry options, they can be priced by using first-order binaries alone.

The next example illustrates the connection between several approaches one can use to price forward start options.

### Example 5.1

Consider the problem of pricing a derivative security with time $T_2$ payoff $V(x_1, x_2, T_2) = x_2/x_1$, where $x_i$ is the price of the asset at time $T_i$ with $T_1 < T_2$. Assume Black–Scholes dynamics on a non-dividend paying asset and price at time $t < T_1$.

We demonstrate three methods for pricing this simple problem. Each method gives a different insight into pricing issues that underlie the EMM approach in the BS framework.

## Method-1

As we have noted previously, over the time interval $(T_1, T_2)$, the asset price $x_1$ is effectively constant. So the payoff at time $T_2$ is just like $V(x_2, T_2) = kx_2$ where $k = x_1^{-1}$. That is $k$ units of the asset. But such a derivative obviously has value at time $T_1$ equal to $kx_1$. But $kx_1 = 1$, so the derivative is equivalent to a $T_1-$maturity ZCB. The pv (at time $t < T_1$) is then given by $V(x,t) = e^{-r\tau_1}$, where as usual, $\tau_i = (T_i - t)$.

## Method-2

This method really just formalizes method-1 using the Tower Law (2.2). We proceed as follows, starting from the FTAP. For $t < T_1 < T_2$, we have

$$
\begin{aligned}
V(x,t) &= e^{-r\tau_2}\mathbb{E}_Q\{X_1^{-1} \cdot X_2 \,|\, \mathcal{F}_t\} \\
&= e^{-r\tau_2}\mathbb{E}_Q\{\mathbb{E}_Q\{X_1^{-1} \cdot X_2 \,|\, \mathcal{F}_1\} \,|\, \mathcal{F}_t\} \\
&= e^{-r\tau_2}\mathbb{E}_Q\{X_1^{-1} \cdot \mathbb{E}_Q\{X_2 \,|\, \mathcal{F}_1\} \,|\, \mathcal{F}_t\} \\
&= e^{-r\tau_2}\mathbb{E}_Q\{X_1^{-1} \cdot X_1 e^{r(T_2 - T_1)} \,|\, \mathcal{F}_t\} \\
&= e^{-r\tau_1}\mathbb{E}_Q\{1 \,|\, \mathcal{F}_t\} \\
&= e^{-r\tau_1}.
\end{aligned}
$$

The second line above is just the Tower Law, with $\mathcal{F}_1$ denoting the filtration up to time $T_1$. The fourth line uses the martingale restriction: $\mathbb{E}_Q\{X_2|\mathcal{F}_1\} = X_1 e^{r(T_2 - T_1)}$.

## Method-3

Using the BS representation $X_i = xe^{(r-\frac{1}{2}\sigma^2)\tau_i + \sigma\sqrt{\tau_i}Z_i}$, where $Z_i \sim N(0,1;\rho)$ with $\rho = \sqrt{\tau_1/\tau_2}$, we obtain

$$
\frac{X_2}{X_1} = e^{(r-\frac{1}{2}\sigma^2)(T_2 - T_1) + \sigma(\sqrt{\tau_2}Z_2 - \sqrt{\tau_1}Z_1)}.
$$

Hence

$$
\begin{aligned}
V(x,t) &= e^{-r\tau_2}\mathbb{E}_Q\{X_1^{-1} \cdot X_2 \,|\, \mathcal{F}_t\} \\
&= e^{-r\tau_1 - \frac{1}{2}\sigma^2(T_2 - T_1)}\mathbb{E}_Q\{e^{\sigma(\sqrt{\tau_2}Z_2 - \sqrt{\tau_1}Z_1)}\} \\
&= e^{-r\tau_1 - \frac{1}{2}\sigma^2(T_2 - T_1)} e^{\frac{1}{2}\sigma^2(T_2 - T_1)} \\
&= e^{-r\tau_1}.
\end{aligned}
$$

The third line above follows from the bivariate GST (3.22) with parameter:

$$
c^2 = \sigma^2[\tau_2 + \tau_1 - 2\rho\sqrt{\tau_1\tau_2}] = \sigma^2(\tau_2 - \tau_1) = \sigma^2(T_2 - T_1).
$$

All three methods naturally give the same result. □

## 5.5 Compound Options

The four basic compound options are the call-on-call, put-on-call, call-on-put, and put-on-put options and were first analyzed by Geske [25]. As the name suggests, a call-on-call option is the right to purchase at time $T_1$ a call option for an agreed price. If the option is exercised at $T_1$, then the holder receives the call option, which expires at the later time, $T_2$. Compound options are clearly examples of dual-expiry options.

The main issue in pricing the call-on-call is the exercise condition at time $T_1$. To determine this condition, let us fix the parameters of the option. Suppose that at $T_1$ we can, if we wish, buy the call option for the agreed price $c_1$, and suppose further the strike price of the option we purchase is $k_2$. The subscripts correspond to the times $T_1$ or $T_2$ at which exercise takes place. Clearly, we should exercise the call-on-call at time $T_1$ iff

$$C_{k_2}(x_1, \tau) > c_1; \qquad \tau = (T_2 - T_1).$$

That is, we exercise at $T_1$ only if the price $c_1$ we agree to pay for the call option is less than the value of the strike $k_2$ call option, with time $\tau = (T_2 - T_1)$ remaining to expiry.

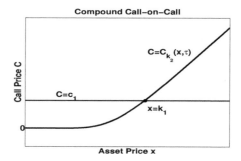

**Compound Call–on–Call**

**FIGURE 5.1:** Graphical representation of the solution $x = k_1$ of the transcendental equation $C_{k_2}(x, \tau) = c_1$

Let $x = k_1$ solve the equation $C_{k_2}(x, \tau) = c_1$. This equation, being transcendental, does not have a closed-form analytic solution. However, since $C_{k_2}(x, \tau)$ is a strictly monotonic increasing function of $x$ taking all values in $[0, \infty)$, the equation is guaranteed to have a unique solution $x = k_1$, for any value of $c_1 > 0$. This unique solution can, in principle, be readily obtained numerically by any of a number of root finding algorithms. Furthermore, for fixed dates $(T_1, T_2)$, and prices $(c_1, k_2)$ the calculation to determine $k_1$ need only be done once. The exercise condition to purchase the call option at $T_1$ is given simply by $x_1 > k_1$, since in this region the call option is worth more than $c_1$. From this discussion, it is clear that the time $T_1$ payoff of the call-on-call option will be

$$V_{cc}(x_1, T_1) = [C_{k_2}(x_1, \tau) - c_1]^+ = [C_{k_2}(x_1, \tau) - c_1]\, \mathbb{I}(x_1 > k_1)$$
$$= [Q^+_{k_2}(x_1, \tau; k_2) - c_1]\, \mathbb{I}(x_1 > k_1). \qquad (5.14)$$

In the last line above, we have expressed the strike $k_2$ call option as a first-order $Q$-option using (4.12). We can now clearly see from this representation, how to write down the pv of call-on-call by inspection. The first term

$$Q^+_{k_2}(x_1, T_1; k_2)\, \mathbb{I}(x_1 > k_1)$$

is by (5.11) a second-order $Q$-option, while the other term $c_1\, \mathbb{I}(x_1 > k_1)$ is the time $T_1$ payoff of an up-type bond binary. Hence, we obtain immediately that for all $t < T_1 < T_2$,

$$\boxed{V_{cc}(x, t) = Q^{++}_{k_1 k_2}(x, t; k_2) - c_1 B^+_{k_1}(x, \tau_1)} \qquad (5.15)$$

where $\tau_1 = (T_1 - t)$. Note once again, the recurring theme in this book, that we have priced the call-on-call option without performing a single integration. Basically, we have only manipulated the time $T_1$ payoff into a form in which we recognize as a portfolio of more elementary contracts. In the present case, a portfolio of a second-order $Q$-option and an up-type bond binary. Both these contracts have been priced earlier in the text. An inspection of these prices shows that the pv of the call-on-call option involves both the univariate and bivariate normal distribution function.

It is instructive to also write down the payoff of the call-on-call option at the second expiry date $T_2$. A little thought will show that the expression

$$V_{cc}(x_1, x_2; T_2) = [(x_2 - k_2)^+ - c_1 e^{r\tau}]\, \mathbb{I}(x_1 > k_1) \qquad (5.16)$$

is the correct result. However, in writing this expression, we should keep in mind that we still had to determine $k_1$ by considering the exercise condition at time $T_1$.

Using a similar notation as above, we can also determine by the same method, the pv's of the other three compound options: the put-on-call, call-on-put and put-on-put. We obtain the following:

$$\boxed{\begin{aligned} V_{pc}(x, t) &= c_1 B^-_{k_1}(x, \tau_1) - Q^{-+}_{k_1 k_2}(x, t; k_2) \\ V_{cp}(x, t) &= Q^{--}_{k_1 k_2}(x, t; k_2) - p_1 B^-_{k_1}(x, \tau_1) \\ V_{pp}(x, t) &= p_1 B^+_{k_1}(x, \tau_1) - Q^{+-}_{k_1 k_2}(x, t; k_2) \end{aligned}} \qquad (5.17)$$

The put options in the final two prices are also assumed to have strike price $k_2$ and trade price $p_1$. The parameter $k_1$ is obtained as the unique solution of the equation $P_{k_2}(x, \tau) = p_1$, provided $p_1 < k_2 e^{-r\tau}$. All these prices also involve both the univariate and bivariate normal distribution functions.

**Parity Relations**

The four compound options just priced satisfy two parity relations, which derive from their time $T_2$ payoffs :

$$V_{cc}(x_1, x_2; T_2) = \left[ (x_2 - k_2)^+ - c_1 e^{r\tau} \right] \mathbb{I}(x_1 > k_1)$$
$$V_{pc}(x_1, x_2; T_2) = \left[ c_1 e^{r\tau} - (x_2 - k_2)^+ \right] \mathbb{I}(x_1 < k_1)$$
$$V_{cp}(x_1, x_2; T_2) = \left[ (k_2 - x_2)^+ - p_1 e^{r\tau} \right] \mathbb{I}(x_1 < k_1)$$
$$V_{pp}(x_1, x_2; T_2) = \left[ p_1 e^{r\tau} - (k_2 - x_2)^+ \right] \mathbb{I}(x_1 > k_1).$$

It follows that

$$[V_{cc} - V_{pc}](x_1, x_2; T_2) = (x_2 - k_2)^+ - c_1 e^{r\tau}$$
$$[V_{cp} - V_{pp}](x_1, x_2; T_2) = (k_2 - x_2)^+ - p_1 e^{r\tau}.$$

Taking the pv of these last two equations gives us the parity relations

$$\boxed{\begin{aligned} V_{cc}(x, t) - V_{pc}(x, t) &= C_{k_2}(x, \tau_2) - c_1 e^{-r\tau_1} \\ V_{cp}(x, t) - V_{pp}(x, t) &= P_{k_2}(x, \tau_2) - p_1 e^{-r\tau_1} \end{aligned}} \qquad (5.18)$$

where $\tau_i = (T_i - t)$; $i = 1, 2$ and $t < T_1 < T_2$.

---

## 5.6  Chooser Options

A chooser option is another example of a dual expiry option, which gives the holder at time $T_1$, the choice of a European call or put option. In the case of a simple chooser, the strike price and expiry dates of the call and put are equal. In a complex chooser, the strikes and expiry dates are generally different. We shall suppose the the expiry dates of the two options are always at the same time $T_2 > T_1$, since nothing is really gained by assuming different dates. Chooser options were first introduced and analyzed by Rubinstein [64].

Simple chooser options can be priced in terms of first-order options only and is left as an Exercise Problem (see Q3) at the end of this chapter.

For the complex chooser we observe that the time $T_1$ payoff will be given by

$$V(x_1, T_1) = \max [C_h(x_1, \tau), \ P_k(x_1, \tau)]$$

where $\tau = (T_2 - T_1)$ and $(h, k)$ are the strikes of the call and put respectively.

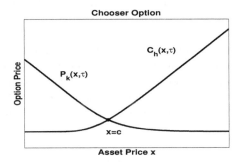

**FIGURE 5.2**: Graphical representation of the solution $x = c$ of the transcendental equation $C_h(x, \tau) = P_k(x, \tau)$

Let $x = c$ solve the transcendental equation $C_h(x, \tau) = P_k(x, \tau)$. This will have unique solution because $C_h(x, \tau)$ is monotonic increasing and $P_k(x, \tau)$ is monotonic decreasing in $x$. As for the compound options, the parameter $c$ will have to be calculated numerically, but for fixed strikes and exercise dates this calculation need only be done once. Then, at time $T_1$ we would choose the call option if $x_1 > c$, since in this region the call price exceeds the put price; and we would choose the put option similarly if $x_1 < c$. It follows that the chooser payoff at time $T_1$ can be written in the equivalent form

$$V(x_1, T_1) = Q_h^+(x_1, T_1; h)\mathbb{I}(x_1 > c) + Q_k^-(x_1, T_1; k)\mathbb{I}(x_1 < c)$$

after we write the call and put prices in terms of first-order $Q$ options. In this representation, we can now write down the pv of the chooser option as the sum of two second-order $Q$-options:

$$\boxed{V(x, t) = Q_{ch}^{++}(x, t; h) + Q_{ck}^{--}(x, t; k)} \tag{5.19}$$

for all $t < T_1 < T_2$. While this price for the chooser option looks quite succinct, it of course is quite a complex expression when written out in terms of its corresponding four bivariate normal distribution functions.

## 5.7    Reset Options

Suppose an investor holds a strike $h$ European call option which expires at time $T_1$. A reset option is another dual expiry option which gives its holder, at time $T_1$, the right to do one of three things: (i) for an agreed fee, extend the life of the option to $T_2 > T_1$ and simultaneously reset the strike from $h$ to $k$; (ii) exercise the original call if it is itm; and (iii) let the option expire worthless if it is otm. Reset options are also called *extendible* options and were first analyzed in detail by Longstaff [55].

Clearly there are many different types of reset options depending on whether the underlying option is a call or a put and depending also on the reset conditions. We shall therefore be content in this section to price the reset call option described above. We do so as another example of pricing by reorganizing the payoff into more basic constituents. Again, no formal integrations are required.

First, observe that the time $T_1$ payoff of the reset call option can be written as

$$V(x_1, T_1) = \max \left[ C_k(x_1, \tau) - c, \ (x_1 - h)^+ \right]$$

where $\tau = (T_2 - T_1)$ and $c$ is the reset fee referred to above. It is then clear we should reset if $C_k(x_1, \tau) - c > (x_1 - h)^+$, otherwise we should exercise the original call if it is itm or let it expire worthless if it is otm.

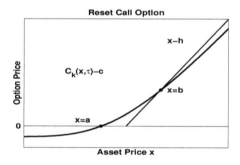

**FIGURE 5.3**: Sketch of reset call option payoff at time $T_1$ with critical asset prices $x = a$ solving $C_k(x, \tau) - c = 0$, and $x = b$ solving $C_k(x, \tau) - c = x - h$.

To find the critical asset prices at $T_1$ that determine which action to take, let $x = a$ and $x = b$ solve the two equations

$$C_k(x, \tau) = c \qquad \text{and} \qquad C_k(x, \tau) = x + c - h.$$

For all three scenarios to be possible, we shall require that the fixed parameters satisfy the condition $0 < c < ke^{-r\tau} - h$. Then the following will apply at time $T_1$:

1. allow the option to expire worthless if $0 < x_1 < a$

2. select the reset option if $a < x_1 < b$; and

3. exercise the original strike $h$ call if $x_1 > b$.

Hence, we may write the time $T_1$ payoff of the reset option as

$$V(x_1, T_1) = [Q_k^+(x_1, T_1; k) - c][\mathbb{I}(x_1 > a) - \mathbb{I}(x_1 > b)] + (x_1 - h)\mathbb{I}(x_1 > b)$$

where we have replaced the call option $C_k(x, \tau)$ by its equivalent $Q$-option representation and used the identity $\mathbb{I}(a < x < b) = \mathbb{I}(x > a) - \mathbb{I}(x > b)$. With this representation we see that $V(x_1, T_1)$ can be replicated by a portfolio of first and second order $Q$-options. The pv is then given by

$$\boxed{\begin{aligned} V(x, t) &= Q_{ak}^{++}(x, t; k) - Q_{bk}^{++}(x, t; k) \\ &\quad - c[B_a^+(x, \tau_1) - B_b^+(x, \tau_1)] + Q_b^+(x, \tau_1; h) \end{aligned}} \tag{5.20}$$

## 5.8  Simple Cliquet Option

Consider a European call option of strike price $k$ and expiry date $T_2$. Suppose the holder of this option is permitted to lock in the option's payoff at a fixed earlier time $T_1$. Such an option is called a *simple cliquet* or *ratchet* option. The advantage of this option is that even if the option finishes otm at time $T_2$, the holder may still receive a positive amount if the option was itm at time $T_1$. Of course this advantage must be reflected in its current price.

Such cliquet options can also be thought of as one-shout options with fixed shout time $T_1$. These options were first analyzed by Thomas [74]. There of course exist in the OTC market more complex cliquets in which locked-in payoffs can occur at multiple fixed times during the life of the option. We develop the technology to handle these more complex options in Chapter 10, but recognize that the simple cliquet above is just another example of a dual expiry option.

To price this option we start from its time $T_2$ payoff

$$V(x_1, x_2, T_2) = \max[(x_1 - k)^+, (x_2 - k)^+]. \tag{5.21}$$

The first term here represents the locked-in value equal to the call option payoff at time $T_1$. The second term is the call option payoff at time $T_2$. Note that for all times $T_1 < t < T_2$, the above payoff keeps $x_1$ constant. We shall exploit this property in pricing the option.

The cliquet option gives its holder the maximum of these two payoffs, which of course can still be zero if both payoffs are otm. As it stands, this payoff

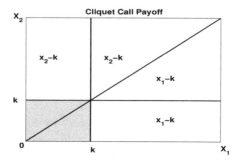

**FIGURE 5.4:** Sketch of cliquet payoff at time $T_2$. Shaded region has zero payoff.

structure does not permit an immediate valuation, but after a some manipulation we shall be able to do precisely that. Though it takes a little thought, the reader should see that the payoff can be written in the equivalent form

$$V(x_1, x_2, T_2) = (x_2 - k)^+ \mathbb{I}(x_1 < k) + [(x_2 - x_1)^+ + (x_1 - k)]\mathbb{I}(x_1 > k). \quad (5.22)$$

Now to price the option, we use a two step process. First we work out the value $V(x_1, T_1)$ at time $T_1$ by keeping $x_1$ constant. Then we go further back in time to find the value $V(x, t)$ for all $t < T_1$. The payoff in (5.22) contains three terms, which we discuss in detail below, at least for all times $T_1 < t < T_2$.

1. The first term $V_1(x_1, x_2, T_2) = (x_2 - k)^+ \mathbb{I}(x_1 < k)$ represent a strike $k$ call option $(x_2 - k)^+$ multiplied by the constant factor $\mathbb{I}(x_1 < k)$. Hence at time $T_1$ its value will be

$$V_1(x_1, T_1) = Q_k^+(x_1, T_1; k) \, \mathbb{I}(x_1 < k)$$

where $Q_k^+(x_1, T_1; k) \equiv C_k(x, \tau)$ with $\tau = (T_2 - T_1)$.

2. From (5.3), the second term $V_2(x_1, x_2, T_2) = (x_2 - x_1)^+ \mathbb{I}(x_1 > k)$ is seen to be the product of a forward start call payoff $(x_2 - x_1)^+$ and a constant term $\mathbb{I}(x_1 > k)$. It follows that at time $T_1$ the value of this second term becomes

$$V_2(x_1, T_1) = g(\tau)x_1 \, \mathbb{I}(x_1 > k); \qquad \tau = (T_2 - T_1),$$

where $g(\tau)$ is defined in (5.1) and $x_1 g(\tau)$ is the time $T_1$ value of the forward start option.

3. The third term, $V_3(x_1, x_2, T_2) = (x_1 - k)\mathbb{I}(x_1 > k)$ depends only on (the constant) $x_1$ and hence has time $T_1$ value

$$V_3(x_1, T_1) = e^{-r\tau}(x_1 - k)\mathbb{I}(x_1 > k); \qquad \tau = (T_2 - T_1).$$

That is, we have simply multiplied the $T_2$ payoff by the discount factor $e^{-r\tau}$ over $(T_1, T_2)$.

Putting all these terms together, we obtain

$$V(x_1, T_1) = Q_k^+(x_1, T_1; k)\,\mathbb{I}(x_1 < k) + [g(\tau)x_1 + e^{-r\tau}(x_1 - k)]\mathbb{I}(x_1 > k) \quad (5.23)$$

It is now a straightforward matter to price the option for any time $t < T_1$. We find

$$\boxed{V(x, t) = Q_{kk}^{-+}(x, t; k) + g(\tau)A_k^+(x, \tau_1) + e^{-r\tau}C_k(x, \tau_1)} \quad (5.24)$$

where as usual $\tau_1 = (T_1 - t)$ is the time remaining to time $T_1$. The price is seen to be a combination of a second-order $Q$-option, an up-type asset binary and a standard European call option.

**REMARK 5.2** The decomposition (5.22) of the payoff at time $T_2$ is not unique. This means there are alternative, but equivalent representations of the price. Indeed, the price given in the Thomas paper looks quite different from Equation (5.24), but the reader can be assured they are equivalent. □

---

## 5.9 Summary

We have studied here, through several different examples, the dual expiry (i.e., single-asset, 2-period) options. Generally, these exotics could be expressed as portfolios of first and second order binary options, and as such their prices in the BS economy, contained one or both univariate and bivariate normal distribution functions. Many exotics in this class have call or put like payoffs and to simplify the notation we introduced here the concept of second order $Q$-options. These are generalizations of the first order $Q$-options introduced in previous chapter. A very useful way of thinking of a second order $Q$-options, is as a binary option at time $T_1$ of a first order $Q$-option which expires at time $T_2$.

In Chapter 10 of this book we shall meet the general multi-period, multi-asset binary option, the so called $\mathbb{M}$-binary, which will include the (1-asset, 2-period) binaries considered in this chapter as special cases.

The next chapter is in a sense the logical sequel to the current one. Here we considered (1-asset, 2-period) options, while the next, on Two-Asset Rainbow options, is concerned with (2-asset, 1-period) options. This topic has been a popular one in the academic literature and includes some very famous results including Margrabe's Formula for an exchange option (the right to exchange one asset for another), and the Stulz formulae for calls and puts on the maximum or minimum of two assets.

## Exercise Problems

Assume standard BS dynamics and notation where appropriate.

1. Let $x_1, x_2$ denote the prices of asset $X$ at time $T_1, T_2$ with $T_1 < T_2$. Find the present values (for $t < T_1$) of binary options with the following time $T_2$ payoffs:

$$\mathbb{I}(x_2 > \alpha x_1); \quad x_1 \mathbb{I}(x_2 > \alpha x_1); \quad x_2 \mathbb{I}(x_2 > \alpha x_1)$$

all with $\alpha > 0$. Hence price a forward start call option with strike price set equal to $\alpha$ times the asset price at time $T_1$.

2. Determine the price of the 2-period derivative with payoff $V(x_1, x_2; T_2) = x_1/x_2$ using the three methods of Example 5.1.

3. A broking firm offers you a *combo-option* on stock $S$. This option pays at time $T_1$, a strike $a$, expiry $T_2$ European put if $0 < x_1 < a$; a strike $c$, expiry $T_2$ European call if $x_1 > c$; and the choice of a one share of $S$ or a ZCB with face-value $b$, otherwise. Here $x_1$ is the share price at time $T_1$. It can be assumed that $t < T_1 < T_2$, where $t$ is the current time and $0 < a < b < c$.

   (a) Determine the current BS price of the combo-option, and

   (b) find the risk-neutral probability that the combo-option finishes out-of-the-money at time $T_2$.

4. A *simple chooser* option gives its holder, at time $T_1$, the right to a call option or a put option, both with the same strike $k$ and expiry date $T_2 > T_1$. Show by a static replication argument, that the price of the simple chooser for $t < T_1$, can be written in terms of a standard European call of strike $k$ and time $\tau_2 = (T_2 - t)$ to expiry, and a standard European put of strike $k' = ke^{-r\tau}$, $(\tau = T_2 - T_1)$ and time $\tau_1 = (T_1 - t)$ to expiry. That is,

$$V(x, t) = C_k(x, \tau_2) + P_{k'}(x, \tau_1).$$

5. Verify that equation (5.19) for the price of a complex chooser option (where the strike prices are different), reduces to the formula in Q4 for the simple chooser, in the limit as $T_2 \to T_1 = T$ and $h \to k$.

6. A contract gives its holder at time $T_1$ the right to swap a strike $k$ call option for a strike $k$ put option, both options expiring at time $T_2 > T_1$. Determine the present value of this dual-expiry swap, assuming the holder receives the call and pays the put, if the swap is exercised.

7. Let $x_1$, $x_2$ denote the price of a given asset at times $T_1$, $T_2$ with $T_1 < T_2$. Determine the price of the 2-period binary options with time $T_2$ payoffs

$$V_1 = x_1 \mathbb{I}(x_1 > x_2) \mathbb{I}(x_1 > k); \qquad V_2 = x_2 \mathbb{I}(x_2 > x_1) \mathbb{I}(x_2 > k)$$

where $k$ is a positive constant. Hence or otherwise, price contracts with payoffs,

$$V_3 = \max(x_1,\, x_2,\, k) \qquad \text{and} \qquad V_4 = [\max(x_1,\, x_2) - k]^+ .$$

8. With $x_1, x_2$ defined as previously, price the two-period derivatives with time $T_2$ payoffs,

$$V_1(x_1, x_2; T_2) = \left( \frac{x_2}{x_1} - k \right)^+ ; \qquad V_2(x_1, x_2; T_2) = \left( \frac{x_1}{x_2} - k \right)^+$$

where $k$ is a positive strike price.

9. An investor is offered a contract, which at the future time $T_1$, gives the investor the choice of $c$ dollars in cash or a strike $k$ put option with expiry dated $T_2 > T_1$. Find the time $t < T_1$ value of the contract.

10. Price a reset put option in terms of the premium $p$, original strike $h$ and reset strike price $k < h$, assuming it is possible at time $T_1$ to do any of: reset, exercise immediately or let the option expire worthless.

—ooOoo—

# Chapter 6

## Two-Asset Rainbow Options

In this chapter we introduce the reader to a special class of multi-asset or rainbow options. Dual-asset options have an expiry $T$ payoff of the form $V(x, y, T) = f(x, y)$, where $x$ and $y$ denote prices of two distinct assets $X$ and $Y$, which are generally correlated. Examples include exchange options and call/put options on the maximum or minimum of two assets. Both these well known examples are analyzed in this chapter.

In the BS framework, future prices of $X$ and $Y$ follow correlated geometric Brownian motions. Thus if the prices of the two assets are observed to be $x$ and $y$ at time $t$, then their future prices at time $T > t$ will have the representation

$$X_T \stackrel{d}{=} x e^{(\mu_1 - \frac{1}{2}\sigma_1^2)\tau + \sigma_1 \sqrt{\tau} Z_1^P}$$
$$Y_T \stackrel{d}{=} y e^{(\mu_2 - \frac{1}{2}\sigma_2^2)\tau + \sigma_2 \sqrt{\tau} Z_2^P}$$

where $\tau = (T - t)$, $(\mu_1, \mu_2)$ are the growth rates of $(X, Y)$ and $(\sigma_1, \sigma_2)$ are their volatilities. Further, $(Z_1^P, Z_2^P)$ are joint standard Gaussian $N(0, 1; \rho)$ random variables with given correlation coefficient $\rho$. The super-script $P$ indicates that these are the asset price representations in the real world measure $P$. Of course, when pricing derivatives on these assets, we shall have to work in the EMM $Q$, and our asset price representations will then be

$$\left. \begin{array}{l} X_T \stackrel{d}{=} x e^{(r - \frac{1}{2}\sigma_1^2)\tau + \sigma_1 \sqrt{\tau} Z_1^Q} \\ Y_T \stackrel{d}{=} y e^{(r - \frac{1}{2}\sigma_2^2)\tau + \sigma_2 \sqrt{\tau} Z_2^Q} \end{array} \right\} \tag{6.1}$$

When using these expressions for the $(X_T, Y_T)$ we shall generally drop the super-script $Q$, it being understood that we are working under the risk-neutral measure.

The price of some derivative $V(x, y, t)$ at any time $t$, will satisfy the two-dimensional BS-pde (refer to Equation (3.31) with $n = 2$)

$$\boxed{V_t = r(V - xV_x - yV_y) - \tfrac{1}{2}(\sigma_1^2 x^2 V_{xx} + \sigma_2^2 y^2 V_{yy} + 2\rho\sigma_1\sigma_2 xy V_{xy})} \tag{6.2}$$

in the domain $t < T$ and $(x, y) \in \mathbb{R}_+^2$, with terminal value $V(x, y, T) = f(x, y)$. Subscripts on $V$ denote partial derivatives.

The corresponding FTAP for this two-asset derivative is given by the formula

$$\boxed{V(x,y,t) = e^{-r\tau}\,\mathbb{E}_Q\{f(X,Y)\,|\,\mathcal{F}_t\}; \qquad \tau = (T-t)}$$ (6.3)

where we have written $(X,Y)$ for $(X_T,Y_T)$. Of course, (6.3) is a formal solution of the pde (6.2).

---

## 6.1 Two-Asset Binaries

Most two-asset exotic options, as for their single asset counterparts, can be expressed as portfolios of binary portfolios. We define and derive prices in this section, for a general class of binary options that have closed-form, analytic expressions in the BS economy. There are four basic types of binary options within this category, and their payoffs and present values are described in the following theorem.

**THEOREM 6.1**

*Consider the four general two-asset binary options, with expiry $T$ payoffs*

$$V_1(x,y,T) = x^p y^q\,\mathbb{I}(s_1 x > s_1 h)\mathbb{I}(s_2 y > s_2 k)$$
$$V_2(x,y,T) = x^p y^q\,\mathbb{I}(sx > sy)$$
$$V_3(x,y,T) = x^p y^q\,\mathbb{I}(s_1 x > s_1 h)\mathbb{I}(sx > sy)$$
$$V_4(x,y,T) = x^p y^q\,\mathbb{I}(s_2 y > s_2 k)\mathbb{I}(sx > sy)$$

*where $(s_1, s_2, s) \in \{\pm\}$, and $(p,q) \in \mathbb{R}$. Then, their present values are given respectively by,*

$$\boxed{\begin{aligned}
V_1(x,y,t) &= x^p y^q e^{\mu\tau}\,\mathcal{N}(s_1 d_1, s_2 d_2;\ s_1 s_2 \rho)\\
V_2(x,y,t) &= x^p y^q e^{\mu\tau}\,\mathcal{N}(sd)\\
V_3(x,y,t) &= x^p y^q e^{\mu\tau}\,\mathcal{N}(s_1 d_1, sd;\ s_1 s \rho_1)\\
V_4(x,y,t) &= x^p y^q e^{\mu\tau}\,\mathcal{N}(s_2 d_2, sd;\ s_2 s \rho_2)
\end{aligned}}$$ (6.4)

*where*

$$\boxed{\begin{aligned}
\mu &= -r + p(r - \tfrac{1}{2}\sigma_1^2) + q(r - \tfrac{1}{2}\sigma_2^2) +\\
&\qquad \tfrac{1}{2}(p^2\sigma_1^2 + q^2\sigma_2^2 + 2pq\rho\sigma_1\sigma_2)\\
\rho_1 &= (\sigma_1 - \rho\sigma_2)/\sigma; \qquad \rho_2 = (\rho\sigma_1 - \sigma_2)/\sigma\\
\sigma^2 &= \sigma_1^2 + \sigma_2^2 - 2\rho\sigma_1\sigma_2
\end{aligned}}$$ (6.5)

*and*

$$d_1(x,\tau) = \frac{\log(x/h)+[r+(p-\frac{1}{2})\sigma_1^2+q\rho\sigma_1\sigma_2]\tau}{\sigma_1\sqrt{\tau}}$$

$$d_2(y,\tau) = \frac{\log(y/k)+[r+(q-\frac{1}{2})\sigma_2^2+p\rho\sigma_1\sigma_2]\tau}{\sigma_2\sqrt{\tau}} \qquad (6.6)$$

$$d(x,y,\tau) = \frac{\log(x/y)+[(p-\frac{1}{2})\sigma_1^2-(q-\frac{1}{2})\sigma_2^2+(q-p)\rho\sigma_1\sigma_2]\tau}{\sigma\sqrt{\tau}}$$

While this theorem seems exceedingly complicated, it nevertheless has the necessary generality to permit pricing of most of the two-asset rainbow options likely to arise in practice. We present below, for purposes of illustration, the proof for the pv of $V_3(x,y,t)$. The proofs for the remaining cases follow similarly. We point out that Chapter 10 analyzes a more general set of binary options, which includes these four binaries as special cases.

**PROOF**   Consider the two asset binary option $V_3(x,y,t)$ for the case $s_1 = s = +$. Then by (6.3), together with (6.1) we obtain:

$$V_3(x,y,t) = e^{-r\tau}\mathbb{E}_Q\{X^pY^q \cdot \mathbb{I}(X>h)\mathbb{I}(X>Y)\}$$
$$= x^p y^q \cdot e^{[-r+p(r-\frac{1}{2}\sigma_1^2)+q(r-\frac{1}{2}\sigma_2^2)]\tau} \cdot$$
$$\mathbb{E}_Q\{e^{\sqrt{\tau}(p\sigma_1 Z_1+q\sigma_2 Z_2)}\mathbb{I}(Z_1>-d_1')\mathbb{I}(Z>-d')\}$$

where

$$d_1'(x,\tau) = [\log(x/h) + (r - \tfrac{1}{2}\sigma_1^2)\tau]/\sigma_1\sqrt{\tau}$$
$$d'(x,y,\tau) = [\log(x/y) + \tfrac{1}{2}(\sigma_2^2 - \sigma_1^2)\tau]/\sigma\sqrt{\tau}.$$

Observe that the condition $X > h$ corresponds to $Z_1 > -d_1'$ and $X > Y$ to $Z > -d'$. To see the latter, note that condition $X > Y$, on taking logs becomes

$$\log x + (r - \tfrac{1}{2}\sigma_1^2)\tau + \sigma_1\sqrt{\tau}Z_1 > \log y + (r - \tfrac{1}{2}\sigma_2^2)\tau + \sigma_2\sqrt{\tau}Z_2$$

which is equivalent to $Z > -d'$ if we define

$$Z = \frac{\sigma_1 Z_1 - \sigma_2 Z_2}{\sigma},$$

with $\sigma^2 = \sigma_1^2 + \sigma_2^2 - 2\rho\sigma_1\sigma_2$. This definition ensures that $Z \sim N(0,1)$ is a standard Gaussian rv, since $\text{corr}(Z_1, Z_2) = \rho$. Then, it is a simple matter to derive:

$$\text{corr}(Z, Z_1) = \frac{\sigma_1}{\sigma}\text{corr}(Z_1, Z_1) - \frac{\sigma_2}{\sigma}\text{corr}(Z_1, Z_2) = \frac{\sigma_1 - \rho\sigma_2}{\sigma} = \rho_1$$

$$\text{corr}(Z, Z_2) = \frac{\sigma_1}{\sigma}\text{corr}(Z_1, Z_2) - \frac{\sigma_2}{\sigma}\text{corr}(Z_2, Z_2) = \frac{\rho\sigma_1 - \sigma_2}{\sigma} = \rho_2.$$

Now, an application of the 2D Gaussian Shift Theorem (3.22), simplifies the expectation to:

$$V_3(x, y, t) = x^p y^q \cdot e^{\mu \tau} \cdot \mathbb{E}_Q\{\mathbb{I}(Z_1 + a' > -d_1') \cdot \mathbb{I}(\frac{\sigma_1(Z_1 + a') - \sigma_2(Z_2 + b')}{\sigma} > -d')\}$$

where $a' = (p\sigma_1 + pq\sigma_2)\sqrt{\tau}$ and $b' = (q\sigma_2 + pp\sigma_2)\sqrt{\tau}$. A little algebra then reduces the result to

$$V_3(x, y, t) = x^p y^q \cdot e^{\mu \tau} \cdot \mathbb{E}_Q\{\mathbb{I}(Z_1 > -d_1) \cdot \mathbb{I}(Z > -d)\},$$

where $d_1 = d_1' + a'$ and $d = d' + b'$. Symmetry of the Gaussian rv's allows us to then write

$$V_3(x, y, t) = x^p y^q \cdot e^{\mu \tau} \cdot \mathbb{E}_Q\{\mathbb{I}(Z_1 < d_1) \cdot \mathbb{I}(Z < d)\}$$

which leads to the final result

$$V_3(x, y, t) = x^p y^q \cdot e^{\mu \tau} \cdot \mathcal{N}(d_1, d; \rho_1)$$

since $\mathrm{corr}(Z, Z_1) = \rho_1$. This completes the proof. □

**REMARK 6.1** The parameters $(p, q)$ that appear in theorem 6.1 permit a wide range of possible two-asset payoff structures. Most values will have $p$ and $q$ equal to zero or one corresponding to two-asset bond or asset binaries respectively. However, other values are possible. For example, options on the geometric mean of $X$ and $Y$ will have payoffs which depend on the expression $\sqrt{xy}$, which corresponds to $p = q = \frac{1}{2}$.

While the theorem refers to four basic types, other binary options can be inferred from them. For example, consider the two-asset binary with payoff $V(x, y, T) = x^p y^q \mathbb{I}(y > k)$. This first-order, two-asset binary option can be obtained from $V_1(x, y, T)$ by taking the limit $h \to 0$ (with $s_1 = +$). But $h$ only appears in the price $V_1(x, y, t)$ in the term $d_1(x, \tau)$, and in the limit, $d_1$ goes to plus infinity. The bivariate normal in this case reduces to a univariate normal, and we actually obtain for this binary option, the price

$$V(x, y, t) = x^p y^q e^{\mu \tau} \mathcal{N}(d_2).$$

□

### Example 6.1
Consider the two asset binary portfolio called a $C$−brick, introduced by Heynen and Kat in [33]. This binary has a time $T$ payoff equal to

$$V(x, t, T) = \mathbb{I}(k_1 < x < k_2) \cdot \mathbb{I}(k_3 < y < k_4).$$

Thus you get one dollar if the asset price pair $(x, y)$ lies inside a rectangular region, and zero otherwise.

To price this derivative, it is only necessary to decompose it into an equivalent portfolio of recognizable binaries. This can be achieved as follows,

$$
\begin{aligned}
V(x, y, T) &= [\mathbb{I}(x > k_1) - \mathbb{I}(x > k_2)] \cdot [\mathbb{I}(y > k_3) - \mathbb{I}(y > k_4)] \\
&= \mathbb{I}(x > k_1)\mathbb{I}(y > k_3) - \mathbb{I}(x > k_1)\mathbb{I}(y > k_4) \\
&\quad - \mathbb{I}(x > k_2)\mathbb{I}(y > k_3) + \mathbb{I}(x > k_2)\mathbb{I}(y > k_4).
\end{aligned}
$$

The present value can then be obtained from derivative $V_1$ in theorem 6.1 with $p = q = 0$ and $s_1 = s_2 = 1$. This gives the result

$$
V(x, y, t) = e^{-r\tau}[\mathcal{N}(d_1, d_3; \rho) - \mathcal{N}(d_1, d_4; \rho) - \mathcal{N}(d_2, d_3; \rho) + \mathcal{N}(d_2, d_4; \rho)]
$$

where

$$
\begin{aligned}
d_i(x, \tau) &= [\log(x/k_i) + (r - \tfrac{1}{2}\sigma_1^2)\tau]/\sigma_1\sqrt{\tau}; & i &= (1, 2) \\
d_i(y, \tau) &= [\log(y/k_i) + (r - \tfrac{1}{2}\sigma_2^2)\tau]/\sigma_2\sqrt{\tau}; & i &= (3, 4).
\end{aligned}
$$

$\square$

## 6.2 The Exchange Option

Probably the first two-asset exotic option to be analyzed was the exchange option: the right to exchange one asset for another at a fixed time $T$ in the future. Assuming we receive asset $X$ and pay asset $Y$ in the exchange, then the payoff at time $T$ will be

$$
V(x, y, T) = (x - y)^+ = (x - y)\mathbb{I}(x > y) \tag{6.7}
$$

Clearly this is equivalent to a portfolio of two-asset binaries: long one contract that pays $x\mathbb{I}(x > y)$ and short a contract that pays $y\mathbb{I}(x > y)$. According to theorem 6.1, the first contract is type $V_2$ with $(p, q, s) = (1, 0, +)$ and the second is also type $V_2$ with $(p, q, s) = (0, 1, +)$. Applying these to the price $V_2(x, y, t)$ in (6.4) we obtain the pv of the exchange option to be

$$
V(x, y, t) = x\mathcal{N}(d) - y\mathcal{N}(d'); \quad [d, d'] = \frac{\log(x/y) \pm \tfrac{1}{2}\sigma^2\tau}{\sigma\sqrt{\tau}} \tag{6.8}
$$

where $\sigma^2 = \sigma_1^2 + \sigma_2^2 - 2\rho\sigma_1\sigma_2$ and $\tau = (T - t)$ is the time remaining to expiry. Equation (6.8) is often referred to as the Margrabe Formula, since it was first derived by Margrabe [57] in 1978. Note that the formula is completely independent of the risk-free rate $r$.

One cannot help but notice the similarity between the Margrabe Formula and the price of a standard European call option (4.7). This is of course no accident, because a standard European call option is in fact an exchange option — it is an exchange of asset $X$ for $k$ units of (riskless) cash. Indeed, we can recover the BS call option price from the Margrabe Formula as follows. The $k$ units of cash represent asset $Y$ and satisfy the ode $dy = rydt$ with $y = k$ at $t = T$. This ode has solution $y = ke^{-r(T-t)} = ke^{-r\tau}$, and can also be considered as an sde with zero volatility, i.e., $\sigma_2 = 0$. Substituting $y = ke^{-r\tau}$ and $\sigma_2 = 0$ into the Margrabe Formula indeed recovers the standard BS call option price.

## PDE Method

It is instructive, from a pedagogical perspective, to see how the Margrabe Formula also derives from the pde method to option pricing. We demonstrate this approach here. Our task is to solve (6.2) subject to the terminal payoff $V(x, y, T) = f(x, y) = (x - y)^+$. Observe first, that the payoff $(x - y)^+$ is homogeneous of degree-1 in $(x, y)$. That is

$$f(\lambda x, \lambda y) = \lambda f(x, y); \qquad \text{for all } \lambda > 0.$$

It follows from purely financial considerations that the price at any time before expiry should also be homogeneous of degree-1. Indeed the BS-pde automatically satisfies this condition, since it is a simple matter to show that if $V(x, y, t)$ satisfies the 2D BS-pde, then so does $V(\lambda x, \lambda y, t)$ for any constant $\lambda > 0$. Hence, if $V(x, y, T)$ is homogeneous of degree-1, then so is $V(x, y, t)$.

Now there is a well-known consequence of homogeneous functions, called Euler's Equation. In its general form it states, that if $V(x, y)$ is homogeneous of degree $n$ so that $V(\lambda x, \lambda y) = \lambda^n V(x, y)$, then $V(x, y)$ satisfies the Euler pde

$$xV_x(x, y) + yV_y(x, y) = nV(x, y), \tag{6.9}$$

where subscripts denote partial derivatives. The proof, which is often set as a question to students studying a course in partial differentiation, is straightforward. Simply differentiate the above homogeneity equation wrt $\lambda$ and then set $\lambda = 1$.

Since our exchange option is homogeneous degree-1 (i.e., $n = 1$ in Euler's pde), we have $xV_x + yV_y = V$ and the BS-pde (6.2) reduces to

$$V_t = -\tfrac{1}{2}(\sigma_1^2 x^2 V_{xx} + \sigma_2^2 y^2 V_{yy} + 2\rho\sigma_1\sigma_2 xy V_{xy}).$$

The dependence on the risk-free rate $r$ disappears completely. Euler's equation also allows us to reduce the BS-pde even further, since we see from $xV_x + yV_y =$

$V$, differentiation wrt both $x$ and $y$, leads to

$$V_x + xV_{xx} + yV_{xy} = V_x \quad \Rightarrow \quad x^2V_{xx} = -xyV_{xy}$$
$$xV_{xy} + V_y + yV_{yy} = V_y \quad \Rightarrow \quad y^2V_{yy} = -xyV_{xy}.$$

Using these in the BS-pde above, reduces it to

$$V_t = -\tfrac{1}{2}\sigma^2 x^2 V_{xx}; \qquad V(x,y,T) = (x-y)^+,$$

where $\sigma^2 = \sigma_1^2 + \sigma_2^2 - 2\rho\sigma_1\sigma_2$. Under this reduction, the pde is seen to be a one-dimensional BS-pde for asset $X$, with volatility $\sigma$ and zero risk-free rate, in which asset price $y$ plays the role of a parameter, appearing only in the payoff function as a strike price. But we have already solved this problem. It obviously represents a European call option of strike price $y$ under zero interest rates. In this scenario, equation (4.7) recovers the Margrabe Formula.

---

## 6.3 Options on the Minimum/Maximum of Two Assets

We consider in this section call and put options on the minimum and maximum of two assets $X$ and $Y$. These options were first considered and analyzed by Stulz [73]. We shall demonstrate here how easy it is to obtain Stulz's formulae using the two-asset binary formulae given in theorem 6.1.

The four payoffs of these two-asset exotic options are respectively

$$V_c^{\min}(x,y,T) = [\min(x,y) - k]^+$$
$$V_c^{\max}(x,y,T) = [\max(x,y) - k]^+$$
$$V_p^{\min}(x,y,T) = [k - \min(x,y)]^+$$
$$V_p^{\max}(x,y,T) = [k - \max(x,y)]^+.$$

Consider first, the call on the minimum. We can write the payoff in the equivalent, more suggestive form,

$$V_c^{\min}(x,y,T) = [\min(x,y) - k]^+$$
$$= x\mathbb{I}(x>k)\mathbb{I}(x<y) + y\mathbb{I}(y>k)\mathbb{I}(x>y) - k\mathbb{I}(x>k)\mathbb{I}(y>k).$$

There are several ways one can demonstrate the equivalence of these payoffs, but the simplest is to sketch them in the positive quadrant of the $xy$−plane. The lines $x = k$, $y = k$ and $x = y$ divide this quadrant into six regions. The payoff in region $[x > k,\ x < y]$ is $(x - k)$, while in region $[y > k,\ x > y]$ it is $(y - k)$. The payoff is zero in the other four regions. Figure 6.1 provides a sketch of this 2D payoff.

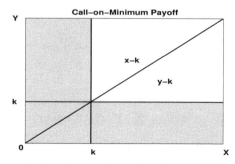

**FIGURE 6.1:** Sketch of Call-on-Minimum of two assets described by $V_c^{\min}(x, y, T)$. Shaded regions have zero payoff.

Now, it is clear that the call on the minimum has a payoff which can be replicated by three two-asset binaries. Using the pricing formulae given in theorem 6.1, we immediately obtain

$$V_c^{\min}(x, y, t) = x\mathcal{N}(d_1, -d; -\rho_1) + y\mathcal{N}(d_2, d'; \rho_2) - ke^{-r\tau}\mathcal{N}(d_1', d_2'; \rho)$$
(6.10)

where

$$[d_1, d_1'] = [\log(x/k) + (r \pm \tfrac{1}{2}\sigma_1^2)\tau]/\sigma_1\sqrt{\tau}$$
$$[d_2, d_2'] = [\log(y/k) + (r \pm \tfrac{1}{2}\sigma_2^2)\tau]/\sigma_2\sqrt{\tau}$$
$$[d, d'] = [\log(x/y) + \pm\tfrac{1}{2}\sigma_2\tau]/\sigma\sqrt{\tau}.$$

The parameters $(\sigma, \rho_1, \rho_2)$ are those previously defined in theorem 6.1.

We can proceed in two different ways to price the call on the maximum of two assets. The first follows a similar procedure used to price the call on the minimum of two assets, while the second method uses parity considerations. Using the first approach, we write the call on the maximum of two assets as

$$V_c^{\max}(x, y, T) = [\max(x, y) - k]^+$$
$$= x\mathbb{I}(x > k)\mathbb{I}(x > y) + y\mathbb{I}(y > k)\mathbb{I}(x < y) - k[1 - \mathbb{I}(x < k)\mathbb{I}(y < k)]$$

This payoff is sketched in Figure 6.2.

Hence, by inspection, we price the call-on-maximum of two assets as the pv of the replicating portfolio of two asset binaries, and obtain

$$V_c^{\max}(x, y, t) = x\mathcal{N}(d_1, d; \rho_1) + y\mathcal{N}(d_2, -d'; -\rho_2)$$
$$-ke^{-r\tau}[1 - \mathcal{N}(-d_1', -d_2'; \rho)]$$
(6.11)

The second method, using parity, derives from the observation that

$$V_c^{\min}(x, y, T) + V_c^{\max}(x, y, T) = (x - k)^+ + (y - k)^+.$$

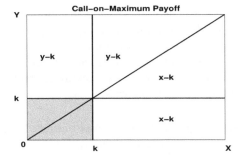

**FIGURE 6.2**: Sketch of Call-on-Maximum of two assets described by $V_c^{\max}(x, y, T)$. Shaded regions have zero payoff.

Hence, for any time $t < T$, we must have

$$\boxed{V_c^{\min}(x, y, t) + V_c^{\max}(x, y, t) = C_k(x, \tau) + C_k(y, \tau)} \qquad (6.12)$$

That is, the sum of the calls on the minimum and maximum of two assets must equal the sum of a standard call on asset $X$ and a standard call on asset $Y$. The two approaches can be reconciled by using the symmetry properties of the bivariate normal presented in Section 3.7. We then find, summing Equations (6.10) and (6.11),

$$\begin{aligned} V_c^{\min}(x, y, t) + V_c^{\max}(x, y, t) &= x\mathcal{N}(d_1) + y\mathcal{N}(d_2) - ke^{-r\tau}[\mathcal{N}(d_1') + \mathcal{N}(d_2')] \\ &= C_k(x, \tau) + C_k(y, \tau) \end{aligned}$$

as required.

We leave the pricing of put options on the minimum and maximum of two assets in the Exercise Problems for this chapter.

---

## 6.4 Product and Quotient Options

These options have found application in the foreign equity and FX markets. Let $X, Y$ denote the two asset prices at expiry $T$. Then the payoffs of product and quotient calls and puts are given in the table below.

| Two-Asset Rainbow Options | | |
|---|---|---|
| | *Product* | *Quotient* |
| *Call* | $(XY - k)^+$ | $(X/Y - k)^+$ |
| *Put* | $(k - XY)^+$ | $(k - X/Y)^+$ |

Observe that the quotient calls and puts are just portfolios of the two-asset binary options considered in Section 6.1 and can be priced by these formulae alone. However, the product calls and puts, which involve exercise conditions like $\mathbb{I}(XY > k)$ and $\mathbb{I}(XY < k)$, are not in the list provided. Still, they are readily priced using similar methods. A general method in Chapter 10 on M-binaries includes such product binaries.

We shall price here the quotient options and relegate the product option to The Exercise Problem for the current chapter.

For the quotient call, the payoff is

$$V_{qc}(x, y, T) = (x/y - k)\mathbb{I}(x > yk) = k(x/z - 1)\mathbb{I}(x > z); \qquad z = ky.$$

If $Y$ follows gBm with volatility $\sigma_2$, then so does $Z = kY$, when $k$ is a positive constant. Thus we only need to substitute $z = ky$ in the two-asset binary formulae for $V_i(x, z, t)$.

For the term $V_a = (x/z)\mathbb{I}(x > z)$

Here we have $p = 1; q = -1$ so we find from $V_2(x, z, t)$ of Equations (6.4), (6.5) and (6.6),

$$V_a(x, z, t) = (x/z)e^{\mu\tau} \mathcal{N}[d(x, z, \tau)]$$

with

$$\mu = -r + \sigma_2(\sigma_2 - \rho\sigma_1); \quad d = \frac{\log(x/z) + (\frac{1}{2}\sigma_2^2 - \frac{1}{2}\sigma_1^2 + \sigma^2)\tau}{\sigma\sqrt{\tau}}.$$

For the term $V_b = \mathbb{I}(x > z)$

Take $p = q = 0$, to obtain

$$V_b(x, z, t) = e^{-r\tau} \mathcal{N}[d'(x, z, \tau)]$$

with

$$d'(x, z, \tau) = \frac{\log(x/z) + \frac{1}{2}(\sigma_2^2 - \sigma_1^2)\tau}{\sigma\sqrt{\tau}} = d(x, z, \tau) - \sigma\sqrt{\tau}.$$

When the two terms are put together, $V_{qc}(x, y, t) = k(V_a - V_b)(x, z, t)$, we obtain

$$\boxed{V_{qc}(x, y, t) = (x/y)e^{\mu\tau} \mathcal{N}[d(x, ky, \tau)] - ke^{-r\tau} \mathcal{N}[d'(x, ky, \tau)]} \qquad (6.13)$$

We observe that this price agrees with a similar formula provided by Zhang [78].

## 6.5 ICIAM Option Competition

The fifth International Congress on Industrial and Applied Mathematics (ICIAM) was held in Sydney, Australia in 2003. The congress included an imbedded colloquium on Financial Mathematics and within this forum an Option Pricing Competition was held. Participants were invited to submit their solutions to pricing the following two-asset binary option, assuming standard 2D, BS dynamics.

$$V(x,y,T) = \mathbb{I}(\sqrt{xy}>c) \cdot \mathbb{I}(x<a) \cdot \mathbb{I}(y<b) \tag{6.14}$$

where $(a,b,c)$ are positive constants satisfying, $ab > c^2$. This condition is required to ensure that the payoff is not identically zero.

The problem has a number of unusual features, which make its pricing quite challenging. First, the exercise condition $\mathbb{I}(\sqrt{xy} > c)$ needs to be addressed, since it is not one of the standard conditions. Second and more subtly, the payoff contains a product of three exercise conditions involving only two asset prices. As we shall explain, payoffs of this kind will require special attention.

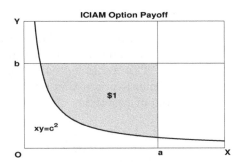

**FIGURE 6.3**: Sketch of ICIAM Option Competition payoff. The binary option pays \$1 in the shaded region and zero everywhere else.

To price this derivative we begin by writing in the usual manner, the asset prices under the EMM $Q$ at time $T$, as

$$X = xe^{\mu_1\tau+\sigma_1\sqrt{\tau}Z_1}; \qquad Y = ye^{\mu_2\tau+\sigma_2\sqrt{\tau}Z_2}$$

where $\mu_i = (r - \frac{1}{2}\sigma_i^2)$, $\tau = (T-t)$ and $(Z_1, Z_2) \sim N(0,1;\rho)$. Then, it follows that the product

$$XY = xye^{(\mu_1+\mu_2)\tau+\sigma\sqrt{\tau}Z}$$

where

$$Z = \frac{\sigma_1 Z_1 + \sigma_2 Z_2}{\sigma}; \quad \sigma^2 = \sigma_1^2 + \sigma_2^2 + 2\rho\sigma_1\sigma_2.$$

It should be clear with $Z$ so defined, that $Z \sim N(0,1)$ with correlation structure

$$\rho_1 = \text{corr}\{Z, Z_1\} = \frac{\sigma_1 + \rho\sigma_2}{\sigma}; \qquad \rho_2 = \text{corr}\{Z, Z_2\} = \frac{\sigma_2 + \rho\sigma_1}{\sigma}.$$

Let us now define the three parameters

$$\left.\begin{array}{l} d_a = [\log(x/a) + \mu_1\tau]/\sigma_1\sqrt{\tau} \\ d_b = [\log(y/b) + \mu_2\tau]/\sigma_2\sqrt{\tau} \\ d_c = [\log(xy/c^2) + (\mu_1+\mu_2)\tau]/\sigma\sqrt{\tau} \end{array}\right\} \qquad (6.15)$$

Then the three conditions $X < a$, $Y < b$ and $XY > c^2$ are readily shown to correspond to $Z_1 < -d_a$, $Z_2 < -d_b$ and $Z > -d_c$ respectively. Furthermore, let $Z_3 = -Z$, so that $Z > -d_c$ corresponds to $Z_3 < d_c$ and for $(i = 1,2)$, $\text{corr}\{Z_i, Z_3\} = -\rho_i$. The present value of the binary can then be determined from

$$\begin{aligned} V(x,y,t) &= e^{-r\tau}\,\mathbb{E}_Q\{\mathbb{I}(X<a)\cdot\mathbb{I}(Y<b)\cdot\mathbb{I}(XY>c^2)\} \\ &= e^{-r\tau}\,\mathbb{E}_Q\{\mathbb{I}(Z_1<-d_a)\cdot\mathbb{I}(Z_2<-d_b)\cdot\mathbb{I}(Z_3<d_c)\}. \quad (6.16) \end{aligned}$$

At this point, one may be tempted to write down the price as

$$V(x,y,t) = e^{-r\tau}\,\mathcal{N}_3(\boldsymbol{d};\,R)$$

in terms of a trinomial normal distribution function, where $\boldsymbol{d}$ denotes the vector with components $[-d_a, -d_b, d_c]$ and $R$ is the correlation matrix for the three Gaussian rv's $(Z_1, Z_2, Z_3)$. This would indeed be correct if the matrix $R$ were non-singular. But the multi-variate Gaussian cdf is undefined for singular correlation matrices. In our problem,

$$R = \begin{bmatrix} 1 & \rho & -\rho_1 \\ \rho & 1 & -\rho_2 \\ -\rho_1 & -\rho_2 & 1 \end{bmatrix}$$

and it is easily demonstrated that $\det R = 0$. Indeed, this must be the case because $Z_3$ is a linear combination of $Z_1$ and $Z_2$. So how are we to proceed?

The trick required to handle the order-3 exercise condition is to use the *Inclusion-Exclusion Principle* to express it as a sum (or union) of order-2 exercise conditions only.

## The Inclusion-Exclusion Principle
Let $A_i$ for $i = 1, 2, \ldots, n$ be a collection of sets such that $\cup_i A_i = \Omega$. Then the *Inclusion-Exclusion Principle* states that

$$\Omega = \sum_i A_i - \sum_{i,j}(A_iA_j) + \sum_{i,j,k}(A_iA_jA_k) + \ldots + (-1)^{n-1}(A_1A_2\ldots A_n) \quad (6.17)$$

In this representation of the principle, sums are over all distinct elements only and products mean "set intersection," i.e., $A_i A_j = A_i \cap A_j$. Furthermore, the sum of two sets means: inclusion of elements of the sets with multiplicity counted; difference of two sets means: exclusion of the elements of the subtracting set, multiplicity counted.

Let $\bar{A} = \Omega - A$ denote the complement of $A$ wrt $\Omega$. That is, the elements of $\bar{A}$ are all those that are in $\Omega$, but not in $A$. Now consider the case of three sets $(A, B, C)$, which is relevant to the ICIAM option pricing problem. The Inclusion-Exclusion Principle then states,

$$\Omega = (A + B + C) - (AB + AC + BC) + ABC$$

or equivalently

$$ABC = \Omega - (A + B + C) + (AB + BC + AC).$$

Now,

$$\bar{A}\bar{B} = (\Omega - A)(\Omega - B) = \Omega - A - B + AB$$
$$\bar{B}C = (\Omega - B)C = C - BC.$$

These follow from the property that $\Omega A = A$ for any set $A$. Hence, we have the interesting decomposition, that for any sets $A, B, C$,

$$ABC = AC - \bar{B}C + \bar{A}\bar{B}. \tag{6.18}$$

This is the result we seek, because it represents the product (or intersection) of three sets in terms of sums of products of only two sets. This decomposition is obviously not unique, since we can interchange any the $(A, B, C)$ without changing the outcome. For instance, interchanging $A$ and $B$, leads to the equivalent decomposition

$$ABC = BC - \bar{A}C + \bar{A}\bar{B}.$$

Now suppose we define the sets: $A = \{x < a\}$, $B = \{x < b\}$ and $C = \{xy > c^2\}$. Then, (6.18) allows us to write the payoff function (6.14) in the equivalent form

$$V(x, y, T) = \mathbb{I}(x < a)\mathbb{I}(xy > c^2) - \mathbb{I}(y > b)\mathbb{I}(xy > c^2) + \mathbb{I}(x > a)\mathbb{I}(y > b).$$

Thus, using the Inclusion-Exclusion Principle we have been able to write down the order-3 binary payoff as a sum of order-2 binary payoffs. The pricing problem now can be carried out without further difficulty, as the sum of these three second order binaries. We obtain

$$\begin{aligned}
V_1(x, y, t) &= e^{-r\tau} \mathbb{E}_Q\{\mathbb{I}(x < a) \cdot \mathbb{I}(xy > c^2)\} \\
&= e^{-r\tau} \mathbb{E}_Q\{\mathbb{I}(Z_1 < -d_a) \cdot \mathbb{I}(Z_3 < d_c)\} \\
&= e^{-r\tau} \mathcal{N}(-d_a, d_c; -\rho_1)
\end{aligned}$$

with similar expressions for the other two terms. Putting these altogether leads to

$$\boxed{V(x, y, t) = e^{-r\tau} \left[ \mathcal{N}(-d_a, d_c; -\rho_1) - \mathcal{N}(d_b, d_c; \rho_2) + \mathcal{N}(d_a, d_b; \rho) \right]} \quad (6.19)$$

for the price of the two asset binary, where $(d_a, d_b, d_c)$ were defined in (6.15).

**REMARK 6.2**    The non-uniqueness of the decomposition (6.18) leads to alternative expressions for the price of the option. These expressions are of course all equivalent to each other and can be derived from the symmetry properties of the bivariate normal distribution function.

This example was included, not so much because of its intrinsic interest, but rather as an illustration of how the Inclusion-Exclusion Principle can be used as an important tool to help price certain classes of exotic options.    ▯

**Postscript**
The ICIAM option pricing competition was won by Pavel Shevchenko of CSIRO, Division of Mathematics, Informatics and Statistics (Sydney, Australia). His approach was to evaluate the integrals corresponding to (6.16), which resulted in quite a different expression for the price of the option. The two prices were eventually reconciled using the identity described in Q10 Exercise Problems of Chapter 3.

---

## 6.6    Executive Stock Option

As a final example of two-asset rainbow options, we consider in this section a simple version of a performance related executive stock option (ESO). Let $X$ denote the company's stock and $Y$ some related stock index. In the ESO we shall price, the executive may exercise call options on the company stock at some fixed date $T$ in the future, but only if at this time, the stock price exceeds the index. This ESO has also been referred to in the literature as an out-performance option.

If $k$ denotes the strike price of the calls, then the value of each at time $T$ is given by the expression

$$V(x, y, T) = (x - k)^+ \mathbb{I}(x > y) = (x - k)\mathbb{I}(x > k)\mathbb{I}(x > y). \quad (6.20)$$

This payoff is equivalent to a portfolio of two-asset binaries. The first component, one long asset binary, has payoff $V_a(x, y, T) = x\mathbb{I}(x > k)\mathbb{I}(x > y)$. The second component has $k$ short bond binaries, each with payoff $V_b(x, y, T) =$

$\mathbb{I}(x > k)\mathbb{I}(x > y)$. Both components are type-3 binaries considered in theorem 6.1, with $[s = s_1 = +, h = k]$ and $(p, q) = (1, 0)$ for the asset binary and $(p, q) = (0, 0)$ for the bond binary.

This leads to the following formula for the value of the ESO at time $t < T$

$$\boxed{V(x, y, t) = x\mathcal{N}(d_k, d; \rho_1) - ke^{-r\tau}\mathcal{N}(d_k', d'; \rho_1)} \qquad (6.21)$$

where

$$d_k(x, \tau) = \frac{\log(x/k) + (r + \frac{1}{2}\sigma_1^2)\tau}{\sigma_1\sqrt{\tau}}; \qquad d_k' = d_k - \sigma_1\sqrt{\tau}$$

$$d(x, y, \tau) = \frac{\log(x/y) + \frac{1}{2}\sigma^2\tau}{\sigma\sqrt{\tau}}; \qquad d' = d - \rho_1\sigma_1\sqrt{\tau}$$

$$\sigma^2 = \sigma_1^2 + \sigma_2^2 - 2\rho\sigma_1\sigma_2; \qquad \rho_1 = \frac{\sigma_1 - \rho\sigma_2}{\sigma}.$$

While this formula has a strong resemblance to the price of a standard European call option, there are significant differences in the detail. In Chapter 10, we shall revisit the pricing of a similar ESO, using the M-binary pricing formulae.

---

## 6.7  Summary

We have studied here a number of different examples of two-asset rainbow (i.e., 1-period, 2-asset) options. Generally, these exotics could be expressed as portfolios of two asset binary options, from a list of the four most common. We have provided prices for these four binaries, so that the pv of a given two-asset option may, in many cases, be determined by simply finding the replicating portfolio. We stress this point because it is a common theme throughout this book. As we move forward, we will find increasingly complex binaries, and the task of replication will not always be immediately obvious.

The Margrabe formula for the price of an option to exchange one asset for another is one of the classical results of exotic option pricing. Due to its importance, we provided both a binary replication method of solution and also a pde method. The pde method demonstrated an interesting concept, namely the reduction of a 2D pde to a 1D pde using Euler's Equation for homogeneous functions. We shall meet this trick again in a later chapter on pricing Asian options.

The ICIAM problem illustrated another very important principle for option pricing, which necessitated using the Inclusion-Exclusion Principle. For a two asset rainbow option, a non-redundant third order exercise condition (product of three 1st order indicators) can always be decomposed into the sum of second order exercise conditions. We have presented the general $n$th order Inclusion-Exclusion Principle in anticipation that it may be required for the higher dimensional M-binaries considered in Chapter 10.

Executive stock options (ESO's), due to their large variety, provide a fruitful source of exotic option pricing problems. We considered a rather simple, performance measured ESO as our last example, as an introduction to several more complex ones considered in later chapters.

---

## Exercise Problems

Assume standard BS dynamics unless otherwise stated.

1. (a) Use the 1D, GST to prove that if $Z_1, Z_2 \sim N(0, 1; \rho)$ are correlated Gaussian rv's with correlation coefficient $\rho$, then

$$\mathbb{E}\{Z_1 e^{aZ_2}\} = \rho\, a e^{\frac{1}{2}a^2}.$$

*Hint*: Write $Z_2 = \rho Z_1 + \sqrt{1 - \rho^2}\, Z$ with $\operatorname{corr}(Z, Z_1) = 0$.

(b) Suppose $X, Y$ are two assets that follow correlated gBm. Show that the price of a derivative with expiry $T$ payoff $V_T = Y_T \log(X_T/a)$ is given, in the usual notation, by

$$V(x, y, t) = y\left[\log(x/a) + (r - \tfrac{1}{2}\sigma_1^2 + \rho\sigma_1\sigma_2)\tau\right].$$

2. A *vulnerable* call option on asset $Y$, tied to another correlated asset $X$, has expiry $T$ payoff

$$V_T = (Y_T - k)^+ \min(X_T/h, 1)$$

where $k$ is the strike price, and $h$ is a "default" level. Offer a financial explanation for this two-asset derivative, and determine its present value under standard BS assumptions.

3. Price correlation calls and puts with expiry $T$ payoffs

$$V_c(x, y, T) = (y - k)^+ \, \mathbb{I}(x > h); \qquad V_p(x, y, T) = (k - y)^+ \mathbb{I}(x > h).$$

Determine the put-call parity relation for these options.

4. Use parity relations to determine the prices of the corresponding two-asset rainbow calls and puts with expiry payoffs

$$V_c(x, y, T) = (y - k)^+ \, \mathbb{I}(x < h); \qquad V_p(x, y, T) = (k - y)^+ \mathbb{I}(x < h).$$

5. Determine the prices of put options on the maximum and minimum of two asset with respective payoffs

$$V_p^{\min}(x, y, T) = [k - \min(x, y)]^+; \qquad V_p^{\max}(x, y, T) = [k - \max(x, y)]^+.$$

Verify these prices satisfy their associated parity relation.

6. Show that the pv of a product call option with expiry payoff $V_T = (X_T Y_T - k)^+$ is given by the expression

$$V(x, y, t) = xy \, e^{(-r + \rho \sigma_1 \sigma_2)\tau} \, \mathcal{N}(d + \sigma\sqrt{\tau}) - ke^{-r\tau} \mathcal{N}(d)$$

where

$$d(x, y, \tau) = \frac{\log(xy/k) + (2r + \frac{1}{2}\sigma_1^2 + \sigma_2^2)\tau}{\sigma\sqrt{\tau}}; \qquad \sigma^2 = \sigma_1^2 + \sigma_2^2 + 2\rho\sigma_1\sigma_2.$$

7. Show that the price of a call option on the geometric mean of two assets, i.e., one with expiry $T$ payoff $V_T = (\sqrt{XY} - k)^+$ is given by

$$V(x, y, t) = \sqrt{xy} \, e^{-\frac{1}{2}\sigma_-^2 \tau} \mathcal{N}(d_1) - ke^{-r\tau} \mathcal{N}(d_2)$$

where

$$d_1(x, y, \tau) = \frac{\log(\sqrt{xy}/k) + (r + \frac{1}{2}\rho\sigma_1\sigma_2)\tau}{\sigma_+\sqrt{\tau}}$$

$$d_2(x, y, \tau) = \frac{\log(\sqrt{xy}/k) + (r - \frac{1}{4}\sigma_1^2 - \frac{1}{4}\sigma_2^2)\tau}{\sigma_+\sqrt{\tau}}$$

and $\quad \sigma_\pm^2 = \frac{1}{4}(\sigma_1^2 + \sigma_2^2 \pm 2\rho\sigma_1\sigma_2).$

8. Determine the pv of a two-asset derivative which gives it holder at time $T$, the right to exchange a strike $k$ call option payoff on asset $X$, with a strike $k$ call option payoff on asset $Y$. Note both calls expire at time $T$.

9. Price the two-asset bond binary which pays one dollar at expiry, provided the asset prices at expiry satisfy: $Y_T > c(X_T)^2$, $X_T > a$ and $Y_T < b$ for any positive values of $(a, b, c)$.

*Hint:* Use the Inclusion-Exclusion Principle.

—ooOoo—

# Chapter 7

## Barrier Options

### 7.1  Introduction

Barrier options are an important class of exotic options which are traded in the FX and OTC markets. They are our first examples of *path dependent* options. Such options have payoffs which depend, at least in part, on some aspect of the actual asset path traced out, and not just on the terminal value of the path. Barrier options have payoffs which depend on whether a given barrier level $x = b$ is crossed or otherwise during the life of the option.

For any given payoff $f(x)$ at time $T$ in the future, there are four standard barrier options. These are referred to as the:

| | | | |
|---|---|---|---|
| 1. | Down-and-out | (D/O) | barrier option |
| 2. | Up-and-out | (U/O) | barrier option |
| 3. | Down-and-in | (D/I) | barrier option |
| 4. | Up-and-in | (U/I) | barrier option |

The D/O barrier option pays $f(x)$ at time $T$, the expiry date of the option, iff during the life of the option, the asset price never falls *below* a given barrier level $x = b$. If the barrier is breached at any time during the interval $(t, T)$, where $t$ denotes current time, the option immediately expires worthless, i.e., it is literally knocked-out at this time. Clearly, a D/O barrier option will be cheaper than a standard option with the same payoff. Indeed, this is one of its main attractions as an investment or hedging instrument. If an investor believes that the asset price is unlikely to fall below the level $b$ during the investment period, then a D/O barrier option may an attractive contract to consider for either speculation or hedging.

The D/I barrier option is the counterpart of the D/O option. If the underlying asset price never hits the barrier $x = b$ (from above) during the period $(t, T)$, then the option expires worthless. If it hits the barrier in this period, then the option expires immediately, and the holder receives a standard option with time $T$ payoff $f(x)$. At the hitting time, the option is literally knocked-in. An investor may use D/I barrier options as an insurance against the underlying asset falling below the barrier level.

U/O and U/I barrier options have similar interpretations as their D/O and D/I cousins. The only real difference is that these options are knocked-out or knocked-in when the barrier is hit or crossed from below. The D/O and U/O options are also called *knock-out options*, while the D/I and U/I contracts are called *knock-in options*. D/O and D/I barrier options exist only in the domain $x > b$, and we refer to this as the *active domain* of the option. Similarly, the active domain of the U/O and U/I barrier options will be $x < b$.

How shall we price these four barrier options in an otherwise standard BS economy? The traditional method obtains prices by considering certain densities of the asset price through the barrier level $x = b$ (see Q2 in the Exercise Problems). One oft quoted paper employing this method is Rich [63]. This is by no means an easy paper to read, involving many pages of complex integrations. Surely, there must be a simpler way to derive the results and indeed there is. Our method, which involves no integrations whatsoever, utilizes the technique (known to physicists and applied mathematicians) as the *Method of Images*. At least in the form presented below, it was first introduced to the Financial Mathematics community by the author in [7].

The Method of Images (or MoI) originates from pde's, so let us begin by writing down the BS-pde's for the four barrier options listed above.

| Option Type | PDE | Active Domain | Bndry. Cond. at $x = b$ | Expiry Cond. at $t = T$ |
|---|---|---|---|---|
| D/O | $\mathcal{L}V_{do} = 0$ | $x > b$ | 0 | $f(x)$ |
| U/O | $\mathcal{L}V_{uo} = 0$ | $x < b$ | 0 | $f(x)$ |
| D/I | $\mathcal{L}V_{di} = 0$ | $x > b$ | $V_0(b,t)$ | 0 |
| U/I | $\mathcal{L}V_{ui} = 0$ | $x < b$ | $V_0(b,t)$ | 0 |

In this table, $\mathcal{L}V = 0$ denotes the BS-pde; the active domain discussed earlier defines the existence domain of the option; the boundary condition at $x = b$ for the knock-out options is obviously zero, while for the knock-in options it is the price of a standard option $V_0(b,t)$ (one without the barrier feature, but with the same payoff $f(x)$ at expiry $T$); note that the expiry condition for the knock-ins is zero, since if the option survives to expiry without hitting the barrier, it expires worthless. The pde's are all TBV problems.

## 7.2 Method of Images

We shall demonstrate first how the Method of Images (MoI) works for a D/O barrier option with general expiry $T$ payoff function $V(x,T) = f(x)$.

The barrier is assumed to be set at $x = b$, and obviously at the current time $t$, we must have $x > b$, otherwise the option would be immediately knocked-out. The TBV problem, repeated here for convenience, is

$$\mathcal{L}V = 0; \qquad V(x, T) = f(x); \qquad V(b, t) = 0 \qquad (7.1)$$

The MoI for solving this pde embeds the problem within a related standard European or simple TV problem of the form $\mathcal{L}U = 0$, $U(x, T) = g(x)$ in $t < T$, but for all $x > 0$. The trick is to find the appropriate function $g(x)$ so that the two prices agree in all respects (including TV and BC) in the active domain of the D/O barrier option, i.e., $x > b$. Then, we can assert that $V(x, t) = U(x, t)$ in $x > b$. In this way, we actually solve the TBV problem describing the D/O barrier option, by pricing the related European option, and generally the latter problem is easier to solve.

**DEFINITION 7.1**    *We say*

$$\overset{*}{F}(x) = \left(\frac{b}{x}\right)^{\alpha} F\left(\frac{b^2}{x}\right); \qquad \alpha = \frac{2r}{\sigma^2} - 1 \qquad (7.2)$$

*is the image of $F(x)$ wrt $x = b$ and the BS differential operator $\mathcal{L}$.*

Sometimes, when the context demands it, we shall write the image of $F(x)$ wrt $x = b$, in operator notation as $\overset{*}{F}(x) = \mathcal{I}_b\{F(x)\}$, where $\mathcal{I}_b$ denotes the *image operator* wrt $x = b$.

The image function has a number of very important properties, given by the following theorem.

**THEOREM 7.1**

1. *The image operator $\mathcal{I}_b$ is an involution, ie $\mathcal{I}_b^2 = I$, where $I$ is the identity operator. This means that for any function $V(x, t)$, we have*

$$\mathcal{I}_b \mathcal{I}_b\{V(x, t)\} = \mathcal{I}_b\{\overset{*}{V}(x, t)\} = V(x, t). \qquad (7.3)$$

2. $\overset{*}{V}(x, t)$ *satisfies the BS-pde, whenever $V(x, t)$ does.*

3. $\overset{*}{V}(x, t) = V(x, t)$ *when $x = b$*

4. *If $x > b$ (resp. $x < b$) is the active domain of $V(x, t)$, then $x < b$ (resp. $x > b$) is the active domain of $\overset{*}{V}(x, t)$.*

If $V(x,t)$ is a derivative price, then we refer to $\overset{*}{V}(x,t)$ as the corresponding *image price*. The proofs of the properties of the image price in theorem 7.1 are straightforward and involve little more than direct calculation.

**PROOF**    Properties 1, 3, and 4 involve little more than elementary algebra, so are left to the reader to confirm.

Property 2 can be demonstrated by showing directly that if $V(x,t)$ satisfies the BS-pde, then so does $\overset{*}{V}(x,t) = (b/x)^\alpha V(y,t)$ where $y = b^2/x$. We first find the partial derivatives

$$\overset{*}{V}_t(x,t) = (b/x)^\alpha V_t(y,t)$$
$$x\overset{*}{V}_x(x,t) = -(b/x)^\alpha [\alpha V(y,t) + yV_y(y,t)]$$
$$x^2\overset{*}{V}_{xx}(x,t) = (b/x)^\alpha [\alpha(\alpha+1)V + 2(\alpha+1)yV_y + y^2V_{yy}].$$

Substitution into the BS-pde and using the identity $(\alpha+1)\sigma^2 = r$ then yields

$$\mathcal{L}\overset{*}{V}(x,t) = \overset{*}{V}_t - r\overset{*}{V} + rx\overset{*}{V}_x + \tfrac{1}{2}\sigma^2 x^2 \overset{*}{V}_{xx}$$
$$= (b/x)^\alpha [V_t - rV + ryV_y + \tfrac{1}{2}\sigma^2 y^2 V_{yy}](y,t) = 0.$$

This completes the proof.    ☐

**REMARK 7.1**    One may well ask where the expression for the image function (7.2) actually comes from. After all, we have simply pulled it "out of the hat," as it were. There are several ways in which it can be obtained, but perhaps the easiest to understand is the following. Suppose we transform the BS-pde to the heat equation $U_\tau = \tfrac{1}{2}\sigma^2 U_{yy}$ for $U = U(y,\tau)$ according to Scheme-1 considered in equation (2.33). Now it is clear that the image solution of the heat equation relative to $y = h$ is given by

$$\overset{*}{U}(y,\tau) = U(2h - y, \tau).$$

Note that this image solution satisfies the same four properties stated in theorem 7.1, relative to the heat equation and $y = h$, as opposed to the BS-pde and $x = b$. If we now simply back-transform $\overset{*}{U}(y,\tau)$ to the original variables, we do indeed obtain (with $h = \log b$) that $\overset{*}{U}(y,\tau) \Rightarrow \overset{*}{V}(x,t)$ as required. Q1 in the Exercise Problems gets you to work out the details.    ☐

We now have all the background needed to state the main result of this section.

**THEOREM 7.2 Method of Images for the D/O Barrier Option**

*Let $V_b^+(x,t)$ solve the TV-problem*

$$\mathcal{L}V_b^+(x,t) = 0; \qquad V(x,T) = f(x)\mathbb{I}(x>b)$$

*in $\{x > 0, \ t < T\}$. Then*

$$\boxed{V_{do}(x,t) = V_b^+(x,t) - \overset{*}{V}_b^+(x,t)} \qquad (7.4)$$

*solves the TBV problem for the D/O barrier option in $\{x > b, \ t < T\}$.*

The proof, which we offer next, relies on the existence and uniqueness (subject to technical conditions) of solutions of the BS-pde.

**PROOF**   We need only show that $V(x,t)$ defined by (7.4) satisfies the BS-pde $\mathcal{L}V = 0$, the TV, $V(x,T) = f(x)$ in $x > b$ and the BC, $V(b,t) = 0$. If these can be demonstrated, then the uniqueness theorem alluded to above, will guarantee that this is the only solution, and hence will represent the D/O barrier option price.

1. Now $V_b(x,t)$ satisfies the BS-pde by assumption and so does its image $\overset{*}{V}_b^+(x,t)$, by the second property of theorem 7.1. But $\mathcal{L}$ is also a linear operator, which means that the difference, $[V_b^+(x,t) - \overset{*}{V}_b^+(x,t)]$ will also satisfy the BS-pde. Hence, $\mathcal{L}V_{do}(x,t) = 0$.

2. $V_{do}(b,t) = 0$ follows directly from the property $V_b = \overset{*}{V}_b$ when $x = b$.

3. Finally, observe that when $t = T$, (7.4) yields, for all $x > 0$,

$$\begin{aligned}
V_{do}(x,T) &= f(x)\mathbb{I}(x>b) - \mathcal{I}_b\{f(x)\mathbb{I}(x>b)\} \\
&= f(x)\mathbb{I}(x>b) - \overset{*}{f}(x)\mathbb{I}(b^2/x>b) \\
&= f(x)\mathbb{I}(x>b) - \overset{*}{f}(x)\mathbb{I}(x<b).
\end{aligned}$$

Thus in the active domain of the D/O barrier option, $x > b$, we do indeed have $V_{do}(x,T) = f(x)$ as required.

This completes the proof.   □

Note that $V_b^+(x,t)$ is just an up-type binary on the standard option which pays $f(x)$ at expiry $T$. Once this price is determined, typically using the FTAP (1.35), we then obtain the corresponding image price using the definition 7.1, and the D/O barrier option price is just their difference. What is quite remarkable about this procedure is that once again, we obtain a complex option price without actually performing a single integration. In the present case, the Method of Images does most of the work, and image prices in the BS world can be calculated essentially as an algebraic operation.

## 7.3 Barrier Parity Relations

It turns out, as we shall demonstrate in this section, that the D/O barrier option is the only one we need to price, since all the others can be derived directly from it using parity relations. To prove these parity relations it is useful to express the payoffs of the four barrier options in terms of future random variables. For this purpose, let us define

$$Y_{t,T} = \min_{t \le s \le T} \{X_s\} \quad \text{and} \quad Z_{t,T} = \max_{t \le s \le T} \{X_s\} \tag{7.5}$$

where $X_s$ denotes the future value of the asset price at time $s$.

The FTAP then determines the four barrier prices as:

$$\left. \begin{array}{l} V_{do}(x,t) = e^{-r\tau} \, \mathbb{E}_Q\{f(X_T)\mathbb{I}(Y_{t,T} > b) \,|\, \mathcal{F}_t\} \\ V_{di}(x,t) = e^{-r\tau} \, \mathbb{E}_Q\{f(X_T)\mathbb{I}(Y_{t,T} < b) \,|\, \mathcal{F}_t\} \\ V_{uo}(x,t) = e^{-r\tau} \, \mathbb{E}_Q\{f(X_T)\mathbb{I}(Z_{t,T} < b) \,|\, \mathcal{F}_t\} \\ V_{ui}(x,t) = e^{-r\tau} \, \mathbb{E}_Q\{f(X_T)\mathbb{I}(Z_{t,T} > b) \,|\, \mathcal{F}_t\} \end{array} \right\} \tag{7.6}$$

It is clear, by simply summing the appropriate pairs of prices that

$$V_{do}(x,t) + V_{di}(x,t) = e^{-r\tau} \, \mathbb{E}_Q\{f(X_T) \,|\, \mathcal{F}_t\}$$
$$V_{uo}(x,t) + V_{ui}(x,t) = e^{-r\tau} \, \mathbb{E}_Q\{f(X_T) \,|\, \mathcal{F}_t\}.$$

But, the rhs of these expressions denotes $V_0(x,t)$, the price of the standard option with the same payoff $f(x)$. We therefore have the two out-and-in parity relations:

$$\boxed{\begin{array}{l} V_{do}(x,t) + V_{di}(x,t) = V_0(x,t) \\ V_{uo}(x,t) + V_{ui}(x,t) = V_0(x,t) \end{array}} \tag{7.7}$$

While the above relations are well-known, the the next pair of parity relations, which are necessary to complete the full cycle, are generally not well-known. They were first derived in Buchen [7].

$$\boxed{\overset{*}{V}_{di}(x,t) = V_{ui}(x,t) \quad \text{and} \quad \overset{*}{V}_{ui}(x,t) = V_{di}(x,t)} \tag{7.8}$$

These two *knock-in* parity relations are not independent of each other, since from the first property in theorem 7.1, the first one implies the second and vice versa.

**PROOF**   The simplest proof of (7.8) involves going back to the pde representations of the D/I and U/I barrier options, listed in the Introduction to this chapter. The TBV-pdes for the D/I and U/I options are:

$$\mathcal{L}V_{di}(x,t) = 0; \text{ in } x > b; \quad V_{di}(b,t) = V_0(b,t); \quad V_{di}(x,T) = 0$$
$$\mathcal{L}V_{ui}(x,t) = 0; \text{ in } x < b; \quad V_{ui}(b,t) = V_0(b,t); \quad V_{ui}(x,T) = 0.$$

It follows, that the image price $\overset{*}{V}_{di}(x,t)$, wrt the barrier level $x = b$, also satisfies the BS-pde (Property 2 of theorem 7.1); the active domain becomes $x < b$ (Property 4); the BC at $x = b$ is $\overset{*}{V}_{di}(b,t) = \overset{*}{V}_0(b,t) = V_0(b,t)$ (Property 3); and $\overset{*}{V}_{di}(x,T) = \mathcal{I}_b(0) = 0$. Thus, $\overset{*}{V}_{di}(x,t)$ satisfies exactly the same TBV problem as $V_{ui}(x,t)$, and by the uniqueness theorem of BS-pde solutions, we may conclude that $\overset{*}{V}_{di}(x,t) = V_{ui}(x,t)$.　　　　　□

Given any one barrier option price, the parity relations (7.7) and (7.8) taken together, allow us to determine the price of the other three. For example, given the D/O barrier price as in (7.4), we have (in order)

$$V_{di}(x,t) = V_0(x,t) - V_{do}(x,t)$$
$$V_{ui}(x,t) = \overset{*}{V}_{di}(x,t) = \overset{*}{V}_0(x,t) - \overset{*}{V}_{do}(x,t)$$
$$V_{uo}(x,t) = V_0(x,t) - V_{uo}(x,t) = [V_0(x,t) - \overset{*}{V}_0(x,t)] + \overset{*}{V}_{do}(x,t).$$

If we substitute, $V_{do}(x,t) = V_b^+(x,t) - \overset{*}{V}_b^+(x,t)$ from (7.4) and use the up-down parity relation $V_b^+(x,t) + V_b^-(x,t) = V_0(x,t)$, we then obtain the full list of barrier option prices in the form

$$
\begin{array}{|l|}
\hline
V_{do}(x,t) = V_b^+(x,t) - \overset{*}{V}_b^+(x,t) \\
V_{di}(x,t) = V_b^-(x,t) + \overset{*}{V}_b^+(x,t) \\
V_{ui}(x,t) = V_b^+(x,t) + \overset{*}{V}_b^-(x,t) \\
V_{uo}(x,t) = V_b^-(x,t) - \overset{*}{V}_b^-(x,t) \\
\hline
\end{array}
\qquad (7.9)
$$

This set of four prices, quite remarkably captures the prices of all four barrier options in terms of just two binary options and their images. The binary options are straightforward European up and down type binaries with the payoffs $f(x)\mathbb{I}(x > b)$ and $f(x)\mathbb{I}(x < b)$ respectively. These can generally be obtained be the methods we advocated in previous chapters of this book.

The price representations are also generic in two different ways. First, they apply for any payoff function $f(x)$, not just calls when $f(x) = (x - k)^+$, and puts when $f(x) = (k-x)^+$. Second, they are generally true for any underlying asset-price process, not just gBm as in the BS-framework. However, in other asset price models, the image operator will be quite different to (7.2), and may not have a simple representation or even exist.

---

## 7.4　Equivalent Payoffs for Barrier Options

Equation (7.9) gives the prices of all four barrier options for an arbitrary payoff function $f(x)$, in terms of European binaries and their images wrt the barrier level $x = b$. This suggests that barrier options have equivalent European payoffs at expiry $T$. Of course these payoffs lead to corresponding European prices for all $t < T$, valid for all $x > 0$. The associated barrier prices agree exactly with these European prices, but only in their active domains. That is, in $x > b$ for the D/O and D/I barrier options, and $x < b$ for the U/O and U/I barrier options. It is in this sense, that barrier option prices have been embedded in an equivalent European pricing framework.

We refer to the equivalent European payoffs at time $T$, simply as *equivalent payoffs*. For the four barrier options, equation (7.9) shows that the equivalent payoffs are in fact the following.

$$\boxed{\begin{aligned}
V_{do}^{eq}(x,T) &= f(x)\mathbb{I}(x>b) - \overset{*}{f}(x)\mathbb{I}(x<b) \\
V_{di}^{eq}(x,T) &= [f(x) + \overset{*}{f}(x)]\mathbb{I}(x<b) \\
V_{ui}^{eq}(x,T) &= [f(x) + \overset{*}{f}(x)]\mathbb{I}(x>b) \\
V_{uo}^{eq}(x,T) &= f(x)\mathbb{I}(x<b) - \overset{*}{f}(x)\mathbb{I}(x>b)
\end{aligned}} \tag{7.10}$$

Observe that the knock-out options have equivalent payoff components in both $x < b$ and $x > b$. However, the knock-in options have non-zero components only their complementary domains: $x < b$ for the D/I (which has active domain $x > b$) and $x > b$ for the U/I (which has active domain $x < b$).

**REMARK 7.2**    The equivalent payoffs clearly show that the MoI for barrier options, involve what might be called *point images* only. That is, for each asset price $x$ in $x \geqslant b$, there is a single image asset price $y = b^2/x$ in $y \leqslant b$ at which the image payoff is calculated. For more complex options such as lookback options for example, we shall see that such point images are insufficient to describe their payoffs. In fact, it is now necessary to consider a continuous distribution of images. $\quad\square$

## 7.5    Call and Put Barrier Options

Suppose the strike price is set at $x = k$ and the barrier level at $x = h$. It turns out that different results for call and put barrier prices occur for the two case $k > h$ and $k < h$. This means there are actually 16 different prices altogether: four barrier types (D/O, D/I, U/O, U/I); times two payoff types (calls/puts); times two ranges $k > h$ and $k < h$. As an illustration of the method, we shall derive here just two of these cases, the D/O call price for $k > h$ and $k < h$. The remaining 14 prices will be listed for completeness, and left for the reader to solve in the Exercise Problems for this chapter.

All 16 prices can be expressed in terms of standard calls/puts, gap calls/puts and their images wrt the barrier level $x = h$. Recall from Section 4.5 that gap calls and puts are like standard call/puts, except they have different strike and exercise prices.

Let the (strike-$k$, exercise price $\xi$) gap call/put prices be denoted by

$$C_\xi(x, \tau; k) = x\mathcal{N}(d_\xi) - ke^{-r\tau}\mathcal{N}(d'_\xi)$$
$$P_\xi(x, \tau; k) = -x\mathcal{N}(-d_\xi) + ke^{-r\tau}\mathcal{N}(-d'_\xi)$$

where $\tau = (T - t)$ and

$$[d_\xi, d'_\xi] = \frac{\log(x/\xi) + (r \pm \frac{1}{2}\sigma^2)\tau}{\sigma\sqrt{\tau}}.$$

Recall that a standard call/put has $\xi = k$.

For a D/O, strike $k$, call barrier option, we have the equivalent payoff at time $T$,

$$\begin{aligned}
V_{do}^{eq}(x, T) &= f(x)\mathbb{I}(x > h) - \overset{*}{f}(x)\mathbb{I}(x < h) \\
&= (x - k)^+ \mathbb{I}(x > h) - (\text{image}) \\
&= \begin{cases} (x - k)\mathbb{I}(x > k) - (\text{image}) & \text{if } k > h \\ (x - k)\mathbb{I}(x > h) - (\text{image}) & \text{if } k < h. \end{cases}
\end{aligned}$$

The term "(image)" means: the image wrt the barrier level $x = h$, of the preceding term. Thus for $k > h$, the equivalent payoff is a standard call minus its image, but for $k < h$ it is a gap call minus its image. The corresponding present values are given by,

$$V_{do}(x, t) = \begin{cases} C_k(x, \tau; k) - \overset{*}{C}_k(x, \tau; k) & \text{if } k > h \\ C_h(x, \tau; k) - \overset{*}{C}_h(x, \tau; k) & \text{if } k > h. \end{cases}$$

Note further, that

$$\overset{*}{C}_k(x, \tau; k) = (h/x)^\alpha C_k(h^2/x, \tau; k); \qquad \overset{*}{C}_h(x, \tau; k) = (h/x)^\alpha C_h(h^2/x, \tau; k)$$

which are readily calculated from the expression for the gap call option above.

The complete list of all call/put barrier options is provided in the following table. The first line of the table was derived above.

| Option Type | Case $k > h$ | Case $k < h$ |
|---|---|---|
| D/O-call | $C_k - \overset{*}{C}_k$ | $C_h - \overset{*}{C}_h$ |
| D/I-call | $\overset{*}{C}_k$ | $C_k - (C_h - \overset{*}{C}_h)$ |
| U/I-call | $C_k$ | $\overset{*}{C}_k + (C_h - \overset{*}{C}_h)$ |
| U/O-call | $0$ | $(C_k - \overset{*}{C}_k) - (C_h - \overset{*}{C}_h)$ |
| D/O-put | $(P_k - \overset{*}{P}_k) - (P_h - \overset{*}{P}_h)$ | $0$ |
| D/I-put | $\overset{*}{P}_k + (P_h - \overset{*}{P}_h)$ | $P_k$ |
| U/I-put | $P_k - (P_h - \overset{*}{P}_h)$ | $\overset{*}{P}_k$ |
| U/O-put | $P_h - \overset{*}{P}_h$ | $P_k - \overset{*}{P}_k$ |

This table of prices shows a very high degree of symmetry, which is rarely seen in other published formulae for call and put barrier options. One of the reasons for this is, that the call and put options satisfy a number of parity relations (apart from the standard put-call parity). One interesting one is the following:

$$C_k(x, \tau; k) - C_h(x, \tau; k) = P_k(x, \tau; k) - P_h(x, \tau; k)$$

which derives from the put-call parity relations $C_k - P_k = C_h - P_h = x - ke^{-r\tau}$.

The existence of such a parity relation means that it is possible to write the call and put barrier options in many different but equivalent ways. As a result, we can write call barriers using put type options and vice versa. Of course, such representations should be discouraged, but the literature is replete with such inelegance.

Note the rather obvious result that the U/O call is worthless if $k > h$ and similarly for the D/O put if $k < h$. In both cases the barrier will be crossed, and hence knocked-out, before the option has a chance to finish in-the-money.

## 7.6  Barrier Option Rebates

Barrier options are often linked with a cash rebate. Holders of a knock-out option may receive a rebate if the option is actually knocked-out before expiry. In this case the rebate may be paid immediately when the option is knocked-out, or payment may be delayed till the expiry date. Knock-in options may sometimes pay a rebate, always at the expiry date, if the option fails to be knocked-in during its life time. The inclusion of a rebate obviously increases the present value of a barrier option, and this increase will be independent of the payoff function $f(x)$. In fact, all rebate terms, regardless of the payoff function, can be expressed as simple portfolios of asset and bond binaries and their images.

The table below lists the pv of the rebate term for the six cases of interest, where $R$ denotes the dollar amount of the cash rebate.

$$
\begin{array}{lll}
V_{do}^R(x,t) = \frac{R}{h}[A_h^-(x,\tau) + \overset{*}{A}_h^+(x,\tau)] & \text{(Paid at KO)} \\
V_{uo}^R(x,t) = \frac{R}{h}[A_h^+(x,\tau) + \overset{*}{A}_h^-(x,\tau)] & \text{(Paid at KO)} \\
V_{do}^R(x,t) = R[B_h^-(x,\tau) + \overset{*}{B}_h^+(x,\tau)] & \text{(Paid at } T) \\
V_{uo}^R(x,t) = R[B_h^+(x,\tau) + \overset{*}{B}_h^-(x,\tau)] & \text{(Paid at } T) \\
V_{di}^R(x,t) = R[B_h^+(x,\tau) - \overset{*}{B}_h^+(x,\tau)] & \text{(Paid at } T) \\
V_{ui}^R(x,t) = R[B_h^-(x,\tau) - \overset{*}{B}_h^-(x,\tau)] & \text{(Paid at } T)
\end{array}
\tag{7.11}
$$

**Rebate Paid at Knock-Out**
To derive these results, consider first the rebate on a D/O barrier option, paid at knock-out (KO) time. The total option value (barrier + rebate) will satisfy the pde

$$\mathcal{L}V = 0; \qquad V(h,t) = R; \qquad V(x,T) = f(x) \quad \text{in } x > h.$$

Since the barrier component satisfies the pde

$$\mathcal{L}V_{do} = 0; \qquad V_{do}(h,t) = 0; \qquad V_{do}(x,T) = f(x) \quad \text{in } x > h$$

it should be clear that the rebate term $V_{do}^R(x,t)$ will satisfy the pde

$$\mathcal{L}V_{do}^R = 0; \qquad V_{do}^R(h,t) = R; \qquad V_{do}^R(x,T) = 0 \quad \text{in } x > h. \tag{7.12}$$

The pde for $V_{do}^R(x,t)$ above, is however, recognized as the pde for a D/I barrier option on a contract with payoff $f(x) = Rx/h$. This can be seen by noting that the associated standard option will then be $V_0(x,t) = Rx/h$ which gives $V_0(h,t) = R$ as required. The corresponding binaries on $f(x)$ will then be $V_b^{\pm}(x,t) = (R/h)A_b^{\pm}(x,\tau)$. Thus the D/I equation (7.9), yields for any time $t < T$, the value of the rebate term as,

$$V_{do}^R(x,t) = (R/h)[A_h^-(x,\tau) + \overset{*}{A}_h^+(x,\tau)].$$

**Rebate Paid at $T$**
If the rebate is, however, paid at expiry $T$, then that is equivalent to receiving $Re^{-r\tau}$ at the knock-out time. In this case, the rebate term satisfies the pde

$$\mathcal{L}V_{do}^R = 0; \qquad V_{do}^R(h,t) = Re^{-r\tau}; \qquad V_{do}^R(x,T) = 0 \quad \text{in } x > h \tag{7.13}$$

and this is recognized as the pde for a D/I option on a contract with $T$ payoff $f(x) = R$. Here the binaries on $f(x)$ will be $V_h^{\pm}(x,t) = R\,B_h^{\pm}(x,\tau)$, and (7.9) then gives the pv of the rebate term as

$$V_{do}^R(x,t) = R[B_h^-(x,\tau) + \overset{*}{B}_h^+(x,\tau)].$$

The other rebate terms follow similarly (see Q5 in Exercise Problems).

## 7.7 Barrier Option Extensions

We present here two straightforward extensions of the barrier options studied so far. The first is to see how the barrier option pricing formulae change if the underlying asset includes a constant dividend yield. The second derives the Method of Images for exponential time varying barriers.

### 1. Dividend Yield

Suppose the underlying asset pays out a continuous dividend at the constant yield $q$. The deliberations of Section 1.12 show how this can be done for vanilla options, by simple adjustments of the non-dividend paying BS formulae. There are just two such adjustments to make for barrier options.

1. Replace $\alpha$ in Equation (7.2) by $\alpha = 2(r - q)/\sigma^2 - 1$; and

2. Replace $x$ in non-image terms by $xe^{-q\tau}$ (an adjustment we have already met) and replace $y = b^2/x$ by $ye^{-q\tau}$ in the image terms.

To see how these adjustments to image terms come about, we shall outline next one approach, using the FTAP. Let $X$ be the future dividend paying asset at time $T$ and let $Y = b^2/X$, where $b$ is the barrier price. Then, since the image price $\overset{*}{V}(x, t)$ is the price of a derivative with $T$ payoff $(b/X)^\alpha F(Y)$, we have by the FTAP,

$$\overset{*}{V}(x,t) = e^{-r\tau}\mathbb{E}_Q\{(b/X)^\alpha F(Y)\}$$
$$X = xe^{(r-q-\frac{1}{2}\sigma^2)\tau + \sigma\sqrt{\tau}Z}; \quad Z \sim N(0,1).$$

This leads, with $y = b^2/x$, to

$$\overset{*}{V}(x,t) = (b/x)^\alpha e^{(-r-\alpha(r-q-\frac{1}{2}\sigma^2))\tau}\mathbb{E}_Q\{e^{-\alpha\sigma\sqrt{\tau}Z}F(ye^{-(r-q-\frac{1}{2}\sigma^2)\tau-\sigma\sqrt{\tau}Z})\}$$
$$= (b/x)^\alpha e^{-\mu\tau}\mathbb{E}_Q\{F(ye^{-(r-q-\frac{1}{2}\sigma^2)\tau-\sigma\sqrt{\tau}(Z-\alpha\sigma\sqrt{\tau})})\}$$

where we have used the GST in the second line, and with $\alpha = 2(r-q)/\sigma^2 - 1$, we find

$$\mu = r + \alpha(r - q - \tfrac{1}{2}\sigma^2) + \tfrac{1}{2}\alpha^2\sigma^2 = r$$
$$\text{and} \quad -(r - q - \tfrac{1}{2}\sigma^2) + \alpha\sigma^2 = (r - q - \tfrac{1}{2}\sigma^2).$$

It follows that

$$\overset{*}{V}(x,t) = (b/x)^\alpha e^{-r\tau}\mathbb{E}_Q\{F(ye^{(r-q-\frac{1}{2}\sigma^2)\tau-\sigma\sqrt{\tau}Z})\}$$
$$= (b/x)^\alpha e^{-r\tau}\mathbb{E}_Q\{F(ye^{(r-q-\frac{1}{2}\sigma^2)\tau+\sigma\sqrt{\tau}Z})\}$$
$$= (b/x)^\alpha V(ye^{-q\tau}, t).$$

In the second line above, we have replaced $Z$ by $-Z$ using the symmetry property of Gaussian rv's, and in the last line, $V(x,t)$ denotes the standard BS price of a European derivative on a non-dividend paying asset, with payoff $F(x)$. This completes the demonstration.

**Example 7.1**
For a D/O call option with $k < h$, for a non-dividend paying asset, the price is

$$V_0(x,t) = C_h(x,t;k) - (b/x)^\alpha C_h(\tfrac{b^2}{x},t;k); \quad \alpha = \frac{2r}{\sigma^2} - 1,$$

while for a dividend paying asset, the price will be

$$V_0(x,t) = C_h(xe^{-q\tau},t;k) - (b/x)^\alpha C_h(\tfrac{b^2 e^{-q\tau}}{x},t;k); \quad \alpha = \frac{2(r-q)}{\sigma^2} - 1.$$

$\square$

## 2. Exponential Barriers
It transpires that the BS framework for barrier options admits exponentially time varying barriers, not just flat, time independent barriers as considered previously. Suppose we have a time varying barrier of the form $b(t) = Be^{\beta t}$, where $B$ is a positive constant, and $\beta$ is the barrier growth rate if $\beta > 0$, or decay rate if $\beta < 0$. Then, all the results we have derived for flat barriers of constant value $b$, also apply to exponential barriers of value $b(t)$, provided we replace the parameter $\alpha = (2r/\sigma^2 - 1)$ by $\alpha = 2(r - \beta)/\sigma^2 - 1$. In particular, we have the image price,

$$\boxed{\overset{*}{V}(x,t) = \left(\frac{b_t}{x}\right)^\alpha V\left(\frac{b_t^2}{x},t\right); \qquad \alpha = \frac{2(r-\beta)}{\sigma^2} - 1} \qquad (7.14)$$

where $b_t = b(t) = Be^{\beta t}$. Obviously, the flat barrier case is recovered when $\beta = 0$.

There are several ways in which this result can be proved, one of which follows a very similar approach to the one used above, to demonstrate the image price of a derivative on a dividend paying asset. See Q6 in Exercise Problems for details. A formal proof can also be found in reference [10].

---

## 7.8 Binomial Model for Barrier Options

Early attempts to price barrier options on a binomial tree ran into considerable difficulties. It was straightforward to execute the backward recursion with the extra boundary condition (for a D/O barrier)

$$V_j = 0 \quad \text{if} \quad X_j > b$$

where $V_j$ is the derivative price, and $X_j$ is the asset price at any node $j$ of the tree. However, this approach led to relative high error rates when compared to the corresponding BS model. These errors were partly mitigated through a continuity correction term analyzed by Broadie et al. in [6]. However, from what we have learned about the method of images for barrier options, it is a simple matter to use the equivalent European payoff at the final nodes of the tree and recurse backwards towards $t = 0$, in the usual way for other European options. The asset tree is constructed as described in Section 4.12, and the backward recursion (4.32),

$$V(x,t) = \rho^{-1} \left[ pV(ux, t + \Delta) + qV(dx, t + \Delta) \right]$$

is started with expiry values, $F(X_j)\mathbb{I}(X_j > b) - \overset{*}{F}(X_j)\mathbb{I}(X_j < b)$, for a D/O barrier option, where $X_j = u^j d^{n-j} x$ for $0 \le j \le n$, denote the asset prices on the expiry nodes of the tree. It should also be fairly clear that the image function $\overset{*}{V}(x,t)$ must satisfy the binomial recursion

$$\overset{*}{V}(x,t) = \rho^{-1} \left[ p\overset{*}{V}(ux, t + \Delta) + q\overset{*}{V}(dx, t + \Delta) \right].$$

The proof of this assertion is also relegated to the Exercise Problems. Using this prescription, there is no need for the continuity correction mentioned above.

To verify this claim, consider a D/O call barrier option with the same parameters as in figure 4.3, and a barrier price set at $b = 9$ dollars. The graphs displayed in figure 7.1 clearly show that the Method of Images applied to the binomial lattice, generates errors relative to the BS price which are comparable to the those of a standard European call option, considered in figure 4.3. The errors generated by the the standard approach, even with the continuity correction applied are much higher. This correction involves moving the barrier from $b$ to $be^{0.5286\sigma\sqrt{\Delta}}$. The numerical factor comes from the expression (see Broadie et al. [6])

$$-\frac{\zeta(\frac{1}{2})}{\sqrt{2\pi}} = 0.5286$$

where $\zeta$ denotes the Riemann zeta function. It is claimed that this is probably the only application in finance where the celebrated Riemann function makes an appearance.

In fact, the continuity correction here only reduces the error by about a factor of two; the Method of Images reduces it by almost a factor of 200.

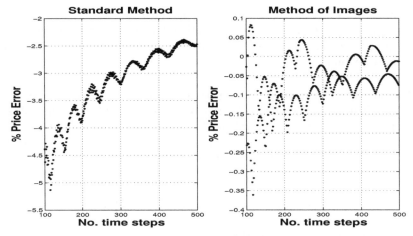

**FIGURE 7.1**: Percentage call price error of the binomial model relative to the BS model for parameters: $x = k = 10$, $T = 0.5$, $b = 9$, $r = 0.10$, $\sigma = 0.25$. Left panel for the standard method with continuity correction; right panel using the Method of Images.

## 7.9 Partial Time Barrier Options

Knock-out barrier options are obviously less expensive than their corresponding standard counterparts (those with the same payoff, but without the barrier feature). However, the reduced premium is offset by the possibility that the option is knocked-out before the expiry date. Investors worried about the possibility of knock-out can partly compensate for this concern by selecting a barrier monitoring window in which they believe knock-out is very unlikely. This idea leads to the class of *partial time barrier options* which were first analyzed in the Black–Scholes framework by Heynen and Kat [31].

There are basically two types of partial time barrier options depending on whether the reduced barrier window lies at the start or at the end of the option's life. The former is called a *start-out*, partial time barrier option; the latter an *end-out*, partial time barrier option. Both options are very similar to the dual expiry options we considered in Chapter 5, and we shall use the pricing methods and notation of that chapter to price these partial time barrier options.

### 7.9.1 Start-out, Partial Time Barrier Options

To fix ideas, let $[t, T_1]$ be the barrier monitoring window, where $t$ is the current time, and let $T = T_2 > T_1$ denote the expiry date of the option. Let $f(x)$

denote the payoff function of the option, provided the option survives to the expiry date. For a standard European option, having no barrier window, with same payoff function $f(x)$, let $V_0(x, \tau)$ denote its value, when time interval $\tau$ remains to expiry.

Consider first a partial time D/O barrier option with barrier price set at $x = h$. Then since there is no barrier over the range $[T_1, T_2]$, the value of the option just after time $T_1$ will be $V_0(x_1, \tau)$, where $x_1$ is the asset price at time $T_1$ and $\tau = (T_2 - T_1)$. This price will then be the time $T_1$ payoff a D/O barrier option, and as we saw in Section 7.4, the associated equivalent European payoff will be given by

$$V_{do}^{eq}(x_1, T_1) = V_0(x_1, \tau)\mathbb{I}(x_1 > h) - \mathcal{I}_h\{\cdots\}$$

where the last term represents the image of the first term relative to the barrier price $h$. In the present case, this will be equal to: $\overset{*}{V}_0(x_1, \tau)\mathbb{I}(x < h)$. It is then just a matter of finding the pv of the equivalent European option with this time $T_1$ payoff.

**Start-out, Partial Time D/O Call**

For a strike $k$ call option with payoff function $f(x) = (x - k)^+$, we obtain

$$V_0(x_1, \tau) = C_k(x_1, \tau) = Q_k^+(x_1, \tau; k)$$

recalling the Definition 4.2 for first order $Q$-options. In this case, the equivalent payoff above will be

$$V_{do}^{eq}(x_1, T_1) = Q_k^+(x_1, \tau; k)\mathbb{I}(x_1 > h) - \mathcal{I}_h\{\cdots\}.$$

From this, and our Definition 5.2 for second order $Q$-options, we can write down, by inspection, the pv of the start-out, partial time, D/O call barrier option, as

$$\boxed{V_{do}(x, t) = Q_{hk}^{++}(x, t; k) - \overset{*}{Q}_{hk}^{++}(x, t; k)} \tag{7.15}$$

**REMARK 7.3** Recall that each second order $Q$-option is the sum of two bivariate normal distribution functions, so that while the formula (7.15) looks simple enough, it is in fact the sum of four such bivariate normals. The image function, is as usual, given by

$$\overset{*}{Q}_{hk}^{++}(x, t; k) = (h/x)^\alpha Q_{hk}^{++}(h^2/x, t; k),$$

where $\alpha = (2r/\sigma^2 - 1)$. Q8 of the Exercise Problems considers the other partial time barrier options, such as U/O, D/I, U/I calls, puts etc. □

## 7.9.2 End-out, Partial Time Barrier Options

Here the barrier monitoring window occupies the range $[T_1, T_2]$ and only kicks in after a time $\tau_1 = (T_1 - t)$ has elapsed. To price an end-out, partial time D/O barrier option, first observe that the option is a standard D/O barrier option for all times in the range $[T_1, T_2]$. Let this price, just after $T_1$, be given by $V_{do}(x_1, \tau)$ where once again $\tau = (T_2 - T_1)$. The option over the time interval $[t, T_1)$, will be a standard European option with payoff $V_{do}(x_1, \tau)$, provided at time $T_1$, the asset price $x_1 > h$. Otherwise, the option will immediately knock-out at time $T_1$ and hence expire worthless. We conclude that the payoff of the end-out, D/O barrier option at time $T_1$ must be

$$V^{eq}(x_1, T_1) = V_{do}(x_1, \tau)\, \mathbb{I}(x_1 > h).$$

The pv of the option is then obtained by finding the pv of an equivalent European option with this time $T_1$ payoff.

### End-out, Partial Time D/O Call
In this case, we have from the table of call and put barrier prices in Section 7.5,

$$V_{do}(x_1, \tau) = C_\ell(x_1, \tau) - \overset{*}{C}_\ell(x_1, \tau); \qquad \ell = \max(h, k).$$

The image is taken wrt the barrier price $h$. Hence, from the above, we have the equivalent European payoff at time $T_1$, given by

$$
\begin{aligned}
V^{eq}_{do}(x_1, T_1) &= [Q^+_\ell(x_1, \tau; k) - \overset{*}{Q}{}^+_\ell(x_1, \tau; k)]\mathbb{I}(x_1 > h)\\
&= Q^+_\ell(x_1, \tau; k)\mathbb{I}(x_1 > h) - \mathcal{I}_h\{Q^+_\ell(x_1, \tau; k)\mathbb{I}(x_1 < h)\}
\end{aligned}
$$

In writing down the second line above, we have used the image property:

$$\mathcal{I}_h\{f(x)\mathbb{I}(x < h)\} = \overset{*}{f}(x)\mathbb{I}(x > h)$$

for any function $f(x)$. This last representation allows us, once again, to write down by inspection the corresponding price of the end-out, partial time, D/O call barrier option, as

$$\boxed{V_{do}(x, t) = Q^{++}_{h\ell}(x, t; k) - \overset{*}{Q}{}^{-+}_{h\ell}(x, t; k)} \qquad (7.16)$$

Following the same procedure as above, it is not difficult, to show that the corresponding end-out, partial time, U/O call barrier option has pv given by the expression

$$\boxed{\begin{aligned} V_{uo}(x, t) = {}&[Q^{-+}_{hk}(x, t; k) - \overset{*}{Q}{}^{++}_{hk}(x, t; k)]-\\ &[Q^{-+}_{hh}(x, t; k) - \overset{*}{Q}{}^{++}_{hh}(x, t; k)] \quad \text{iff } h > k \end{aligned}} \qquad (7.17)$$

The option has zero value if $h < k$.

**REMARK 7.4**   Heynen and Kat [31] consider a second type of end-out, partial time barrier option (their type B1), in which the barrier monitoring occurs in the interval $[T_1, T_2]$. This type depends on the asset price level at time $T_1$. If the asset price is above the barrier price, i.e., $x_1 > h$, then the option becomes a D/O, partial time barrier option. If the asset price is below the barrier price, i.e., $x_1 < h$, then the option becomes an U/O, partial time barrier option. In other words, this option will not automatically be knocked out at time $T_1$. The equivalent payoff at time $T_1$, will then be given by the expression

$$V(x_1, T_1) = V_{do}(x_1, \tau)\mathbb{I}(x_1 > h) + V_{uo}(x_1, \tau)\mathbb{I}(x_1 < h).$$

It follows, that the pv of this end-out, partial time barrier option (for a call payoff) will just be the sum of the prices of the D/O and U/O, end-out, call barrier options derived in (7.16) and (7.17).

These expressions can become very complicated. For example, in the case $h > k$, the price will be the sum of six second-order $Q$ options and their images. Since each such $Q$ option is itself a sum of two bi-variate normals, the price will contain twelve such bi-variate normal distribution functions. □

---

## 7.10   Double Barriers

Double barrier options, also known as *corridor* options, are derivatives whose payoff depends on the underlying asset hitting either an upper or a lower barrier. The distinction between up and down types is lost for double barrier options; we only have knock-out and knock-in types. We saw that single barrier options have just one image price associated with the given barrier. Unfortunately, this is not the case for double barrier options. Here we need to consider a doubly-infinite sequence of image prices, which in turn leads to a doubly-infinite series to represent the price of any double barrier option.

Kunimoto and Ikeda [47] have analyzed the case of double barrier calls and puts with exponential barriers using mostly integration methods. A more general solution, for arbitrary payoffs, using methods espoused in this book can be found in Buchen and Konstandatos [10]. We present here the latter case for flat barriers only, in order to keep the exposition as simple as possible.

Let $a$ and $b$, independent of time $t$ with $0 < a < b$, denote the lower and upper (flat) barriers respectively. Then the following theorem (similar to theorem 7.2) gives the Method of Images for knock-out, double barrier options, in the Black–Scholes framework.

**THEOREM 7.3**

*Let $\mathcal{L}$ denote the BS-pde operator and let $V_{ab}(x,t)$ solve the TV problem*

$$\mathcal{L}V_{ab}(x,t) = 0; \quad V_{ab}(x,T) = f(x)\mathbb{I}(a < x < b) \qquad (7.18)$$

*in $x > 0$, $t < T$. Then the price of the knock-out, double barrier option, for all $t < T$ and $a < x < b$, satisfying the TBV problem*

$$\mathcal{L}V(x,t) = 0; \quad V(x,T) = f(x); \quad V(a,t) = V(b,t) = 0 \qquad (7.19)$$

*is given by the doubly-infinite sum*

$$V(x,t) = \sum_{n=-\infty}^{\infty} \lambda^{n\alpha} \left[ V_{ab}(\lambda^{2n}x,t) - \overset{*}{V}_{ab}(\lambda^{2n}x,t) \right] \qquad (7.20)$$

*where $\lambda = b/a$, $\alpha = 2r/\sigma^2 - 1$ and*

$$\overset{*}{V}_{ab}(\lambda^{2n}x,t) = \left(\frac{a}{x}\right)^{\alpha} V_{ab}\left(\frac{\lambda^{2n}a^2}{x},t\right) \quad \text{or (resp.)} \quad \left(\frac{b}{x}\right)^{\alpha} V_{ab}\left(\frac{\lambda^{2n}b^2}{x},t\right)$$

*is the image function wrt $x = a$ or (resp.) $x = b$.*

**REMARK 7.5**    Note that because of the identities

$$\mathbb{I}(a < x < b) \equiv \mathbb{I}(x > a) - \mathbb{I}(x > b) \equiv \mathbb{I}(x < b) - \mathbb{I}(x < a)$$

the European derivative $V_{ab}(x,t)$ can be expressed either as the difference of two up-type or two down-type binaries,

$$V_{ab}(x,t) = V_a^+(x,t) - V_b^+(x,t) = V_b^-(x,t) - V_a^-(x,t).$$

It is also important to realize that the last line of the theorem does not imply that the image functions with respect to $a$ and $b$ are equal, but rather that the doubly-infinite sum of these images are equal.

Theorem 7.3 states a general result for arbitrary payoff $V(x,T) = f(x)$. To get a representation of the price, albeit in terms of a doubly-infinite sum, we need only determine the the price of the binary option $V_{ab}(x,t)$. This being a TV or standard European option (without any barriers), it can generally be obtained, either by static replication methods, or from the FTAP.

If $Y_{t,T}$ and $Z_{t,T}$ represent the minimum and maximum asset prices over the interval $[t,T]$, then in the spirit of (7.6) for single barriers, we can write the double KO-barrier price as

$$V(x,t) = e^{-r\tau}\mathbb{E}_Q\{f(X_T)\mathbb{I}(Y_{t,T} > a)\mathbb{I}(Z_{t,T} < b) \,|\, \mathcal{F}_t\} \qquad (7.21)$$

where $Q$ denotes the EMM. ▯

The proof of theorem 7.3, which is presented next, avoids any integrations (the standard approach as described in Q2 of the Exercise Problems), but instead exploits a variety of image operator properties.

### 7.10.1 Proof of MoI for Double Barrier Options

First, to simplify notation, let the operator $\mathcal{I}_{ab}$ denote the double image sequence: $\mathcal{I}_a\mathcal{I}_b$. The image wrt $b$ is performed first and is then followed by the image wrt $a$. We attach a similar interpretation to the multiple image operators: $\mathcal{I}_{abc}$ and $\mathcal{I}_{abcd}$ for possibly other barrier prices: $c, d > 0$.

Define the doubly-infinite sequence of image operators, by

$$\mathcal{K}_a^b = I - \mathcal{I}_a + \mathcal{I}_{ba} - \mathcal{I}_{aba} + \mathcal{I}_{baba} - \cdots$$
$$-\mathcal{I}_b + \mathcal{I}_{ab} - \mathcal{I}_{bab} + \mathcal{I}_{abab} - \cdots$$

where $I$ is the identity operator and the $\mathcal{I}$'s are image operators. Observe the symmetry: $\mathcal{K}_a^b = \mathcal{K}_b^a$. Then, as we shall demonstrate, the double barrier price satisfying (7.19) is given, at least formally, by the expression

$$V(x,t) = \mathcal{K}_a^b\{V_{ab}(x,t)\}. \tag{7.22}$$

We need to verify that this expression satisfies three things: first that it satisfies the BS-pde; second that $V(x,T) = f(x)$ for $a < x < b$; and third that $V(x,t)$ vanishes at $x = a$ and $x = b$. We proceed to verify these three conditions.

1. Now $V_{ab}(x,t)$ satisfies the BS-pde by definition, and by Part 2. of theorem 7.1, so does any image of $V_{ab}(x,t)$, such as $\mathcal{I}_a V_{ab}(x,t)$ or $\mathcal{I}_{ab}V_{ab}(x,t)$ etc. Further, since the BS-pde is linear, any sum of such images of $V_{ab}(x,t)$ will also satisfy the BS-pde. It follows, that the expression (7.22), subject to convergence issues, will also satisfy the BS-pde.

2. To verify the expiry condition, observe that from (7.22), evaluated at time $T$,

$$V(x,T) = V_{ab}(x,T) + \text{image sequence}\{V_{ab}(x,T)\}.$$

The image sequence contains images wrt $a$ and $b$ only and for $a < x < b$, these images will all be valued at prices outside the interval $[a,b]$, where $V_{ab}(x,T)$ vanishes. For example, $\mathcal{I}_a V_{ab}(x,T)$ is valued at the price $a^2/x$, which for $x > a$, lies in the domain $x < a$. A similar outcome applies to all the images in the sequence. Hence, we are left with $V(x,T) = V_{ab}(x,T) = f(x)$, when $x$ is restricted to the range $a < x < b$.

3. Observe next, that after a little rearrangement, we can write $\mathcal{K}_a^b$ in either of the two factorized forms

$$\mathcal{K}_a^b = (I - \mathcal{I}_a)\mathcal{H}_{ba} = (I - \mathcal{I}_b)\mathcal{H}_{ab} \qquad (7.23)$$

where $\mathcal{H}_{ba} = I - \mathcal{I}_b + \mathcal{I}_{ba} - \mathcal{I}_{bab} + \mathcal{I}_{baba} - \cdots$ and $\mathcal{H}_{ab}$ is obtained from $\mathcal{H}_{ba}$ by interchanging $a$ and $b$. However, by part 3. of theorem 7.1, we know that $(I - \mathcal{I}_a)U(x,t)$ vanishes at $x = a$ for any function $U(x,t)$; and similarly for $(I - \mathcal{I}_b)U(x,t)$ at $x = b$. Hence, (7.23) shows that $\mathcal{K}_a^b\{V_{ab}(x,t)\}$ vanishes at both $x = a$ and $x = b$.

The intuition behind the operator $\mathcal{K}_a^b$ should not escape the reader. If $x = a$ is the first barrier hit, we subtract $\mathcal{I}_a V_{ab}$ from $V_{ab}$ to satisfy the BC at $x = a$; then we add a term $\mathcal{I}_{ba}V_{ab}$ to satisfy the BC at $x = b$; but now the BC is no longer satisfied at $x = a$, so we subtract a further term $\mathcal{I}_{aba}V_{ab}$. We continue this process indefinitely to obtain the first infinite sequence of images of $V_{ab}$. The second infinite sequence starting with $-\mathcal{I}_b V_{ab}$ follows a similar pattern, corresponding to $x = b$ as the first barrier hit, satisfying the BC at $x = b$, then $x = a$, back to $x = b$ and so on.

The next part of the proof relies on a number of properties of image sequences. It is useful to collect these properties under a single heading and to think of these properties as representing the *algebra* of image operators.

## The Algebra of Image Operators

1. $\mathcal{I}_a^2 = \mathcal{I}_{aa} = I$ for any $a > 0$

2. $\mathcal{I}_a \mathcal{I}_{(\lambda b)} = \mathcal{I}_{(a/\lambda)}\mathcal{I}_b \quad$ for any $\lambda > 0$

3. For any $(a, b, c) > 0, \quad \mathcal{I}_{abc} = \mathcal{I}_a \mathcal{I}_b \mathcal{I}_c = \mathcal{I}_{(ac/b)}$

4. Define $\mathcal{H}_{ab}^0 = I$ and for positive integer $n$, define $\mathcal{H}_{ab}^n = \mathcal{I}_{ab}\mathcal{I}_{ab}\cdots\mathcal{I}_{ab}$ with $n$ copies of $\mathcal{I}_{ab} = \mathcal{I}_a \mathcal{I}_b$. Then

$$H_{ab}^n = \mathcal{I}_b \mathcal{I}_{(b^{n+1}/a^n)} \qquad \text{and} \qquad H_{ab}^{-n} = H_{ba}^n. \qquad (7.24)$$

## Proofs

1. Although we have already met this involution property in part 1. of theorem 7.1, we present the proof here. Applying the image operator $\mathcal{I}_a$ twice to an arbitrary function $V(x,t)$, we find:

$$\begin{aligned}
\mathcal{I}_a^2 V(x,t) = \mathcal{I}_a \overset{*}{V}(x,t) &= \mathcal{I}_a \left\{ (a/x)^\alpha V(a^2/x,t) \right\} \\
&= (a/x)^\alpha \overset{*}{V}(a^2/x,t) \\
&= (a/x)^\alpha \cdot (a/(a^2/x))^\alpha \, V(a^2/(a^2/x),t) \\
&= V(x,t).
\end{aligned}$$

Hence $\mathcal{I}_a^2 = \mathcal{I}_{aa}$, for any $a > 0$, is the identity operator.

2. Consider the double image $\mathcal{I}_{ab}$. We find explicitly,

$$
\begin{aligned}
\mathcal{I}_{ab} V(x,t) &= \mathcal{I}_a \, (b/x)^\alpha \, V(b^2/x, t) \\
&= (a/x)^\alpha \cdot (b/(a^2/x))^\alpha \, V(b^2/(a^2/x), t) \\
&= (b/a)^\alpha \, V(b^2 x/a^2, t).
\end{aligned}
\tag{7.25}
$$

It follows from this representation, that replacing $b$ by $\lambda b$ is equivalent to replacing $a$ by $a/\lambda$. That is, $\mathcal{I}_a \mathcal{I}_{(\lambda b)} = \mathcal{I}_{(a/\lambda)} \mathcal{I}_b$.

3. Here we compute the product of three images, using the previous result for double images.

$$
\begin{aligned}
\mathcal{I}_{abc} V(x,t) &= \mathcal{I}_a \cdot \mathcal{I}_{bc} V(x,t) \\
&= \mathcal{I}_a \cdot (c/b)^\alpha \, V(c^2 x/b^2, t) \\
&= (a/x)^\alpha \, (c/b)^\alpha \, V(c^2(a^2/x)/b^2, t) \\
&= (d/x)^\alpha \, V(d^2/x, t) \quad \text{with} \quad d = ac/b.
\end{aligned}
$$

It follows that $\mathcal{I}_{abc} = \mathcal{I}_d$ where $d = ac/b$.

4. We prove that $H_{ab}^n = \mathcal{I}_b \mathcal{I}_{(b^{n+1}/a^n)}$, by induction on $n$. For $n = 0$, the formula reduces to $H_{ab}^0 = \mathcal{I}_b \mathcal{I}_b = I$, by virtue of the involution property of $\mathcal{I}_b$. Now, assume the formula is true for $n$, and let us compute, using the triple image above,

$$
\begin{aligned}
H_{ab}^{n+1} &= H_{ab}^n \mathcal{I}_{ab} \\
&= \mathcal{I}_b \cdot \mathcal{I}_{(b^{n+1}/a^n)} \mathcal{I}_a \mathcal{I}_b \\
&= \mathcal{I}_b \cdot \mathcal{I}_{(b^{n+2}/a^{n+1})}.
\end{aligned}
$$

Since this last expression agrees with the formula for $H_{ab}^n$ with $n$ replaced by $(n+1)$, the result is proved.

The formula for $H_{ab}^n$ allows us to define $H_{ab}^n$, for negative $n$ as well. We formally replace $n$ by $-n$ in the formula. This gives, with $\lambda = b/a$,

$$
\begin{aligned}
H_{ab}^{-n} &= \mathcal{I}_b \mathcal{I}_{(b^{-n+1}/a^{-n})} \\
&= \mathcal{I}_b \mathcal{I}_{(a^n/b^{n-1})} = \mathcal{I}_b \mathcal{I}_{(\lambda a^{n+1}/b^n)} \\
&= \mathcal{I}_{(b/\lambda)} \mathcal{I}_{(a^{n+1}/b^n)} \\
&= \mathcal{I}_a \mathcal{I}_{(a^{n+1}/b^n)} = \mathcal{H}_{ba}^n.
\end{aligned}
$$

This completes the proofs of the algebraic properties of image operators. $\square$

We have spent some considerable effort establishing important properties of image operators. The next result brings us a step closer to establishing Equation (7.20) in theorem 7.3.

**LEMMA 7.1**

*The image sequence $\mathcal{K}_a^b$ has any of the following four equivalent representations,*

$$\mathcal{K}_a^b = (I - \mathcal{I}_a) \sum_{n=-\infty}^{\infty} \mathcal{H}_{ab}^n = (I - \mathcal{I}_a) \sum_{n=-\infty}^{\infty} \mathcal{H}_{ba}^n$$

$$= (I - \mathcal{I}_b) \sum_{n=-\infty}^{\infty} \mathcal{H}_{ba}^n = (I - \mathcal{I}_b) \sum_{n=-\infty}^{\infty} \mathcal{H}_{ab}^n. \qquad (7.26)$$

**PROOF**   It is only necessary to prove one of the four, since the symmetry properties $\mathcal{K}_a^b = \mathcal{K}_b^a$ and $\mathcal{H}_{ab}^{-n} = \mathcal{H}_{ba}^n$ will then establish the other three. So, starting from (7.23), we find, after some rearrangement of terms,

$$\begin{aligned}
\mathcal{H}_{ba} &= I - \mathcal{I}_b + \mathcal{I}_{ba} - \mathcal{I}_{bab} + \mathcal{I}_{baba} - \mathcal{I}_{babab} + \cdots \\
&= [I + \mathcal{I}_{ba} + \mathcal{I}_{baba} + \cdots] - \mathcal{I}_b[I + \mathcal{I}_{ab} + \mathcal{I}_{abab} + \cdots] \\
&= \sum_{n=0}^{\infty} \mathcal{H}_{ba}^n - \mathcal{I}_b \sum_{n=0}^{\infty} \mathcal{H}_{ab}^n \\
&= \sum_{n=0}^{\infty} \mathcal{H}_{ba}^n - \mathcal{I}_a \mathcal{I}_{ab} \sum_{n=0}^{\infty} \mathcal{H}_{ab}^n \qquad \text{since } \mathcal{I}_a^2 = I \\
&= \sum_{n=0}^{\infty} \mathcal{H}_{ba}^n - \mathcal{I}_a \sum_{n=1}^{\infty} \mathcal{H}_{ab}^n.
\end{aligned}$$

Next, we calculate, again using $\mathcal{I}_a^2 = I$ in the first line,

$$\begin{aligned}
(I - \mathcal{I}_a)\mathcal{H}_{ba} &= \sum_{n=0}^{\infty} \mathcal{H}_{ba}^n - \mathcal{I}_a \sum_{n=0}^{\infty} \mathcal{H}_{ba}^n - \mathcal{I}_a \sum_{n=1}^{\infty} \mathcal{H}_{ab}^n + \sum_{n=1}^{\infty} \mathcal{H}_{ab}^n \\
&= (I - \mathcal{I}_a) \left( \sum_{n=0}^{\infty} \mathcal{H}_{ba}^n + \sum_{n=1}^{\infty} \mathcal{H}_{ab}^n \right) \\
&= (I - \mathcal{I}_a) \left( \sum_{n=0}^{\infty} \mathcal{H}_{ba}^n + \sum_{n=1}^{\infty} \mathcal{H}_{ba}^{-n} \right) \\
&= (I - \mathcal{I}_a) \sum_{n=-\infty}^{\infty} \mathcal{H}_{ba}^n.
\end{aligned}$$

This completes the proof of Lemma 7.1.                                         □

The final step in the proof of the Method of Images for double barrier options is now at hand. For any function $U(x,t)$, using the double image

(7.25), we obtain

$$\mathcal{H}^n_{ba} U(x,t) = \mathcal{I}_b \mathcal{I}_{(b^{n+1}/a^n)} U(x,t)$$
$$= (b/a)^{n\alpha} U(b^{2n}x/a^{2n},t)$$
$$= \lambda^{n\alpha} U(\lambda^{2n}x,t) \qquad \text{where } \lambda = b/a.$$

So, putting everything together, the final result for the price of a knock-out, double barrier option with barriers at $x = a, b$ is given by

$$V(x,t) = \mathcal{K}^b_a V_{ab}(x,t)$$
$$= (I - \mathcal{I}_a) \sum_{n=-\infty}^{\infty} \mathcal{H}^n_{ba} V_{ab}(x,t)$$
$$= (I - \mathcal{I}_a) \sum_{n=-\infty}^{\infty} \lambda^{n\alpha} V_{ab}(\lambda^{2n}x,t)$$
$$= (I - \mathcal{I}_b) \sum_{n=-\infty}^{\infty} \lambda^{n\alpha} V_{ab}(\lambda^{2n}x,t).$$

The last two lines above give equivalent representations for $V(x,t)$, based on the symmetries stated in Lemma 7.1. □

## 7.10.2  Double Barrier Calls and Puts

Theorem 7.3 provides a prescription how to price knock-out, double barrier options within the BS economy, for essentially an arbitrary payoff function $f(x)$. When considering specifically call and put options, we need only apply this method with $f(x) = (x - k)^+$ and $f(x) = (k - x)^+$ respectively, where $k$ denotes the strike price.

### Double Barrier Call Options

Assume that relative to the barrier prices $(a, b)$, the strike price satisfies $k < b$, otherwise the knock-out option will always expire worthless. Let $\ell = \max(a, k)$. Then, in theorem 7.3, we calculate,

$$V_{ab}(x, T) = (x - k)^+ [\mathbb{I}(x > a) - \mathbb{I}(x > b)] = (x - k)[\mathbb{I}(x > \ell) - \mathbb{I}(x > b)]$$

Hence, the corresponding pv of $V_{ab}$ will be given, in terms of $Q-$options, by

$$V_{ab}(x, t) = Q^+_\ell(x, t; k) - Q^+_b(x, t; k)$$

for all $t < T$. Equation (7.20) then gives the following formula for the price of a knock-out, double barrier call option:

$$V^{ko}_c(x,t) = \sum_{n=-\infty}^{\infty} \lambda^{n\alpha} \left[ Q^+_\ell(\lambda^{2n}x, t; k) - Q^+_b(\lambda^{2n}x, t; k) \right]$$
$$- \sum_{n=-\infty}^{\infty} \lambda^{n\alpha} \left[ \overset{*}{Q}{}^+_\ell(\lambda^{2n}x, t; k) - \overset{*}{Q}{}^+_b(\lambda^{2n}x, t; k) \right]$$

$$(7.27)$$

The images $\overset{*}{Q_\ell^+}$ and $\overset{*}{Q_b^+}$ appearing above can be taken either wrt the lower barrier $a$ or the upper barrier $b$.

The corresponding price $V_c^{ki}(x,t)$, of a knock-in, double barrier call option is easily found using the (general) parity relation,

$$V^{ko}(x,t) + V^{ki}(x,t) = V_0(x,t) \qquad (7.28)$$

where $V_0(x,t)$ denotes the price of a standard European option (i.e., one without any barriers) with the same payoff function $f(x)$. In the present case of a call option, this yields,

$$V_c^{ki}(x,t) = C_k(x,t) - V_c^{ko}(x,t). \qquad (7.29)$$

## Double Barrier Put Options

For put options, assume $k > a$, otherwise the knock-out version of the option will always expire worthless, and let $\ell = \min(b,k)$. Then, following a similar analysis for the call option above, we find

$$V_{ab}(x,T) = (k-x)\left[\mathbb{I}(x<\ell) - \mathbb{I}(x<a)\right]$$

with pv, for all $t < T$,

$$V_{ab}(x,t) = Q_\ell^-(x,t;k) - Q_a^-(x,t;k)$$

The corresponding price of the knock-out, double barrier put option is then

$$
\begin{aligned}
V_p^{ko}(x,t) = & \sum_{n=-\infty}^{\infty} \lambda^{n\alpha} \left[Q_a^-(\lambda^{2n}x,t;k) - Q_a^+(\lambda^{2n}x,t;k)\right] \\
& - \sum_{n=-\infty}^{\infty} \lambda^{n\alpha} \left[\overset{*}{Q_\ell^-}(\lambda^{2n}x,t;k) - \overset{*}{Q_b^-}(\lambda^{2n}x,t;k)\right]
\end{aligned}
\qquad (7.30)
$$

Using (7.28), we also obtain the associated knock-in, double barrier put price, as

$$V_p^{ki}(x,t) = P_k(x,t) - V_p^{ko}(x,t). \qquad (7.31)$$

**REMARK 7.6** The above prices for double barrier calls and puts do agree with the published results of Kunimoto and Ikeda [47]. Issues of convergence of the doubly-infinite sums in which the prices are given, has been addressed in Buchen and Konstandatos [10].

Extensions of the double-barrier analysis given above, including: effects of a dividend paying asset, exponential barriers, and partial time, double barrier windows are readily handled using methods we have addressed in previous calculations. They can also be found in Konstandatos [45], [46] and the cited literature. $\Box$

## 7.11   Sequential Barrier Options

Sequential barrier options are another variation of the standard knock-out and knock-in barrier options, which have become popular in the Japanese OTC equity and FX markets. They are similar to compound options, in that they are barrier options, which if not knocked-out, deliver a contract which is another barrier option with a different barrier price. However, in the next section, we shall consider another family of barrier options, which are indeed called *compound barrier options* and these differ in detail to the sequential barrier options considered here.

Thus, in order to avoid confusion, a great deal of care must be taken to clarify their precise payoffs. The first comprehensive analysis of sequential barrier options options in a BS framework, using integration methods, can be found in Pfeffer [61]. We present below, our approach to pricing them, and as usual no integrations are required at all. Furthermore, our method gives a prescription for an arbitrary payoff function, not just for calls and puts.

There are essentially eight different sequential barrier options for any given expiry $T$ payoff function $f(x)$. They are

| Sequential Barrier Options | |
| --- | --- |
| Knock-in Sequential Barriers | Knock-out Sequential Barriers |
| UI/DI | UO/DI |
| UI/DO | UO/DO |
| DI/UI | DO/UI |
| DI/UO | DO/UO |

The UI/DI option is an up-and-in barrier option with payoff equal to a down-in-barrier option set at a lower barrier price. That is, if the UI barrier is knocked in at time $t_b$ before $T$, the holder gets a DI barrier option with life $(T-t_b)$ remaining. The others are interpreted the same way. There are always two barrier levels for sequential barrier options, and we shall always take the lower one to be $x = a$ and the higher one to be $x = b$, with $a < b$. Note that other potential types, such as UI/UI, UI/UO etc. don't appear in the list, as they do not really make much sense in the classification of sequential barrier options.

The two columns representing knock-in and knock-out sequential barriers have some important basic differences. For the knock-in types: monitoring of the payoff barrier (e.g., the DI-barrier in UI/DI) only starts once the option has been knocked-in. What the asset price did before the knock-in time is irrelevant. For the knock-out types: monitoring of the payoff barrier (e.g., the

UI-barrier in DO/UI) starts from the current time $t < T$. These differences have profound effects on their respective prices.

Perhaps the simplest way to see how these features affect the prices of sequential barriers is to express them using their FTAP representations. To this end, let $Y_{t,T}$, $Z_{t,T}$ denote the minimum and maximum asset prices over $[t, T]$ and let $t_a$, $t_b$ denote the first passage times to the price levels $x = a$ and $x = b$, given we know the asset price at time $t$. Then the eight sequential barrier options have prices given by the following.
For the KI-sequential barriers:

$$V_{ui/di}(x,t) = e^{-r\tau}\mathbb{E}_Q\{f(X_T)\mathbb{I}(Y_{t_b,T} > a)\mathbb{I}(t_b < T)|\mathcal{F}_t\}$$
$$V_{ui/do}(x,t) = e^{-r\tau}\mathbb{E}_Q\{f(X_T)\mathbb{I}(Y_{t_b,T} < a)\mathbb{I}(t_b < T)|\mathcal{F}_t\}$$
$$V_{di/ui}(x,t) = e^{-r\tau}\mathbb{E}_Q\{f(X_T)\mathbb{I}(Y_{t_a,T} > b)\mathbb{I}(t_a < T)|\mathcal{F}_t\}$$
$$V_{di/uo}(x,t) = e^{-r\tau}\mathbb{E}_Q\{f(X_T)\mathbb{I}(Y_{t_a,T} < b)\mathbb{I}(t_a < T)|\mathcal{F}_t\}.$$

Note that the condition $\mathbb{I}(t_b < T)$ is equivalent to $\mathbb{I}(Z_{t,T} > b)$.
For the KO-sequential barriers:

$$V_{uo/di}(x,t) = e^{-r\tau}\mathbb{E}_Q\{f(X_T)\mathbb{I}(Y_{t,T} < a)\mathbb{I}(Z_{t,T} < b)|\mathcal{F}_t\}$$
$$V_{uo/do}(x,t) = e^{-r\tau}\mathbb{E}_Q\{f(X_T)\mathbb{I}(Y_{t,T} > a)\mathbb{I}(Z_{t,T} < b)|\mathcal{F}_t\}$$
$$V_{do/ui}(x,t) = e^{-r\tau}\mathbb{E}_Q\{f(X_T)\mathbb{I}(Y_{t,T} > a)\mathbb{I}(Z_{t,T} > b)|\mathcal{F}_t\}$$
$$V_{do/uo}(x,t) = e^{-r\tau}\mathbb{E}_Q\{f(X_T)\mathbb{I}(Y_{t,T} > a)\mathbb{I}(Z_{t,T} < b)|\mathcal{F}_t\}.$$

Several things immediately result from these representations.

1. First, we have two parity relations

$$\begin{aligned}
V_{ui/di}(x,t) + V_{ui/do}(x,t) &= V_{ui}(x,t); \quad (x < b)\\
V_{di/ui}(x,t) + V_{di/uo}(x,t) &= V_{di}(x,t); \quad (x > a)
\end{aligned} \tag{7.32}$$

where $V_{ui}(x,t)$ and $V_{di}(x,t)$ are the prices of standard up-and-in, and down-and-in barrier options with the same payoff $f(x)$.

2. From (7.21) we have

$$\boxed{V_{uo/do}(x,t) = V_{do/uo}(x,t) = V_{db}^{ko}(x,t); \quad (a < x < b)} \tag{7.33}$$

where $V_{db}^{ko}(x,t)$ denotes the price of a double barrier knock-out option, and is given explicitly by (7.20).

3. Since $\mathbb{I}(Y_{t,T} < a) = 1 - \mathbb{I}(Y_{t,T} > a)$ and $\mathbb{I}(Z_{t,T} > b) = 1 - \mathbb{I}(Z_{t,T} < b)$ we also have

$$\boxed{\begin{aligned}
V_{uo/di}(x,t) &= V_{uo}(x,t) - V_{db}^{ko}(x,t); \quad (x < b)\\
V_{do/ui}(x,t) &= V_{do}(x,t) - V_{db}^{ko}(x,t); \quad (x > a)
\end{aligned}} \tag{7.34}$$

where $V_{uo}(x,t)$ and $V_{do}(x,t)$ are the prices of standard up-and-out, and down-and-out barrier options with the same payoff $f(x)$.

From these observations, it is clear we need only price the two sequential barrier options, $V_{ui/di}(x,t)$ and $V_{di/ui}(x,t)$ and the list is complete.

**Pricing the UI/DI Sequential Barrier Option**
We price this option by finding its equivalent European payoff at time $T$. Since this option is first an UI barrier (relative to $x = b$), we have from (7.10)

$$V^{eq}(x,T) = g(x)\mathbb{I}(x>b) + \mathcal{I}_b[g(x)\mathbb{I}(x<b)]$$

where $\mathcal{I}_b$ is the image operator wrt $x = b$, and we have used the property

$$\mathcal{I}_b[g(x)]\mathbb{I}(x>b) = \mathcal{I}_b[g(x)\mathbb{I}(x<b)].$$

In the above, $g(x)$ denotes the payoff of the UI barrier option, which in the present case is that of a DI-barrier option relative to $x = a$. Thus, also from (7.10), we have

$$g(x) = f(x)\mathbb{I}(x<a) + \mathcal{I}_a[f(x)\mathbb{I}(x>a)].$$

Hence, putting the two expression together leads to:

$$V^{eq}(x,T) = f(x)\mathbb{I}(x<a)\mathbb{I}(x>b) + \mathcal{I}_a[f(x)\mathbb{I}(x>a)]\mathbb{I}(x>b)$$
$$+ \mathcal{I}_b[f(x)\mathbb{I}(x<a)\mathbb{I}(x<b)] + \mathcal{I}_b[\mathcal{I}_a\{f(x)\mathbb{I}(x>a)\}\mathbb{I}(x>b)].$$

However, the first term vanishes since $a < b$; the second term also vanishes as

$$\text{2nd-term} = \mathcal{I}_a[f(x)\mathbb{I}(x>a)\mathbb{I}(x<a^2/b)] = 0.$$

In the third term, $\mathbb{I}(x<a)\mathbb{I}(x<b) = \mathbb{I}(x<a)$; and for the fourth term

$$\text{4th-term} = \mathcal{I}_b\mathcal{I}_a[f(x)\mathbb{I}(x>a)\mathbb{I}(x<a^2/b)] = \mathcal{I}_{ba}[f(x)\mathbb{I}(x>a)].$$

Hence
$$V^{eq}(x,T) = \mathcal{I}_b[f(x)\mathbb{I}(x<a)] + \mathcal{I}_{ba}[f(x)\mathbb{I}(x>a)]. \tag{7.35}$$

This is the equivalent European payoff for the sequential UI/DI barrier option, we are seeking. To get the price of the option is now straightforward. Let $V_a^{\pm}(x,t)$ denote the prices of up and down binaries, which pay at time $T$, respectively the amounts $f(x)\mathbb{I}(x>a)$ and $f(x)\mathbb{I}(x<a)$. Then the pv is given by,

$$\boxed{V_{ui/di}(x,t) = \mathcal{I}_b[V_a^-(x,t)] + \mu^\alpha V_a^+(\mu^2 x, t); \qquad (\mu = a/b)} \tag{7.36}$$

We have used (7.25) in writing down the expression for the double image.

**Pricing the DI/UI Sequential Barrier Option**

A similar analysis, which is not repeated here, for the sequential DI/UI barrier option, yields

$$V_{di/ui}(x,t) = \mathcal{I}_a[V_b^+(x,t)] + \lambda^\alpha V_b^-(\lambda^2 x, t); \qquad (\lambda = b/a) \qquad (7.37)$$

**Sequential UI/DI Calls**

We use a similar notation for gap calls as in Section 7.5. Thus

$$C_\xi(x,t;k) = x\mathcal{N}(d_1) - ke^{-r\tau}\mathcal{N}(d_2); \quad d_{1,2} = \frac{\log(x/\xi) + (r \pm \frac{1}{2}\sigma^2)\tau}{\cdot}\sigma\sqrt{\tau}$$

Then applying Equation (7.36), with $f(x) = (x-k)^+$ and $k < b$, for the price of the sequential UI/DI call, we obtain

|  | $k > a$ | $k < a$ |
|---|---|---|
| $V_a^-(x,t)$ | $0$ | $C_k(x,t;k) - C_a(x,t;k)$ |
| $V_a^+(x,t)$ | $C_k(x,t;k)$ | $C_a(x,t;k)$ |

This leads to the prices:

$$V_{ui/di}(x,t) = \begin{cases} \mu^\alpha C_k(\mu^2 x, t; k) & (k > a) \\ \overset{*}{C}_k(x,t;k) - \overset{*}{C}_a(x,t;k) + \mu^\alpha C_a(\mu^2 x, t; k) & (k < a) \end{cases}$$

$$(7.38)$$

where $\overset{*}{C}$ denotes the image wrt $x = b$.

**Sequential DI/UI Calls**

Here we find, since $k < b$,

$$V_b^+(x,t) = C_b(x,t;k)$$
$$V_b^-(x,t) = C_k(x,t;k) - C_b(x,t;k)$$

and (7.37) gives the price

$$V_{di/ui}(x,t) = \overset{*}{C}_b(x,t;k) + \lambda^\alpha[C_k(\lambda^2 x, t; k) - C_b(\lambda^2 x, t; k)] \qquad (7.39)$$

where now, $\overset{*}{C}$ denotes the image wrt $x = a$.

**REMARK 7.7**   While the formula (7.39) agrees with Pfeffer's result, our formula (7.38) for the sequential UI/DI call price, contains extra terms (the image terms for the case $k < a$) not appearing in Pfeffer's paper.   □

## 7.12   Compound Barrier Options

The barrier options we analyze in this section are more akin to the compound vanillas (call/puts) considered in Section 5.5. They differ from the sequential barrier options of the previous section by having two *fixed* expiry dates $T_1$ and $T_2$ with $T_1 < T_2$. Over the interval $[t, T_1]$ we have a barrier option (of any of the four types) with barrier level $x = a$; the payoff of this barrier option at time $T_1$ is another barrier option (also of any of the four types) with different barrier level $x = b$. There is no predefined relationship between $a$ and $b$. The payoff of this second barrier option at time $T_2$ will be designated by an arbitrary function $f(x)$.

Clearly, these compound barrier options are different from the sequential barrier options of the previous section, but there are strong similarities. However, while there were only eight different sequential barrier options for any given payoff, there will be the full complement of sixteen different compound barrier options. Many of these will of course be linked through parity relations.

We shall analyze in detail just one: namely, the DO/UO compound barrier option, and present a table of results for all sixteen types. To be clear, the DO/UO compound barrier option is a DO barrier option over $[t, T]$ with barrier level $x = a$, the time $T_1$ payoff being an UO barrier option with barrier level $x = b$ and payoff function $f(x)$. For this option to finish itm at time $T_2$ (i.e., to pay $f(x)$), the underlying asset price must not fall below $x = a$ during $[t, T_1]$, and not rise above $x = b$ during $[T_1, T_2]$.

### The DO/UO Compound Barrier Option
We determine the price of this compound option by finding its equivalent European payoffs: first at time $T_1$ and then at time $T_2$. Now at time $T_1$, the payoff will be an UO barrier $b$ option with time $\tau = (T_2 - T_1)$ remaining to expiry, but only if the asset price $x_1$ at $T_1$ is above $a$, i.e., $x_1 > a$. If in fact $x_1 < a$, the option gets knocked out instantly at time $T_1$. Hence with the equivalent DO payoff (7.10) we can write

$$V_{do/uo}^{eq}(x_1, T_1) = V_{uo}^{eq}(x_1, \tau)\mathbb{I}(x_1 > a) - \mathcal{I}_a[V_{uo}^{eq}(x_1, \tau)\mathbb{I}(x_1 > a)].$$

But, also from (7.10), we have

$$V_{uo}^{eq}(x_2, T_2) = f(x_2)\mathbb{I}(x_2 < b) - \mathcal{I}_b[f(x_2)]\mathbb{I}(x_2 > b)$$

where, for clarity, we have written the lhs at the fixed time $T_2$, rather than the relative time we employed at time $T_1$. Since $x_1$ is fixed for all times in

$[T_1, T_2]$, we can put the two expressions above together, to yield

$$V_{do/uo}^{eq}(x_1, x_2; T_2) = [f(x_2)\mathbb{I}(x_2 < b) - \mathcal{I}_b[f(x_2)]\mathbb{I}(x_2 > b)]\mathbb{I}(x_1 > a) -$$
$$\mathcal{I}_a \{[f(x_2)\mathbb{I}(x_2 < b) - \mathcal{I}_b[f(x_2)]\mathbb{I}(x_2 > b)]\mathbb{I}(x_1 > a)\}.$$

Now to keep things manageable, let us introduce some simplifying notation: define the second order $f(x)-$binary

$$f_{a,b}^{s_1 s_2}(x_1, x_2) = f(x_2)\,\mathbb{I}(s_1 x_1 > s_1 a)\,\mathbb{I}(s_2 x_2 > s_2 b) \tag{7.40}$$

where $(s_1, s_2)$ are plus or minus signs. Then using, by now, standard properties of image operators and how they operate on functions like $f(x)\mathbb{I}(x > a)$, we can write the above equivalent DO/UO payoff in the form,

$$\boxed{V_{do/uo}^{eq}(x_1, x_2; T_2) = f_{a,b}^{+-} - \mathcal{I}_a f_{a,b}^{+-} - \mathcal{I}_b f_{\bar{a},b}^{--} + \mathcal{I}_{ab} f_{\bar{a},b}^{--}} \tag{7.41}$$

where $\bar{a} = b^2/a$. This is the $T_2$ equivalent payoff we seek, and is seen to be a portfolio of second order $f(x)-$binaries and their images wrt both $x_1 = a$ and $x_2 = b$. Hence if we can price the second order binary on $f(x)$, we can price the corresponding DO/UO compound barrier option.

### Table of Compound Barrier Equivalent Payoffs

It transpires, and indeed it should be clear, that each of the 16 compound barrier option payoffs has the same structure as (7.41). That is to say, each belongs to one of:

$$V^{eq}(x_1, x_2; T_2) = \pm f_{a,b}^{\pm\pm} \pm \mathcal{I}_a f_{a,b}^{\pm\pm} \pm \mathcal{I}_b f_{\bar{a},b}^{\pm\pm} \pm \mathcal{I}_{ab} f_{\bar{a},b}^{\pm\pm}.$$

Thus each of the four terms has a triple $(s_1, s_2, s_3)$ of signs: $s_1$ being the sign in front of term; $(s_2, s_3)$ being the pair of superscripts. The table below simply lists this triple of signs for each of the sixteen compound barrier options.

| | Compound Barrier | $f_{a,b}$ | $\mathcal{I}_a f_{a,b}$ | $\mathcal{I}_b f_{\bar{a},b}$ | $\mathcal{I}_{ab} f_{\bar{a},b}$ |
|---|---|---|---|---|---|
| 1 | DO/DO | $(+,+,+)$ | $(-,+,+)$ | $(-,-,+)$ | $(+,-,+)$ |
| 2 | DI/DO | $(+,-,+)$ | $(+,+,+)$ | $(-,+,+)$ | $(-,-,+)$ |
| 3 | UI/DO | $(+,+,+)$ | $(+,-,+)$ | $(-,-,+)$ | $(-,+,+)$ |
| 4 | UO/DO | $(+,-,+)$ | $(-,-,+)$ | $(-,+,+)$ | $(+,+,+)$ |
| 5 | DO/DI | $(+,+,-)$ | $(-,+,-)$ | $(+,-,+)$ | $(-,-,+)$ |
| 6 | DI/DI | $(+,-,-)$ | $(+,+,-)$ | $(+,+,+)$ | $(+,-,+)$ |
| 7 | UI/DI | $(+,+,-)$ | $(+,-,-)$ | $(+,-,+)$ | $(+,+,+)$ |
| 8 | UO/DI | $(+,-,-)$ | $(-,-,-)$ | $(+,+,+)$ | $(-,+,+)$ |
| 9 | DO/UI | $(+,+,+)$ | $(-,+,+)$ | $(+,-,-)$ | $(-,-,-)$ |
| 10 | DI/UI | $(+,-,+)$ | $(+,+,+)$ | $(+,+,-)$ | $(+,-,-)$ |
| 11 | UI/UI | $(+,+,+)$ | $(+,-,+)$ | $(+,-,-)$ | $(+,+,-)$ |
| 12 | UO/UI | $(+,-,+)$ | $(-,-,+)$ | $(+,+,-)$ | $(-,+,-)$ |
| 13 | DO/UO | $(+,+,-)$ | $(-,+,-)$ | $(-,-,-)$ | $(+,-,-)$ |
| 14 | DI/UO | $(+,-,-)$ | $(+,+,-)$ | $(-,+,-)$ | $(-,-,-)$ |
| 15 | UI/UO | $(+,+,-)$ | $(+,-,-)$ | $(-,-,-)$ | $(-,+,-)$ |
| 16 | UO/UO | $(+,-,-)$ | $(-,-,-)$ | $(-,+,-)$ | $(+,+,-)$ |

## Example 7.2

As an example, consider the case of a DO/UO compound call, with time $T_2$ payoff given by (7.41), with $f(x) = (x-k)^+$ and assume $a < k < b$. In this case, we obtain the following:

$$f_{a,b}^{+-}(x_1, x_2) = (x_2 - k)\mathbb{I}(x_1 > a)[\mathbb{I}(x_2 > k) - \mathbb{I}(x_2 > b)]$$
$$= Q_{ak}^{++}(x_1, x_2; k) - Q_{ab}^{++}(x_1, x_2; k).$$
$$f_{\bar{a},b}^{--}(x_1, x_2) = (x_2 - k)\mathbb{I}(x_1 < \bar{a})[\mathbb{I}(x_2 > k) - \mathbb{I}(x_2 > b)]$$
$$= Q_{\bar{a}k}^{-+}(x_1, x_2; k) - Q_{\bar{a}b}^{-+}(x_1, x_2; k).$$

These $T_2$ payoffs are just the second order $Q$–options of (5.12). Hence, the pv of the DO/UO compound call option will be

$$
V_{do/uo}(x, t) = \left\{
\begin{array}{l}
[Q_{ak}^{++}(x, t; k) - Q_{ab}^{++}(x, t; k)] \\
-[\hat{Q}_{ak}^{++}(x, t; k) - \hat{Q}_{ab}^{++}(x, t; k)] \\
-[\overset{*}{Q}_{\bar{a}k}^{-+}(x, t; k) - \overset{*}{Q}_{\bar{a}b}^{-+}(x, t; k)] \\
+[\bar{Q}_{\bar{a}k}^{-+}(x, t; k) - \bar{Q}_{\bar{a}b}^{-+}(x, t; k)]
\end{array}
\right.
\tag{7.42}
$$

where, recalling the image functions,

$$\hat{Q}(x, t) = \mathcal{I}_a Q(x, t) = (a/x)^\alpha Q(a^2/x, t)$$
$$\overset{*}{Q}(x, t) = \mathcal{I}_b Q(x, t) = (b/x)^\alpha Q(b^2/x, t)$$
$$\bar{Q}(x, t) = \mathcal{I}_{ab} Q(x, t) = (b/a)^\alpha Q(b^2 x/a^2, t).$$

We are content to leave the price of this compound call barrier as in (7.42), since when expressed in terms of the associated bi-variate normals (see Equations (5.9), (5.10) and (5.13)), the eight terms above become sixteen. ☐

### Another ESO Example

Regulatory bodies require that ESO's be valued for compliance and tax issues. They permit a range of possible pricing methodologies including Monte Carlo simulation, binomial lattice methods and also BS modeling. There is an enormous range of ESO's in current use, and many contain a wide range of special features concerning: vesting periods, performance hurdles, performance roll-overs, exercise scenarios, survival statistics and others.

We considered a simple performance based ESO (Executive Stock Option) in Section 6.6 of the previous chapter. Here we price a more complex ESO which includes performance monitoring up to a vesting date $T_1$, followed by an exercise period up to a maturity date $T_2$. This ESO uses a DO barrier to monitor performance in the first interval, and exercise of the options will be governed by an UO barrier in the second interval. Hence, this ESO will include a compound DO/UO barrier option as part of its pricing regime.

The precise details of the ESO we shall price are described next. It was analyzed, principally by the methods advocated here, in the PhD thesis of Kyng [49]. Suppose the ESO is granted as an atm call option today, represented by time $t$. The company's share price is monitored over the interval $[t, T_1]$, during which time the executive is not permitted to exercise. If at any time during this interval, the share price drops below a minimum performance level $x = a$, the ESO immediately expires worthless (and the exec is shown the door). If the ESO survives the performance period, then it vests at time $T_1$, signalling that the executive may now exercise the option at any time up to the maturity date $T_2$. It has been argued, that many executives don't exercise their options optimally, but wait till they are suitably deep itm. Hence the way these options are exercised, they are not really American, but rather European with an UO barrier feature. To explain this last point, suppose our executive decides to exercise if and when the share price reaches some fraction $\gamma$ of the strike price $k$, where obviously $\gamma > 1$. For example, they might exercise when $\gamma$ equals 120% . In this case, the executive will receive a dollar amount, $R = (x - k) = (\gamma - 1)k > 0$. This is obviously equivalent to an UO barrier option with barrier level $b = \gamma k$ and a rebate $R$. We considered such rebates in Section 7.6. If the share price is at or above $b = \gamma k$ at time $T_1$, then the executive will exercise immediately. We assume the strike price $k$ satisfies $a < k < b$.

*Payoff at $T_1$*
Since the ESO is a DO barrier option prior to $T_1$, with barrier level $a$, we have

its equivalent payoff as

$$V^{eq}(x_1, T_1) = U(x_1, \tau)\mathbb{I}(x_1 > a) - \mathcal{I}_a[U(x_1, \tau)\mathbb{I}(x_1 > a)] \tag{7.43}$$

where $U(x_1, \tau)$ denotes the payoff of a contract with time $\tau = (T_2 - T_1)$ remaining to maturity. This payoff consists of three terms:

$$U_1(x_1, \tau) = (x_1 - k)^+ \, \mathbb{I}(x_1 > b)$$

corresponding to immediate exercise of the option when $x_1 > b$, i.e., above the executive's exercise threshold;

$$U_2(x_1, \tau) = V_{uo}(x_1, \tau)\mathbb{I}(x_1 < b)$$

the UO call barrier option with barrier level $b$, which kicks in at time $T_1$ if $x_1 < b$, i.e., below the exercise threshold; and

$$U_3(x_1, \tau) = V_R(x_1, \tau)\mathbb{I}(x_1 < b)$$

corresponding to the rebate term of the UO barrier option, which in this case is the call option payout if the executive decides to exercise before maturity. Putting these all together leads to:

$$V^{eq}(x_1, T_1) = (x_1 - k)\mathbb{I}(x_1 > b) + V_{uo}(x_1, \tau)\mathbb{I}(a < x_1 < b)$$
$$+ V_R(x_1, \tau)\mathbb{I}(a < x_1 < b) \quad - \quad \mathcal{I}_a\{\cdots\cdots\} \tag{7.44}$$

where $\mathcal{I}_a\{\cdots\}$ represents the image wrt $x_1 = a$ of the preceding three terms. When we include the expressions:

$$V_{uo}(x_1, \tau) = [Q_k^+(x_1, \tau; k) - \overset{*}{Q}{}_k^+(x_1, \tau; k)] - [Q_b^+(x_1, \tau; k) - \overset{*}{Q}{}_b^+(x_1, \tau; k)]$$

and

$$V_R(x_1, \tau) = R[B_b^+(x_1, \tau) + \overset{*}{B}{}_b^-(x_1, \tau)]$$

we are finally in a position to price the ESO. The result is:

$$V_{ESO}(x, t) = \begin{cases} Q_b^+(x, \tau_1; k) \\ +[Q_{ak}^{++} - \overset{*}{Q}{}_{\bar{a}k}^{-+}](x, t; k) \\ -[Q_{ab}^{++} - \overset{*}{Q}{}_{\bar{a}b}^{-+}](x, t; k) \\ -[Q_{bk}^{++} - \overset{*}{Q}{}_{bk}^{-+}](x, t; k) \\ +[Q_{bb}^{++} - \overset{*}{Q}{}_{bb}^{-+}](x, t; k) \\ +R[B_{ab}^{++} + \overset{*}{B}{}_{\bar{a}b}^{--}](x, t) \\ -R[B_{bb}^{++} + \overset{*}{B}{}_{bb}^{--}](x, t) \\ -\mathcal{I}_a\{\cdots\cdots\} \end{cases} \tag{7.45}$$

Here $\overset{*}{Q}$ denotes the image wrt $x = b$. The first term $Q_b^+(x, \tau_1; k)$ where $\tau_1 = (T_1 - t)$, is a first order gap call and represents (with its image wrt $a$)

the immediate $T_1$ exercise component of the ESO, if $x_1 > b$. If you count the number of normal distribution functions in this expression, you will find a total of 44 of them, which explains why we have left the price in the above, somewhat, abstract form.

We offer a final word of caution: due to the various parity relations for both first order and second order $Q$−options and bond binaries, the price (7.45) has many different but equivalent representations. Matching results (in general) with other published ones can be a nightmare, as the author could testify on many occasions.

---

## 7.13 Outside Barrier Options

Consider a standard European option, on asset $X$, with expiry $T$ payoff function $f(x)$. The related down-and-out barrier option also pays $f(x)$ at time $T$, but only if the underlying asset price of $X$ stays above the barrier price $h$ throughout the life of the option. If it hits the barrier at any time before expiry, the option expires worthless. Outside barrier options are similar, but with one important change. The option is knocked-out, not if asset $X$ hits the barrier, but if a different asset $Y$ hits the barrier. Outside barrier options were first analyzed by Heynen and Kat in [30].

Outside barrier options are two-dimensional rainbow options and their prices will be governed by the 2D Black–Scholes pde (6.2), repeated here for convenience

$$V_t = r(V - xV_x - yV_y) - \tfrac{1}{2}(\sigma_1^2 x^2 V_{xx} + \sigma_2^2 y^2 V_{yy} + 2\rho\sigma_1\sigma_2 xy V_{xy}).$$

$V = V(x, y, t)$ is the pv at time $t < T$ and the TV or payoff condition is $V(x, y, T) = f(x)$, independent of $y$. For the D/O, outside barrier option with barrier price at $y = h$, the domain of the pde is $(x > 0, \ y < h)$, and the BC at $y = h$ is $V(x, h, t) = 0$. For this pde, it has been assumed that assets $X$ and $Y$ follow correlated gBm's with volatilities $\sigma_1$ for $X$ and $\sigma_2$ for $Y$. The correlation coefficient of their log-returns is assumed to be constant and equal to $\rho$.

The solution to the above 2D pde can be obtained by an appropriate Method of Images, which depends on the following result.

**THEOREM 7.4**
*Let $V(x, y, t)$ be any solution of the 2D, BS-pde. Then its image wrt $y = h$,*

*is given by*

$$\overset{*}{V}(x,y,t) = \left(\frac{h}{y}\right)^\alpha V\left[\left(\frac{h}{y}\right)^\beta x, \left(\frac{h^2}{y}\right), t\right] \qquad (7.46)$$

*where $\alpha = 2r/\sigma_2^2 - 1$ and $\beta = 2\rho\sigma_1/\sigma_2$.*

**REMARK 7.8**    The 2D image function $\overset{*}{V}(x,y,t)$ satisfies all the properties listed in theorem 7.1 for 1D images. In particular, it also satisfies the 2D, BS-pde, as can be demonstrated by a rather tedious exercise in partial differentiation. However, we give a proof of this result in Chapter 10, where the general multi-asset image function is derived.

Note that for uncorrelated assets, $\rho = 0$ and the 2D image reduces essentially to the 1D image. ⬜

Everything we learned about 1D images, including the Method of Images, carries over with obvious adjustment to 2D images. In particular, for all four types of outside barriers, the equivalent 2D European payoffs (see Equation (7.10) for the 1D case) will be

$$\begin{aligned}
V_{do}^{eq}(x,y,T) &= f(x)\mathbb{I}(y>h) - \overset{*}{f}(x)\mathbb{I}(y<h) \\
V_{di}^{eq}(x,y,T) &= [f(x) + \overset{*}{f}(x)]\mathbb{I}(y<h) \\
V_{ui}^{eq}(x,y,T) &= [f(x) + \overset{*}{f}(x)]\mathbb{I}(y>h) \\
V_{uo}^{eq}(x,y,T) &= f(x)\mathbb{I}(y<h) - \overset{*}{f}(y)\mathbb{I}(x>h)
\end{aligned} \qquad (7.47)$$

where $\overset{*}{f}(x) = \mathcal{I}_{\{y=h\}} f(x) = (h/y)^\alpha f[(h/y)^\beta x]$.

We calculate next a few of the more common examples of outside barriers.

**Outside Call Barrier Options**
The payoff function is $f(x) = (x - k)^+$, so the equivalent payoff for the D/O case can be written as

$$V_{do}^{eq}(x,y,T) = (x - k)\mathbb{I}(x>k)\mathbb{I}(y>h) - \{\text{image wrt } y = h\}$$

However, we recognize the first term above as a portfolio of two-asset, binary options considered in Section 6.1. Theorem 6.1 gives the pv of this first term as $\mathcal{A}_{kh}^{++}(x,y,t) - k\mathcal{B}_{kh}^{++}(x,y,t)$, where

$$\mathcal{A}_{kh}^{++}(x,y,t) = x\mathcal{N}(d_{1k}, d_{2h} + \rho\sigma_1\sqrt{\tau}; \rho); \qquad \mathcal{B}_{kh}^{++}(x,y,t) = e^{-r\tau}\mathcal{N}(d_{1k}', d_{2h}'; \rho)$$

and

$$[d_{1k}, d_{1k}'] = \frac{\log(x/k) + (r \pm \frac{1}{2}\sigma_1^2)\tau}{\sigma_1\sqrt{\tau}}; \qquad [d_{2h}, d_{2h}'] = \frac{\log(y/h) + (r \pm \frac{1}{2}\sigma_2^2)\tau}{\sigma_2\sqrt{\tau}}.$$

With these definitions, we can therefore write the pv of the outside, D/O call barrier option as

$$V_{do}^c(x,y,t) = \begin{aligned}&[A_{kh}^{++}(x,y,t) - \overset{*}{A}_{kh}^{++}(x,y,t)]\\&-k[B_{kh}^{++}(x,y,t) - \overset{*}{B}_{kh}^{++}(x,y,t)]\end{aligned} \tag{7.48}$$

The three independent parity relations (7.7) and (7.8) still apply for outside barriers, so now that we have priced the D/O outside call barrier option, the other three follow immediately.

$$V_{di}^c(x,y,t) = C_k(x,t) - V_{do}^c(x,y,t)$$
$$V_{ui}^c(x,y,t) = \overset{*}{V}_{di}^c(x,y,t)$$
$$V_{uo}^c(x,y,t) = C_k(x,t) - V_{ui}^c(x,y,t).$$

**Outside Put Barrier Options**
Since the analysis is very similar, we are content to simply state the results. For the D/O outside put barrier option,

$$V_{do}^p(x,y,t) = \begin{aligned}&-[A_{kh}^{-+}(x,y,t) - \overset{*}{A}_{kh}^{-+}(x,y,t)]\\&+k[B_{kh}^{-+}(x,y,t) - \overset{*}{B}_{kh}^{-+}(x,y,t)]\end{aligned} \tag{7.49}$$

and, from the parity relations,

$$V_{di}^p(x,y,t) = P_k(x,t) - V_{do}^p(x,y,t)$$
$$V_{ui}^p(x,y,t) = \overset{*}{V}_{di}^p(x,y,t)$$
$$V_{uo}^p(x,y,t) = P_k(x,t) - V_{ui}^p(x,y,t).$$

In Equation (7.49), an obvious adjustment needs to made for the different (super-scripted) signs.

---

## 7.14  Reflecting Barriers

All the barriers we have looked at so far in this chapter are examples of what are called *absorbing barriers*, where the variable of interest vanishes. When the derivative of the variable, wrt the asset price, vanishes we have a *reflecting barrier*. It is also possible to have more complicated BC's, called *mixed BC's*, at which a linear combination of the variable and its derivative, vanishes. Let the barrier be located at $x = b$. Then, the three basic types of BC's are listed in the following table.

| Boundary Type | BC at $x = b$ |
|---|---|
| Absorbing | $V(b, t) = 0$ |
| Reflecting | $V_x(b, t) = 0$ |
| Mixed | $V_x(b, t) + \beta V(b, t) = 0$ |

There exist derivative securities where reflecting BC's for the BS-pde are the operative type. These include the value of lease contracts, when regarded as real options. See Dixit and Pindyck [20] for examples. A comprehensive analysis of such contracts, using the methods of this book, can be found in the PhD thesis of Ho-Shon [34].

Consider the following BS-pde, terminal, mixed-BV problem for $V(x, t)$ in the domain $(t < T, \ x > b)$,

$$\mathcal{L}V(x, t) = 0; \quad V(x, T) = f(x); \quad V_x + \beta V = 0 \text{ at } x = b$$

for some real constant $\beta$. This more general case includes problems with reflecting boundaries when $\beta = 0$.

In order to solve this problem, we shall utilize the theory of log-volutions and Mellin Transforms introduce in Section 2.12. In this formalism, the BS-pde operator $\mathcal{L}$ can be expressed as

$$\mathcal{L}V = V_t - rV + (r - \tfrac{1}{2}\sigma^2)DV + \tfrac{1}{2}\sigma^2 D^2 V; \qquad D = x\frac{d}{dx}$$

and the BC at $x = b$, can be written as $\mathcal{B}V = DV + \beta b V = 0$.

### PROPOSITION 7.1

*The solution of the mixed TBV problem $\mathcal{L}V = 0$, $V(x, T) = f(x)$, $\mathcal{B}V = 0$ at $x = b$, for $V(x, t)$ in the domain $(t < T, \ x > b)$ is given, in operational form, by the expression*

$$V(x, t) = V_b^+(x, t) - (\mathcal{B}^{-1}\mathcal{I}_b\mathcal{B}) V_b^+(x, t) \tag{7.50}$$

*where $V_b^+(x, t)$ solves the (standard European) TV problem $\mathcal{L}V_b^+ = 0$ in $(t < T; x > 0)$, with payoff condition $V_b^+(x, T) = f(x)\mathbb{I}(x > b)$.*

**REMARK 7.9**  Proposition 7.1 basically states the Method of Images for mixed BV problems of the BS-pde. $\mathcal{I}_b$ is our old friend the image operator wrt $x = b$. Note that if $\mathcal{B} = I$, the identity operator, then (7.50) reduces precisely to the Method of Images (7.4) for absorbing barriers. As expected, $\mathcal{B}^{-1}$ is the inverse of the first-order differential operator $\mathcal{B}$. It should therefore be some kind of integration operator. Its precise representation is given below. ☐

**PROOF**   The proof of the proposition requires us only to verify (assuming uniqueness of solutions of the mixed TVB problem for the BS-pde) that the expression for $V(x,t)$ satisfies the BS-pde, the TV, $V(x,T) = f(x)$ in $x > b$, and the BC, $\mathcal{B}V = 0$ at $x = b$.

1. Now $V_b^+(x,t)$ satisfies the BS-pde by definition. Also for $D = x\frac{d}{dx}$, so does $\mathcal{B}(D)V_b^+(x,t)$, for any operator depending only on $D$. This follows from the observation that when the BS-pde is expressed in terms of the operator $D$, it is linear and has constant coefficients.

   The image function $\mathcal{I}_b U(x,t)$, also satisfies the BS-pde whenever $U(x,t)$ does, so clearly the term $(\mathcal{B}^{-1}\mathcal{I}_b\mathcal{B})\,V_b^+(x,t)$ must satisfy the BS-pde. Hence $V(x,t)$ defined by (7.50) is indeed a solution of the BS-pde $\mathcal{L}V = 0$.

2. Now at expiry, $V_b^+(x,T) = f(x)\mathbb{I}(x > b)$ which vanishes for all $x < b$. Similarly, $\mathcal{B}V_b^+(x,T)$ will also vanish for all $x < b$, and the image $\mathcal{I}_b\mathcal{B}\,V_b^+(x,T)$ will vanish for all $x$ in the complementary domain, $x > b$, as also will $(\mathcal{B}^{-1}\mathcal{I}_b\mathcal{B})\,V_b^+(x,T)$. Hence, $V(x,T) = f(x)$ for all $x > b$, as required.

3. Now (7.50) can be written in the equivalent form

$$\mathcal{B}V(x,t) = U(x,t) - \mathcal{I}_b U(x,t); \qquad U(x,t) = \mathcal{B}V_b^+(x,t).$$

   But from property 3 of theorem 7.1, $U(x,t) - \mathcal{I}_b U(x,t) = 0$ at $x = b$. Hence, $\mathcal{B}V(x,t)$ vanishes at $x = b$.

The above three items complete the proof of proposition 7.1.   ⬚

But of course, the operational form of the solution (7.50) is useless until we can evaluate it explicitly. This can be done by taking the Mellin Transform, of the equivalent representation

$$\mathcal{B}V(x,t) = \mathcal{B}V_b^+(x,t) - \mathcal{I}_h\mathcal{B}V_b^+(x,t). \tag{7.51}$$

Let $\hat{V}(s,t)$ denote the Mellin Transform of $V(x,t)$. Then M5. of the table of Mellin Transforms in Section 2.12.1, gives the Mellin Transform of $\mathcal{B}(D)V(x,t)$ as $\mathcal{B}(-s)\hat{V}(s,t) = (\beta b - s)\hat{V}(s,t)$. But what about the Mellin Transform of the image function $\mathcal{I}_b U(x,t)$? The next lemma gives the required result.

*LEMMA 7.2*
*The Mellin Transform of $V(x,t) = \mathcal{I}_b U(x,t) = (b/x)^\alpha U(b^2/x,t)$, for any function $U(x,t)$, for which $\hat{U}(s,t)$ exists, is given by the expression*

$$\boxed{\hat{V}(s,t) = b^{2s-\alpha}\,\hat{U}(\alpha - s, t)} \tag{7.52}$$

**PROOF**  The proof depends only on properties listed in the table of Mellin Transforms. We find $\hat{V}(s,t)$ in four easy steps (in an obvious notation).

Using M2. $\qquad\qquad\qquad U(b^2x,t) \rightarrow b^{-2s}\,\hat{U}(s,t)$

Using M3. $\qquad\qquad\qquad U(b^2/x,t) \rightarrow b^{2s}\,\hat{U}(-s,t)$

Using M4. $\qquad\qquad x^{-\alpha}U(b^2/x,t) \rightarrow b^{2(s-\alpha)}\,\hat{U}(\alpha-s,t)$

Using M1. $\qquad\quad (b/x)^\alpha U(b^2/x,t) \rightarrow b^{2s-\alpha}\,\hat{U}(\alpha-s,t)$

This completes the proof. $\qquad\qquad\qquad\qquad\qquad\qquad\qquad\qquad$ ▯

Let $\gamma = \beta b$ and $U(x,t) = V_b^+(x,t)$. Taking the Mellin Transform of (7.51) now yields,

$$(\gamma-s)\hat{V}(s,t) = (\gamma-s)\hat{U}(s,t) - [\gamma-(\alpha-s)]\,b^{2s-\alpha}\hat{U}(\alpha-s,t)$$

$$\therefore \quad \hat{V}(s,t) = \hat{U}(s,t) + \left[1 + \frac{2\gamma-\alpha}{s-\gamma}\right]b^{2s-\alpha}\hat{U}(\alpha-s,t).$$

Using Mellin Transform properties M6 and M8 and (7.52), we obtain the inverse Mellin Transform of $V(x,t)$ as

$$V(x,t) = U(x,t) + \overset{*}{U}(x,t) + (2\gamma-\alpha)\left[x^{-\gamma}\mathbb{I}(x<1) \star \overset{*}{U}(x,t)\right]$$

$$= U(x,t) + \overset{*}{U}(x,t) + (2\gamma-\alpha)\int_0^\infty \overset{*}{U}(y,t)\left(\frac{y}{x}\right)^\gamma \mathbb{I}(\tfrac{x}{y}<1)\frac{dy}{y}$$

where $\overset{*}{U}(x,t) = \mathcal{I}_b U(x,t)$ denotes the image of $U(x,t)$ wrt $x = b$.

In conclusion, the solution of the mixed BV problem for the BS-pde is given by:

$$\boxed{V(x,t) = V_b^+(x,t) + \overset{*}{V}_b^+(x,t) + (2\gamma-\alpha)\int_x^\infty \overset{*}{V}_b^+(y,t)\left(\frac{y}{x}\right)^\gamma \frac{dy}{y}} \qquad (7.53)$$

and for a reflecting boundary (case $\gamma = 0$), the solution simplifies to

$$\boxed{V(x,t) = V_b^+(x,t) + \overset{*}{V}_b^+(x,t) - \alpha\int_x^\infty \overset{*}{V}_b^+(y,t)\frac{dy}{y}} \qquad (7.54)$$

**REMARK 7.10**  These equations for solving mixed and reflecting BV problems for the BS-pde are quite remarkable. They differ from the absorbing boundary solution $V(x,t) = V_b^+(x,t) - \overset{*}{V}_b^+(x,t)$, in one very important respect. This, of course, is the addition of the integrals over the image function $\overset{*}{V}_b^+(y,t)$. The interpretation of this observation is as follows: we need only a single point image for an absorbing boundary, but we require a continuous distribution of images for mixed and reflecting boundaries. ▯

**Equivalent European Payoffs**
If we set $t = T$ in the above expressions, we generate the corresponding equivalent European payoffs for mixed and reflecting BV problems. The results are:
For the mixed BV problem in $x > b$,

$$\boxed{\begin{aligned} V^{eq}(x,T) &= f(x)\mathbb{I}(x>b) + \overset{*}{f}(x)\mathbb{I}(x<b) \\ &+ (2\gamma - \alpha)\left[\int_x^b \overset{*}{f}(y)\left(\frac{y}{x}\right)^\gamma \frac{dy}{y}\right]\mathbb{I}(x<b) \end{aligned}}$$ (7.55)

and for a reflecting BV problem in $x > b$,

$$\boxed{V(x,T) = f(x)\mathbb{I}(x>b) + \overset{*}{f}(x)\mathbb{I}(x<b) - \alpha\left[\int_x^b \overset{*}{f}(y)\frac{dy}{y}\right]\mathbb{I}(x<b)}$$ (7.56)

The corresponding equivalent payoffs for mixed and reflecting BV problems in the domain $x < b$, are given by:

For the mixed BV problem in $x < b$,

$$\boxed{\begin{aligned} V^{eq}(x,T) &= f(x)\mathbb{I}(x<b) + \overset{*}{f}(x)\mathbb{I}(x>b) \\ &- (2\gamma - \alpha)\left[\int_b^x \overset{*}{f}(y)\left(\frac{y}{x}\right)^\gamma \frac{dy}{y}\right]\mathbb{I}(x>b) \end{aligned}}$$ (7.57)

and for a reflecting BV problem in $x < b$,

$$\boxed{V(x,T) = f(x)\mathbb{I}(x<b) + \overset{*}{f}(x)\mathbb{I}(x>b) + \alpha\left[\int_b^x \overset{*}{f}(y)\frac{dy}{y}\right]\mathbb{I}(x>b)}$$ (7.58)

To illustrate this theory, we give just one example of a reflecting barrier problem.

**Pricing a Real Estate Lease**
This "real options" example has its origins in the book by Dixit and Pindyck [20]. Let $V(x,t)$ denote the value of a lease on a property, which expires at time $T$ in the future, and where $x$ denotes the market rent at time $t$. Then, assuming the rent process is gBm with volatility $\sigma$ and dividend yield $q$, it can be shown that $V(x,t)$ satisfies the BS, TBV problem

$$V_t = rV - (r-q)xV_x - \tfrac{1}{2}\sigma^2 x^2 V_{xx}; \qquad (t < T; \ x < b)$$

with terminal value $V(x,T) = H(x)$, and reflecting BC, $V_x(b,t) = 0$. Here

$$H(x) = \frac{b}{q}\left[\frac{x}{b} - \frac{1}{\beta}\left(\frac{x}{b}\right)^\beta\right]$$

represents the equilibrium value of the property, which happens to satisfy the time-independent BS-pde. For an interpretation of the parameters $(q, b, \beta)$, all of which are positive, the reader is referred to the above reference, or the paper by Grenadier [27]. This paper also prices a range of other, more complex lease contracts and does so using first passage densities and, by the authors' own admission, many "grueling integrations."

Our Method of Images for reflecting boundaries, on the other hand, considerably reduces the complexity of the calculations and does so by eschewing all but one elementary integration in the equivalent European payoff. A comprehensive treatment of pricing lease contracts by these methods can be found in the PhD thesis of Ho-Shon [34].

Let $U_k(x, t)$ solve the above BS, pde with reflecting BC at $x = b$, and TV, $U_k(x, T) = (x/b)^k$. Then, by linearity of the BS-pde, the solution of the lease problem will be

$$V(x, t) = \frac{b}{q} \left[ U_1(x, t) - \frac{1}{\beta} U_\beta(x, t) \right].$$

We use Equation (7.58) to find the equivalent European payoff. For the elementary integration,

$$\int_b^x \overset{*}{f}(y) dy/y = \int_b^x (b/y)^\alpha (b/y)^k dy/y = \frac{1}{\alpha + k} [1 - (b/x)^{\alpha + k}].$$

Hence

$$U_k^{eq}(x, T) = \left( \frac{x}{b} \right)^k \mathbb{I}(x < b) + \frac{\alpha}{\alpha + k} \mathbb{I}(x > b) + \frac{k}{\alpha + k} \left( \frac{b}{x} \right)^{\alpha + k} \mathbb{I}(x > b)$$

$$= \left( \frac{x}{b} \right)^k \mathbb{I}(x < b) + \frac{\alpha}{\alpha + k} \mathbb{I}(x > b) + \frac{k}{\alpha + k} \mathcal{I}_b \left[ \left( \frac{x}{b} \right)^k \mathbb{I}(x < b) \right]$$

where $\mathcal{I}_b$ denotes the image wrt $x = b$. This payoff is recognized as a portfolio of turbo binaries which we considered in section 4.8. Hence, we have without further calculation, the following representation for the pv of $U_k(x, t)$,

$$U_k(x, t) = b^{-k} P_b^-(x, t; k) + \frac{\alpha}{\alpha + k} B_b^+(x, t) + \frac{k b^{-k}}{\alpha + k} \overset{*}{P}_b^-(x, t; k)$$

where the formal expression for $P_b^-(x, t; k)$ is given by Equation (4.17).

The corresponding price of the lease contract can then be written as

$$\boxed{\begin{aligned} V(x, t) = \frac{1}{q} &\left[ A_b^-(x, t) + \frac{1}{\alpha + 1} \overset{*}{A}_b^-(x, t) + \frac{\alpha b}{\alpha + 1} B_b^+(x, t) \right] \\ &- \frac{b}{q} \left[ b^{-\beta} P_b^-(x, t; \beta) + \frac{b^{-\beta}}{\alpha + \beta} \overset{*}{P}_b^-(x, t; \beta) + \frac{\alpha}{\beta(\alpha + \beta)} B_b^+(x, t) \right] \end{aligned}} \tag{7.59}$$

Thus even for a simple fixed term lease, the price is a complex portfolio of asset, bond and turbo binaries and their images. Things become even more complicated when other features, like reset clauses are included. The reader is referred to the Grenadier paper and Ho-Shon thesis for analyses of these contracts.

---

## 7.15 Summary

This lengthy chapter on barrier options in many ways epitomizes the essence of this book. The traditional way to price these path-dependent options (see Exercise Problem 2) is a two-step process. In Step 1, calculate the first passage density or joint asset price & running max/min through a given barrier; and in Step 2 compute the discounted expectation of the option payoff with respect to these joint densities. While these methods ultimately work, there is a heavy toll in terms efficiency and mathematical elegance.

Better methods do exist, as the techniques described in this chapter hopefully demonstrate. Boundary value problems in many branches of Theoretical Physics employ a technique called the Method of Images (MoI). For example, if you wanted to solve the heat equation in the domain $x > 0$, with an absorbing BC, $u(0,t) = 0$ and IC $u(x,0) = f(x)$, the MoI does it for you very quickly. Here is how it can be done. Find the solution $u_0(x,t)$ for the unbounded domain $x \in \mathbb{R}$, with IC $u_0(x,0) = f(x)\mathbb{I}(x>0)$, then the required solution is just

$$u(x,t) = u_0(x,t) - u_0(-x,t).$$

The problem for $u_0(x,t)$, being just an IV-problem (and not one with a boundary condition) is generally easy to solve, and once this is obtained, the solution to the the BV-problem can be written down by inspection. The key observation is that $u_0(-x,t)$ is a special symmetry solution of the heat equation, called the *image solution* and it satisfies all the properties of images described in this chapter.

Basically the MoI for the BS-pde, although it looks a little more complicated, follows exactly this prescription. Indeed, one way of deriving the MoI for the BS-pde, is to transform the latter into the heat equation and use the simple heat equation MoI (see Exercise Problems Q1). Undoing these transformations then leads to the MoI for Black–Scholes option pricing. It's that simple! We even make it easier to apply, by deriving what we call the equivalent European payoff of the option. In the heat equation example above, the solution to the BV problem can obviously be obtained by solving the IV problem, with IV

$$u_{eq}(x,0) = f(x)\mathbb{I}(x>0) - f(-x)\mathbb{I}(x<0).$$

This is the equivalent IV (translate to equivalent European value in the BS context) that solves the absorbing BV-problem. Few seem to have noticed that you can solve path-dependent barrier options, by embedding them into an associated European option pricing problem. And in the option pricing arena, European options are almost always easier to price.

We use the MoI exclusively to price a range of barrier options in this chapter. Given the D/O and U/O barrier prices, two well known parity relations have been traditionally been used to price the corresponding D/I and U/I options. These are basically: (knock-out + knock-in = standard option). But the MoI also provides a third parity relation, also not recognized by traditional methods. This is the property: image(D/I)=U/I and vice versa, image(U/I)=D/I. With this third parity relation, we only ever need to price one of the four standard barrier options: the other three then follow from parity relations.

These methods are so powerful, that we are able to extend the analysis in many directions. These include application of the MoI to partial time barrier options, double barrier options, sequential barriers, compound barrier options and even two-asset barrier options (and extended to the general multi-asset case in Chapter 10). We also apply the MoI to pricing barrier options on a binomial lattice. It was thought that since lattice nodes don't generally line up with the barrier, that a continuity correction had to be made to get agreement with BS barrier prices. But use of the MoI, as we demonstrate, eschews the need for such continuity corrections. Indeed, we get just as good agreement with BS as for simple European options. And of course this comes about, because barrier options can be expressed in terms of equivalent European options!

Our final section of this chapter addresses the problem of pricing a barrier option with a reflecting BC, rather than an absorbing BC. Such problems are rare in finance, but they do exist in pricing property leases, for example. No doubt, other applications will emerge, with the proliferation of option theoretic ideas penetrating the field of real options. We derived the MoI for the reflecting barrier problem, and saw that it is much more complex than the MoI for absorbing barriers. In place of a single image function, we now have a continuous distribution of images. Nevertheless, we showed how the MoI in this situation can be effectively used, but confess, we now are required to do some integrations, albeit only simple algebraic ones. The use of Mellin Transforms and log-volutions plays an important role in the development of this theory.

The next chapter addresses the other well-known path-dependent class of options know as lookback options. Here too, we develop new methods for pricing these complex options. In fact, we point out a little known relationship which shows that lookback options are nothing more than integrated barrier options. We can then bring the powerful methods of the current chapter to bear on the pricing technology for lookback options. That is exactly how we shall proceed.

---

## Exercise Problems

1. Transform the BS-pde $\mathcal{L}V(x,t) = 0$, for the domain $x > b$, $t < T$, to the heat equation $U_{yy} = \frac{1}{2}\sigma^2 U_\tau$ where

$$y = \log(x/b); \qquad V(x,t) = e^{-\frac{1}{2}\alpha y - \beta\tau}U(y,\tau); \qquad \tau = T - t$$

(see Remark 7.1), in the domain $y > 0$, $\tau > 0$. Show that the image of $U(y,\tau)$ relative to the boundary $y = 0$ is given by $\overset{*}{U} = U(-y,0)$, and hence obtain the image function $\overset{*}{V}(x,t) = (b/x)^\alpha V(b^2/x,t)$ for the BS-pde relative to the barrier $x = b$.

2. Let the underlying asset price at time $T$, under the EMM, $Q$, be denoted by $X_T = xe^{Z_T}$, where $Z_T \sim N(\mu\tau, \sigma^2\tau)$ and $\mu = (r - \frac{1}{2}\sigma^2)$, $\tau = (T-t)$ and $x = X_t$ is the current asset price. Then the pdf of $Z_T$ is given by

$$f(z) = \frac{1}{\sigma\sqrt{\tau}}\phi\left(\frac{z - \mu\tau}{\sigma\sqrt{\tau}}\right); \qquad \phi(z) = \frac{1}{\sqrt{2\pi}}e^{-\frac{1}{2}z^2}; \quad z \in \mathbb{R}.$$

It can also be shown that the joint pdf

$$g(z) = \frac{\partial}{\partial z}\mathbb{P}\{Z_T > z; \min_{t \le s \le T}(Z_s) < h\} = \frac{\partial}{\partial z}\mathbb{P}\{Z_T < z; \max_{t \le s \le T}(Z_s) > h\}$$

where

$$g(z) = \frac{e^{2h\mu/\sigma^2}}{\sigma\sqrt{\tau}}\phi\left(\frac{z - 2h - \mu\tau}{\sigma\sqrt{\tau}}\right).$$

(a) Use this information to show that the price of a D/O barrier option, with payoff function $F(X_T)$ and barrier price $b$, is given by

$$V_{do}(x,t) = \int_h^\infty F(xe^z)f(z)\,dz - \int_h^\infty F(xe^z)g(z)\,dz; \qquad (x > b)$$

where $h = \log(b/x)$.

(b) Write down similar expressions for D/I, U/O and U/I barrier options and hence verify the three parity relations

$$V_{do} + V_{di} = V_0; \quad V_{uo} + V_{ui} = V_0; \quad \overset{*}{V}_{do} = V_{ui}$$

where $V_0(x,t)$ denotes the price of a European option with payoff $F(X_T)$.

*Note:* This question illustrates the "standard" approach to pricing barrier options.

3. Given the price of the D/O call barrier option as

$$V_{do}(x,t) = C_\ell(x,t;k) - \overset{*}{C}_\ell(x,t;k); \quad \ell = \max(h,k),$$

use parity relations to derive the remaining three: D/I, U/I and U/O call barrier option, for the two cases $(k > h)$ and $(k < h)$, as given in the table of call barrier prices in the text.

4. Derive the corresponding prices for the four put barrier prices, by first deriving the formula

$$V_{uo}(x,t) = P_\ell(x,t;k) - \overset{*}{P}_\ell(x,t;k); \quad \ell = \min(h,k)$$

for the U/O put barrier price.

5. Derive the prices in Equation (7.11) for the rebate terms:

(a) $V_{uo}^R(x,t)$ paid at knock-out; and

(b) $V_{uo}^R(x,t)$, $V_{di}^R(x,t)$, $V_{ui}^R(x,t)$ paid at expiry $T$.

6. (a) Show formally, by direct differentiation that the image function

$$\overset{*}{V}(x,t) = \left(\frac{b_t}{x}\right)^\alpha V\left(\frac{b_t^2}{x},t\right); \quad \alpha = \frac{2(r-\beta)}{\sigma^2} - 1$$

where $b_t = Be^{\beta t}$, satisfies the standard BS-pde.

(b) Derive this expression for the image function, relative to an exponential barrier $b_t$, by first transforming the BS-pde to the heat equation.

7. Show that the binomial recursion

$$\overset{*}{V}(x,t) = \rho^{-1}\left[p\overset{*}{V}(ux,t+\Delta) + q\overset{*}{V}(dx,t+\Delta)\right]$$

where $\overset{*}{V} = (b/x)^\alpha V(b^2/x,t)$, reduces to the BS-pde for $V(x,t)$ in the limit $\Delta \to 0$.

8. (a) Derive parity relations connecting the four start-out, partial time, call barrier options and use these to price the corresponding D/I, U/I, and U/O call barrier options.

   (b) Show that the price of the D/O start-out, partial time, put barrier options can be expressed in the form

$$V_{do}(x,t) = Q_{hk}^{+-}(x,t;k) - \overset{*}{Q}_{hk}^{+-}(x,t;k).$$

   Hence, derive prices for the D/I, U/I, and U/O put barrier options.

9. (a) Derive parity relations for end-out, partial time barrier options.

   (b) Use these parity relations to price end-out, partial time D/I, U/I and U/O call barrier options.

   (c) Do the same for end-out, partial time put barrier options.

10. Show that in the notation of the text of Section 7.11, that the pv of the DI/UI sequential barrier option is given by (see (7.37))

$$V_{di/ui}(x,t) = \mathcal{I}_a[V_b^+(x,t)] + \lambda^\alpha V_b^-(\lambda^2 x,t); \qquad (\lambda = b/a).$$

11. (a) Derive the sign signature displayed in the table of compound barrier options of Section 7.12 for the UO/DI compound barrier option.

   (b) Hence determine the pv of both an UO/DI compound call barrier option, and a corresponding UO/DI compound put barrier option.

12. Determine the price of a contract that pays one unit of asset $X$ at expiry $T$, provided the price of a correlated asset $Y$, stays above the constant barrier level $h$, throughout the life of the contract.

13. Consider a barrier option with current price $V(x,t;h)$ where $h$ denotes the barrier level. In a published article, it was suggested that the function

$$U(x,t;a,b) = \frac{1}{b-a} \int_a^b V(x,t;h)\,dh; \qquad (0 < a < b)$$

represents the price of a *soft* barrier option.

   (a) Show that for a D/I call of strike price $k > b$,

$$U(x,t;a,b) = \mathcal{J}_a^b \left\{ \frac{A_k^+(x,\tau)}{\alpha+3} - \frac{kB_k^+(x,\tau)}{\alpha+1} + \frac{2kP_k^+(x,\tau;\beta)}{(\alpha+1)(\alpha+3)} \right\}$$

   where $\mathcal{J}_a^b$ is the image operator $\frac{b\mathcal{I}_b - a\mathcal{I}_a}{b-a}$ and $P_k^+(x,\tau;\beta)$ is the turbo binary with payoff $(x/k)^\beta \, \mathbb{I}(x > k)$, and $\beta = -\frac{1}{2}(\alpha+1)$.
   *Hint*: Integrate the expiry $T$ payoff rather than the time $t$ price.

190


(b) It was argued that a soft barrier option is knocked-in or out proportionally to the depth of penetration of the barrier range $[a, b]$. Show that such a claim cannot possibly be correct.

14. A one-rung *ladder option*, with ladder price $h$ and monotonic increasing payoff function $f(x)$, has the expiry payoff

$$V(x, T) = \max[f(x), \ f(h)\mathbb{I}(z > h)]$$

where $z$ denotes the running maximum asset price over $[t, T]$. Thus if the asset price ever reaches the ladder price $h$ during the life of the option, the minimum payoff is locked in at the value $f(h)$.

(a) Show that the ladder payoff can also be written in the form

$$V(x, T) = f(x) + [f(h) - f(x)]\mathbb{I}(x < h)\mathbb{I}(z > h).$$

(b) For the case of a strike $k < h$ call option, where $f(x) = (x - k)^+$, show that the ladder payoff can be replicated by a portfolio containing: a long strike $k$ European call option; a short U/I put barrier option with strike price $k$ and barrier price $h$; and a long U/I put barrier option with strike price $h$ and barrier price $h$.

15. Multi-rung ladder options lock in successive payoff values $f_i$, where $f_i = f(h_i)$ provided $h_i < z < h_{i+1}$. Sketch in the $xz$−plane the ladder payoff

$$V(x, T) = \max[f(x), \ f_i\mathbb{I}(h_i < z < h_{i+1})]; \quad (i = 1, 2, \ldots, n)$$

where $h_{n+1} = \infty$.

—ooOoo—

# Chapter 8

## Lookback Options

### 8.1 Introduction

In general terms, lookback options are path-dependent exotic options whose payoffs depend not only on the underlying asset price at expiry, but also on the maximum or minimum asset price over some pre-defined monitoring window. Lookback options are sometimes called *hindsight* options or options of *least regret*. The latter name comes from the observation that the payoffs of these options can be tailored to minimize the chance that they finish out-of-the money. Lookback options may also be used to hedge against large fluctuations in the underlying, or to optimize timing into or out of the market. For example, a floating strike lookback call option (which we shall define presently) effectively allows the holder to buy the underlying asset at its minimum price over the lifetime of the option. Similarly, a floating strike put option allows an investor to sell the underlying at the maximum price over this period. Thus a knowledgeable trader can effectively "buy at the low and sell at the high."

Standard methods of pricing lookback options were first explored in Goldman et al. (1979) [26] and later by Conze and Viswanathan (1991) [16]. The approach we use here, as expected by now, will follow a very different path, and was first presented in Buchen and Konstandatos (2005) [9]. This method was derived from an important, but apparently little known relationship between lookback options and barrier options. Thus much of the power developed for barrier options in the previous chapter can be brought to bear on the pricing of lookback options. While we shall present most of the details in this chapter, further aspects can be found in the PhD thesis of Konstandatos (2003) [45] and Konstandatos (2008) [46].

To fix the notation, let $x = X_t$ denote the current asset price at time $t$, such that $0 \leq t \leq T$, and define two new state variables $(y, z)$ by

$$y = Y_t = \min_{0 \leq s \leq t} X_s \qquad \text{and} \qquad z = Z_t = \max_{0 \leq s \leq t} X_s. \qquad (8.1)$$

Thus $y$ and $z$ denote the running minimum and running maximum of the underlying asset price over the continuous time window $[0, t]$. Hence, time

191

zero defines the start time for monitoring the underlying asset price. Note that this feature is already somewhat different from the barrier monitoring window, which for the standard type, starts at the current time $t$ and finishes at the expiry date $T$.

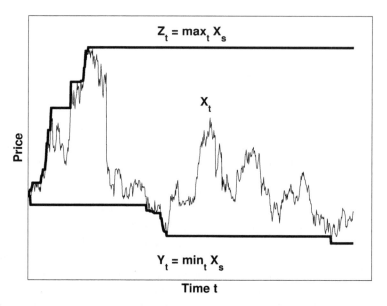

**FIGURE 8.1:** Comparison of the running minimum $y = Y_t$ and running maximum $z = Z_t$ for a given underlying asset process $x = X_t$.

Figure 8.1 displays a sample of what a running minimum and running maximum asset price looks like in practice, relative to the asset price itself. Observe that at $t = 0$, we have $X_0 = Y_0 = Z_0$ and that for long stretches of time, both $y = Y_t$ and $z = Z_t$ are constant. In fact, changes to $y$ and $z$ only occur when a new minimum or maximum asset price is encountered.

This last point is very important because it impacts significantly on how lookback options are priced using the BS-pde. The result is captured in the following theorem.

### THEOREM 8.1

*Let $V(x, y, t)$ and $V(x, z, t)$ denote the prices of minimum and maximum type lookback options. Then, $V(x, y, t)$ and $V(x, z, t)$ both satisfy the standard BS-pde*

$$V_t = rV - rxV_x - \tfrac{1}{2}\sigma^2 x^2 V_{xx}$$

*in the respective domains $x > y$ and $x < z$; $t < T$ and satisfy the BC's*

$$\boxed{V_y(x, y, t) = 0 \text{ at } x = y \qquad and \qquad V_z(x, z, t) = 0 \text{ at } x = z} \qquad (8.2)$$

*where subscripts denote partial derivatives.*

Before we offer an explanation (rather than a formal proof) of theorem 8.1, the reader should be aware of an important feature of the above theorem. This is the observation that both $V(x, y, t)$ and $V(x, z, t)$ satisfy the standard BS-pde with respect to the variables $(x, t)$, and that the new variables $(y, z)$ act only as parameters in the formulation. We shall exploit this property to the hilt in deriving our approach to pricing lookback options.

**Explanation of Theorem 8.1**

The BS-pde for minimum type lookback options[1] can be derived by setting up a self-financing, dynamic hedge portfolio, consisting of a long position in one lookback option, and a short position in $h$ units of the underlying asset. This portfolio has current value $P(x, y, t) = V(x, y, t) - hx$. After a small time interval $dt$, the change in value of the lookback option will by Itô's Lemma, be given by the expression

$$dV(x, y, t) = (V_t + \tfrac{1}{2}\sigma^2 x^2 V_{xx})dt + V_x dx + V_y dy$$

But the last term $V_y dy$ is equal to zero for all times over which there is no change to the running minimum, i.e., when $dy = 0$. Arbitrage considerations require that $dV$ cannot have finite jumps, since they cannot be hedged away. Thus if $V_y dy = 0$ before the asset price hits a new minimum, it must remain zero afterwards as well. But since $dy$ differs from zero only when $x = y$, i.e., at a new minimum, we conclude that $V_y$ must vanish at $x = y$. As remarked in the footnote, we also conclude that $V_z = 0$ at $x = z$ for a maximum type lookback option. Furthermore, since $V_y dy = 0$ for all conditions, the expression for $dV(x, y, t)$ above is the usual one for standard European options, and the hedge portfolio will therefore lead to the standard BS-pde, independent of the variable $y$ (or $z$). □

---

## 8.2  Equivalent Payoffs for Lookback Options

We saw in the previous chapter in Section 7.4 that barrier options could be priced by finding their equivalent European payoffs. These equivalent payoffs,

---

[1] A similar explanation follows with appropriate changes from $y$ to $z$, for maximum type lookback options.

derived from the Method of Images, depend on both the original payoff function $f(x)$, and its image wrt the barrier price. It is possible to derive similar equivalent payoffs for minimum and maximum type lookback options as well. However, there is an important difference. Barrier options require just a single image function, while lookback options require a continuous distribution of image functions. The following theorem provides the details.

### THEOREM 8.2

1. Let $f(x, y)$ denote the time $T$ payoff of a minimum type lookback option. Then the equivalent European payoff is given by the pair of equations

$$
\begin{aligned}
V_{\min}^{eq}(x, y, T) &= f(x, y)\mathbb{I}(x > y) + g(x, y)\mathbb{I}(x < y) \\
\text{where} \quad g(x, y) &= f(x, x) - \int_x^y \overset{*}{f}_\xi(x, \xi)\, d\xi
\end{aligned}
\tag{8.3}
$$

2. Let $F(x, z)$ denote the time $T$ payoff of a maximum type lookback option. Then the equivalent European payoff is given by the pair of equations

$$
\begin{aligned}
V_{\max}^{eq}(x, z, T) &= F(x, z)\mathbb{I}(x < z) + G(x, z)\mathbb{I}(x > z) \\
\text{where} \quad G(x, z) &= F(x, x) + \int_z^x \overset{*}{F}_\xi(x, \xi)\, d\xi
\end{aligned}
\tag{8.4}
$$

**REMARK 8.1**   Theorem 8.2 expresses a surprising result. The traditional way of pricing lookback options utilizes the joint density of the underlying asset's gBm and its corresponding running minimum (resp. maximum). The option payoff must then be integrated over this joint density to obtain its price. Needless to say the calculation is both lengthy and complicated. However, the lookback theorem above permits the calculation to be done virtually by inspection — generally only a simple algebraic integration is required.

Two words of caution are necessary in order to interpret theorem 8.2 correctly.

1. In the interest of expediency, we have been a little carefree with our notation in equations (8.3) and (8.4). In both equations the variable $x$ does indeed represent the asset price at time $T$. However, the variables $(y, z)$ must be taken as parameters, and in fact represent the running minimum and maximum asset price up to the current time $t$ only, and not as may be expected, up to the expiry date date $T$. Perhaps, the notation $V_{\min}(X, y, T)$ and $V_{\max}(X, z, T)$ would be more appropriate, where upper-case variables denote prices at time $T$ and lower-case ones at the current time $t$. However, we are content to stick with the

present notation, with this interpretation being understood. Curiously, the equivalent European payoff depends not only on what is happening at the expiry date $T$, but also on the what is happening, *vis-a-vie* parameters $(y, z)$, at the current time $t$.

2. We must also be careful how we interpret the terms $\overset{*}{f}_\xi(x, \xi)$ and $\overset{*}{F}_\xi(x, \xi)$ in the theorem. These terms contain a partial derivative wrt $\xi$ and an image, also wrt $\xi$. However, differentiation and imaging are operators which do not commute. That is, the result depends on the order in which they are done. In the theorem, the differentiation is performed first, then the imaging. Hence,

$$\overset{*}{f}_\xi(x, \xi) = \mathcal{I}_\xi[f_\xi(x, \xi)] = (\xi/x)^\alpha f_\xi(\bar{x}, \xi); \quad \bar{x} = \xi^2/x$$

with a similar expression for $\overset{*}{F}_\xi(x, \xi)$. The partial derivative wrt $\xi$ is carried out with respect to the second variable in $f(\bar{x}, \xi)$ only, even though $\bar{x}$ depends on $\xi$.

☐

The proof of theorem 8.2, which is presented next, depends on a neat connection between lookback options and knock-out barrier options.

**PROOF**   We begin with the minimum type lookback option, which satisfies the Black–Scholes, TBV problem

$$\mathcal{L}V = 0; \quad V(x, y, T) = f(x, y); \quad V_y(x, y, t) = 0 \text{ at } x = y$$

in the domain $t, T$; $x > y$. Let $U(x, y, t) = V_y(x, y, t)$, the partial derivative of $V(x, y, t)$ wrt $y$. Then since $y$ appears only as a parameter in the above BS-pde, we see that $U(x, y, t)$ solves the TBV problem

$$\mathcal{L}U = 0; \quad U(x, y, T) = f_y(x, y); \quad U(x, y, t) = 0 \text{ at } x = y$$

in the same domain. However, we recognize this problem for $U(x, y, t)$ as a down-and-out barrier option with payoff function $f_y(x, y)$ and barrier price $y$. The equivalent European payoff for this D/O option is given by the first equation in (7.10). That is

$$U^{eq}(x, y, T) = f_y(x, y)\mathbb{I}(x > y) - \overset{*}{f}_y(x, y)\mathbb{I}(x < y)$$

where $y$ denotes a "fixed" parameter, representing the running minimum asset price up to time $t$. Integration wrt $y$ then recovers the equivalent payoff

$V^{eq}(x, y, t)$ for the minimum type lookback option. This yields, the result

$$V^{eq}(x, y, T) - V^{eq}(x, 0, T)$$

$$= \int_0^y [f_\xi(x, \xi)\mathbb{I}(x > \xi) - \overset{*}{f_\xi}(x, \xi)\mathbb{I}(x < \xi)]d\xi$$

$$= \int_0^{\min(x,y)} f_\xi(x, \xi)d\xi - \mathbb{I}(x < y) \int_x^y \overset{*}{f_\xi}(x, \xi)d\xi$$

$$= [f(x, \min(x, y)) - f(x, 0)] - \mathbb{I}(x < y) \int_x^y \overset{*}{f_\xi}(x, \xi)d\xi$$

$$= [f(x, y)\mathbb{I}(x > y) - f(x, 0)] + g(x, y)\mathbb{I}(x < y).$$

The identification, $V^{eq}(x, 0, T) = f(x, 0)$ then establishes Equation (8.3). The proof of Equation (8.4) for the maximum type lookback option follows a very similar analysis, so we leave the details to the reader. 

**REMARK 8.2**

1. Note that while the physical domains of the minimum and maximum type lookback options are $x > y$ and $x < z$ respectively, no such constraints are imposed in the equivalent payoffs for these options. Indeed, the image part of the equivalent payoffs are specifically located in the complementary domains, $x < y$ and $x > z$.

2. As for the barrier options, our equivalent payoffs for lookback options are for arbitrary payoff functions of the form $f(x, y)$ and $F(x, z)$. The case of a payoff function $H(x, y, z)$ which depends simultaneously on both the running minimum $y$ and running maximum $z$, is more complicated, and as might be expected, will generally depend on an associated double barrier equivalent payoff. An example is provided by Q12 in the Exercise Problems.

---

## 8.3    The Generic Lookback Options $m(x, y, t)$ and $M(x, z, t)$

Consider two elementary lookback options: the first is a contract that pays at expiry $T$ the minimum asset price over $[0, T]$; the second is a contract that pays at expiry $T$ the maximum asset price over $[0, T]$. We refer to these two special derivatives as *generic* lookback options. While they are interesting contracts in their own right, there is another reason for studying them: all the standard lookback call and put options can be expressed in terms of these generic lookbacks.

## Generic Minimum Lookback

Let $m(x, y, t)$ denote the present value of the generic minimum contract. Then the payoff at time $T$ is given simply by $f(x, y) = y$, independent of $x$. In this case,

$$f(x, \xi) = \xi; \qquad f_\xi(x, \xi) = 1; \qquad \overset{*}{f_\xi}(x, \xi) = (\xi/x)^\alpha$$

According to theorem 8.2, the corresponding equivalent European payoff can be calculated as follows:

$$m^{eq}(x, y, T) = y\mathbb{I}(x > y) + \left[x - \int_x^y (\xi/x)^\alpha \, d\xi\right] \mathbb{I}(x < y)$$

$$= y\mathbb{I}(x > y) + x\mathbb{I}(x < y) - \beta[y(y/x)^\alpha - x]\mathbb{I}(x < y)$$

$$= (1 + \beta)x\mathbb{I}(x < y) + y[\mathbb{I}(x > y) - \beta\overset{*}{\mathbb{I}}(x > y)]$$

where $\beta = (\alpha + 1)^{-1} = \frac{\sigma^2}{2r}$ and the image in the final line, is taken wrt $y$. Observe that in calculating this equivalent payoff, only an elementary integration was needed.

The equivalent payoff is seen to be that of a portfolio of asset and bond binaries: in particular, this portfolio contains a long position in $(1 + \beta)$ down-type asset binaries; a long position in $y$ up-type bond binaries; and a short position in $\beta y$ (imaged) up-type bond binaries. No further calculation is therefore necessary. We can write down the price $m(x, y, t)$ for all $0 < t < T$, by inspection. Using the asset and bond binary notation introduce in Chapter 4, we obtain,

$$\boxed{m(x, y, t) = (1 + \beta) A_y^-(x, \tau) + y[B_y^+(x, \tau) - \beta \overset{*}{B_y^+}(x, \tau)]} \qquad (8.5)$$

where $\tau = (T - t)$ denotes the time remaining to expiry.

Observe, that despite the complexity of this path-dependent option, we have, with the valuation technology at our disposal, obtained its price by little more than an elementary integration and some clever mathematical manipulations.

## Generic Maximum Lookback

Let $M(x, z, t)$ denote the present value of the generic maximum contract. Then the payoff at time $T$ is given simply by $F(x, z) = z$, independent of $x$. A similar calculation to the one above, yields

$$M^{eq}(x, z, T) = z\mathbb{I}(x < z) + \left[x + \int_z^x (\xi/x)^\alpha \, d\xi\right] \mathbb{I}(x > z)$$

$$= z\mathbb{I}(x < z) + x\mathbb{I}(x > z) + \beta[x - z(z/x)^\alpha l]\mathbb{I}(x > z)$$

$$= (1 + \beta)x\mathbb{I}(x > z) + z[\mathbb{I}(x < z) - \beta\overset{*}{\mathbb{I}}(x < z)].$$

This too is a portfolio of asset and bond binaries, and once again we can write down by inspection the current value as

$$M(x, z, t) = (1 + \beta)\, A_z^+(x, \tau) + z[B_z^-(x, \tau) - \beta \overset{*}{B}_z^-(x, \tau)] \qquad (8.6)$$

One cannot help but notice the strong symmetry between the two prices for $m(x, y, t)$ and $M(x, z, t)$. The only differences are: $z$ replaces $y$, and the superscripted signs change parity, i.e., up-binaries become down-binaries and vice versa.

**REMARK 8.3**

1. Although the complementary image domains, $(x < y)$ for $m(x, y, t)$ and $(x > z)$ for $M(x, z, t)$, appear in the equivalent payoffs for these generic lookbacks, the final expressions (8.5) and (8.6) are valid only in the *physical* domains $(x > y)$ and $(x < z)$ respectively.

2. We have presented the theory of lookback options on a non-dividend paying asset only. However, using the barrier option extension for dividends considered in Section 7.7, it is a straightforward matter to include their effect on lookback options. In particular, the parameter $\beta = \frac{\sigma^2}{2r}$ will need to be replaced by $\beta = \frac{\sigma^2}{2(r-q)}$, where $q$ is the constant dividend yield of the underlying asset.

$\Box$

## 8.4 The Standard Lookback Calls and Puts

There are six standard lookback calls and puts, which are listed in the following table.

| Standard Lookback Payoffs | | | |
|---|---|---|---|
| | Floating Strike | Normal Fixed Strike | Reverse Fixed Strike |
| Call | $(X_T - Y_T)^+$ | $(Z_T - k)^+$ | $(Y_T - k)^+$ |
| Put | $(Z_T - X_T)^+$ | $(k - Y_T)^+$ | $(k - Z_T)^+$ |

Observe that, since $X_T \geq Y_T$ and $X_T \leq Z_T$, floating strike calls and puts are always exercised at expiry (unless of course the equality sign occurs). Thus for floating strike calls and puts, the plus function indicator can be dropped altogether. This makes their pricing very straightforward.

## Floating Strike Lookback Call

The floating strike lookback call option is like a standard call option except that the strike price $k$ is replaced by $Y_T$, the running minimum asset price over $[0, T]$. In one sense, this is optimal for the holder, since the lower the strike, the bigger the payoff. For this reason, the floating strike call is sometimes referred to as an option of *least regret*.

The payoff at expiry is $f(x, y) = (x - y)^+ = (x - y)$, which from Equation (8.5), and the result $\text{pv}(X_T) = x$, gives the following value for the floating strike call:

$$\boxed{V_c(x, y, t) = x - m(x, y, t)} \tag{8.7}$$

A useful alternative expression, which is the one more often found in the literature, can be derived using the specific representation (8.5) for $m(x, y, t)$ in terms of asset and bond binaries.

$$
\begin{aligned}
V_c(x, y, t) &= x - (1 + \beta) A_y^-(x, \tau) - y[B_y^+(x, \tau) - \beta \overset{*}{B}_y^+(x, \tau)] \\
&= [A_y^+(x, \tau) - y B_y^+(x, \tau)] + \beta[y \overset{*}{B}_y^+(x, \tau) - A_y^-(x, \tau)] \\
&= C_y(x, \tau) + L_c(x, y, \tau)
\end{aligned}
$$

where $C_y(x, \tau)$ is the price of a standard European call option of strike price $y$, and

$$L_c(x, y, \tau) = \beta[y \overset{*}{B}_y^+(x, \tau) - A_y^-(x, \tau)] \tag{8.8}$$

is the so-called *lookback premium*. This is the extra price the lookback call option attracts over and above the standard European price. In deriving the above we have used the asset binary parity relation $A_y^+ + A_y^- = x$. Generally, we would expect floating strike lookback calls to be more expensive than their European counterparts, so that we should have $L_c(x, y, \tau) > 0$.

## Floating Strike Lookback Put

The floating strike lookback put option is also an option of least regret, since the strike price is set at $k = Z_T$, the maximum asset price over $[0, T]$. Thus the holder gets to sell the asset at the highest price attained over the life of the option. The payoff at expiry is $F(x, z) = (z - x)^+ = (z - x)$, so the current value is

$$\boxed{V_p(x, z, t) = M(x, z, t) - x} \tag{8.9}$$

The lookback premium $L_p(x, z, \tau)$ is calculated, using (8.6) and the parity relation $A_z^+ + A_z^- = x$ as follows

$$
\begin{aligned}
V_p(x, z, t) &= (1 + \beta) A_z^+(x, \tau) + z[B_z^-(x, \tau) - \beta \overset{*}{B}_z^-(x, \tau)] - x \\
&= [z B_z^-(x, \tau) - A_z^-(x, \tau)] + \beta[A_z^+(x, \tau) - z \overset{*}{B}_z^-(x, \tau)] \\
&= P_z(x, \tau) + L_p(x, z, \tau)
\end{aligned}
$$

where

$$L_p(x, z, \tau) = \beta[A_z^+(x, \tau) - z\overset{*}{B_z}(x, \tau)]. \tag{8.10}$$

### Normal Fixed Strike Lookback Put

We know that $m(x, y, t)$ solves the BS-pde problem:

$$\mathcal{L}m = 0; \quad m(x, y, T) = y; \quad m_y(x, y, t) = 0 \text{ at } x = y$$

in the domain $(t < T; \ x > y)$. One might be tempted to conclude since after all, $y$ is only a parameter in this system, that $V(x, y, t) = m(x, f(y), t)$ then solves the BS-pde with expiry payoff $V(x, y, T) = f(y)$. This is of course incorrect, but it is instructive to see why it is wrong. While this $V(x, y, t)$ indeed satisfies the BS-pde and the expiry payoff condition, it does not satisfy the BC at $x = y$. In fact

$$V_y(x, y, t) = f'(y)m_y(x, f(y), t) \neq 0 \text{ at } x = y.$$

Does this mean that the only solution of the form $m(x, f(y), t)$ occurs for $f(y) = y$ ? The answer is a somewhat surprising no, as demonstrated by the next result.

### PROPOSITION 8.1

$V(x, y, t) = m(x, f(y), t)$ *solves the BS-pde problem*

$$\mathcal{L}V = 0; \quad V(x, y, T) = f(y); \quad V_y(x, y, t) = 0 \quad at \ x = y$$

*in the domain* $(t < T; \ x > y)$, *iff* $f(y) = y$ *or* $f(y) = \min(y, k)$, *for any positive constant* $k$.

**PROOF**  As argued above, it is only necessary to verify the BC at $x = y$. That is, we require

$$V_y(x, y, t) = f'(y)m_y(x, f(y), t) = 0 \text{ at } x = y$$

given that $m_y(x, y, t) = 0$ at $x = y$. Clearly, $f(y) = y$ satisfies the condition, but so does $f(y) = \min(y, k)$, since, then $f'(y) = \mathbb{I}(y < k)$, and

$$V_y(x, y, t) = \mathbb{I}(y < k)\, m_y(x, \min(y, k), t) = \mathbb{I}(y < k)\, m_y(x, y, t)$$

which vanishes at $x = y$.  ∎

Consider now the normal fixed strike lookback put option, whose expiry payoff (from the table) is given by $f(x, y) = (k - y)^+$, independent of $x$. This payoff can be written in the equivalent form

$$V_p(x, y, T) = (k - y)^+ = k - \min(y, k).$$

Hence, from proposition 8.1 and the observation that $pv(k) = ke^{-r\tau}$, we have

$$\boxed{V_p(x, y, t) = ke^{-r\tau} - m(x, y \wedge k, t)} \qquad (8.11)$$

where $(y \wedge k)$ is a common alternative expression for $\min(y, k)$.

**REMARK 8.4** Observe that if the current running minimum asset price $y$ is above the strike price $k$, i.e., it is currently otm, then the option, has price $V_p(x, y, t) = ke^{-r\tau} - m(x, k, t)$ independent of $y$. When the option is currently itm, so that $y < k$, the price is $V_p(x, y, t) = ke^{-r\tau} - m(x, y, t)$, which now depends on $y$. ▯

**Normal Fixed Strike Lookback Call**
Here the payoff is given by $V_c(x, z, t) = (z - k)^+ = \max(z, k) - k$. The corresponding counterpart to proposition 8.1 is the following.

**PROPOSITION 8.2**
$V(x, y, t) = M(x, f(z), t)$ *solves the BS-pde problem*

$$\mathcal{L}V = 0; \quad V(x, z, T) = f(z); \quad V_z(x, z, t) = 0 \ \ at \ x = z$$

*in the domain* $(t < T; \ x < z)$, *iff* $f(z) = y$ *or* $f(z) = \max(z, k)$, *for any positive constant* $k$.

We forego the proof, which is virtually identical to that for $m(x, f(y), t)$. Hence, with proposition 8.2, we obtain

$$\boxed{V_c(x, z, t) = M(x, z \vee k, t) - ke^{-r\tau}} \qquad (8.12)$$

where $(z \vee k)$ is a common alternative expression for $\max(z, k)$.

**Reverse Fixed Strike Lookback Call/Put**
The payoffs of reverse fixed strike lookback calls and puts are respectively,

$$V_c(x, y, T) = (y - k)^+ = y - \min(y, k)$$
$$V_p(x, z, T) = (k - z)^+ = \max(z, k) - z.$$

From propositions 8.1 and 8.2, and formulae (8.5) and (8.6), we obtain their current prices,

$$\boxed{\begin{aligned} V_c(x, y, t) &= m(x, y, t) - m(x, y \wedge k, t) \\ V_p(x, z, t) &= M(x, z \vee k, t) - M(x, z, t) \end{aligned}} \qquad (8.13)$$

An interesting feature of these formulae is that both vanish identically for all times up to expiry $T$, if at any current time $t$ the options are out-of-the-money.

That is, $V_c(x, y, t) = 0$ for all times up to expiry, if $y < k$, and $V_p(x, z, t) = 0$ for all times up to expiry, if $z > k$. If we think about it, this is exactly what we should expect, since once these options are otm, they must remain otm for the rest of their existence.

## 8.5   Partial Price Lookback Options

Lookback options are generally quite expensive, relative to their standard counterparts. One way of reducing premium costs is by changing the floating strikes: increasing them for floating strike calls and reducing them for floating strike puts. The payoffs of these partial price, floating strike options are, respectively,

$$V_c(x, y, T) = (x - \lambda y)^+ \qquad \text{for } \lambda \geq 1 \tag{8.14}$$
$$V_p(x, z, T) = (\mu z - x)^+ \qquad \text{for } \mu \leq 1. \tag{8.15}$$

Conze and Viswanathan [16] also managed to price these derivatives by standard methods. Our approach, on the other hand, using theorem 8.2 for the equivalent European payoffs, gives the results in quick fashion using only simple algebraic integrations.

For the partial price, floating strike call option, our calculations (with some elementary details omitted) produce the following:

$$f(x, y) = (x - \lambda y)^+; \quad f(x, x) = 0;$$
$$f_\xi(x, \xi) = -\lambda \mathbb{I}(x > \lambda \xi); \quad \overset{*}{f}_\xi(x, \xi) = -\lambda(\xi/x)^\alpha \mathbb{I}(x < \xi/\lambda);$$
$$V_c^{eq}(x, y, T) = (x - \lambda y)^+ + \lambda \mathbb{I}(\lambda x < y) \int_{\lambda x}^{y} (\xi/x)^\alpha \, d\xi$$
$$= (x - \lambda y)^+ + \lambda^{\alpha+2} \beta[(y'(y'/x)^\alpha - x] \mathbb{I}(x < y'); \quad y' = y/\lambda$$
$$= (x - \lambda y)^+ + \lambda^{\alpha+2} \beta[(y' \overset{*}{\mathbb{I}}(x > y') - x \mathbb{I}(x < y')].$$

The corresponding present value can then be expressed in the form

$$\boxed{V_c(x, y, t) = C_{\lambda y}(x, \tau) + \lambda^{\alpha+2} \beta[y' \overset{*}{B}{}_{y'}^+(x, \tau) - A_{y'}^-(x, \tau)]} \tag{8.16}$$

where the image above is taken relative to $y' = y/\lambda$. The second term above is obviously the lookback premium and can be expressed as $\lambda^{\alpha+2} L_c(x, y', \tau)$ where $L_c(x, y, \tau)$ was defined earlier in Equation (8.8). The partial price, floating strike, lookback call value clearly reduces to our previous floating strike price, when $\lambda = 1$.

A similar calculation for the partial time, floating strike lookback put option, leads to the current value,

$$\boxed{V_p(x, z, t) = P_{\mu z}(x, \tau) + \mu^{\alpha+2}\beta[A_{z'}^+(x, \tau) - z'\overset{*}{B}_{z'}^-(x, \tau)]} \qquad (8.17)$$

where the image above is taken relative to $z' = z/\mu$ and the associated lookback premium (the second term above) can be expressed as $\mu^{\alpha+2}L_p(x, z', \tau)$, where $L_p(x, z, t)$ was defined in Equation (8.10).

---

## 8.6  Partial Time Lookback Options

Just as we could divide the time horizon into a standard window and a fixed monitoring window for partial time barrier options, we can do the same for lookback options. The first BS analysis of partial time lookback options was carried out by Heynen and Kat [32] in 1995. The analysis using the approach advocated by theorem 8.2, has been solved in the PhD thesis [45] of Konstandatos (2003).

There exists a fairly large menagerie of such partial time options, depending on (a) start-out or end-out monitoring windows, and (b) the lookback payoff. We shall, however, be content to price just two examples from this family: a start-out, floating strike, lookback call; and an end-out, fixed strike, lookback call.

**Start-out, Floating Strike, Lookback Call**
The lookback monitoring window is $[0, T_1]$, the expiry date is $T_2 > T_1$ and the payoff at time $T_2$ will be $V_c(x, y, T_2) = (x - y)^+$ where $y = \min\limits_{0 \le s \le T_1} X_s$. We wish to price the derivative for all $t$ in $0 < t < T_1$. Observe first, that over the the interval $[T_1, T_2]$ the parameter $y$, having already been observed at $T_1$, is fixed. Hence, over this interval, the derivative is a standard European call option of strike price $y$. It follows that the lookback payoff at time $T_1$ is simply $V_c(x, y, T_1) = C_y(x, \tau)$ where $\tau = (T_2 - T_1)$.

Using the equivalent European payoff formula (8.3), we compute the following (see the Remark 8.5) below for an explanation of the details).

$$f(x, \xi) = C_\xi(x, \tau); \quad f_\xi(x, \xi) = -B_\xi^+(x, \tau);$$
$$f(x, x) = g(\tau)x; \quad \text{(price of a forward start call, see Eq(5.1))}$$
$$V_c^{eq}(x, y, T) = C_y(x, \tau)\mathbb{I}(x > y) + \left[g(\tau)x + \int_x^y \overset{*}{B}_\xi^+(x, \tau)d\xi\right]\mathbb{I}(x < y).$$

The integral above looks difficult, but a simple trick allows its calculation in algebraic terms alone.

**LEMMA 8.1**

For any $\tau \geq 0$, let $I_{(x,y)}(x,\tau) = \int_x^y \overset{*}{B}_\xi^+(x,\tau)d\xi$. Then

$$I_{(x,y)}(x,\tau) = \beta[y\overset{*}{B}_y^+(x,\tau) - A_y^-(x,\tau) - g'(\tau)x]$$
$$g'(\tau) = e^{-r\tau}\mathcal{N}(a'\sqrt{tau}) - \mathcal{N}(-a\sqrt{\tau})$$

where $(a, a') = \frac{r}{\sigma} \pm \frac{1}{2}\sigma$.

**PROOF** Since the integrand is a solution of the BS-pde, so is $I_{(0,y)}(x,\tau)$ and the corresponding expiry value at $\tau = 0$, is

$$I_{(0,y)}(x,0) = \int_0^y \overset{*}{\mathbb{I}}(x>\xi)\,d\xi = \beta[y\overset{*}{\mathbb{I}}(x>y) - x\mathbb{I}(x<y)].$$

This integral is elementary. Hence, we can write down the value of the integral for any $\tau > 0$, as the pv of this binary option portfolio. We obtain,

$$I_{(0,y)}(x,\tau) = \beta[y\overset{*}{B}_y^+(x,\tau) - A_y^-(x,\tau)]$$

which happens to be equal to the floating strike lookback call premium (8.8). From this last result, with $y = x$, we then obtain $I_{(0,x)}(x,\tau) = \beta g'(\tau)x$. So finally, since $I_{(x,y)}(x,\tau) = I_{(0,y)}(x,\tau) - I_{(0,x)}(x,\tau)$, we find

$$I_{(x,y)}(x,\tau) = \beta[y\overset{*}{B}_y^+(x,\tau) - A_y^-(x,\tau) - g'(\tau)x].$$

This completes the proof. $\square$

Our equivalent European payoff is therefore given by the expression,

$$V_c^{eq}(x,y,T) = C_y(x,\tau)\mathbb{I}(x>y) + [g(\tau)x + I_{(x,y)}(x,\tau)]\,\mathbb{I}(x<y)$$

This is the form in which we can immediately write down its present value in terms of second-order $Q$-options and second-order asset and bond binaries. The result, with $\tau_1 = (T_1 - t)$, is

$$\boxed{\begin{aligned} V_c(x,y,t) &= Q_{yy}^{++}(x,t;y) + \hat{g}(\tau)A_y^-(x,\tau_1) \\ &\quad + \beta[y\overset{*}{B}_{yy}^{++}(x,t) - A_{yy}^{--}(x,t)] \end{aligned}}$$

$$(8.18)$$

where

$$\hat{g}(\tau) = g(\tau) - \beta g'(\tau) = 1 - (1-\beta)\mathcal{N}(-a\sqrt{\tau}) - (1+\beta)e^{-r\tau}\mathcal{N}(a'\sqrt{\tau}).$$

**REMARK 8.5** The following provides some of the missing detail above.

- The derivative of a call option wrt its strike is its *kappa*. This is where the term $f_\xi(x,\xi) = -B_\xi^+(x,\tau)$ comes from. A simple way to derive it, starts from the expiry value of a call option: $C_\xi(\dot{x},0) = (x-\xi)^+$. Its derivative wrt $\xi$ is just $-\mathbb{I}(x>\xi)$, and the pv of this payoff is $-B_\xi^+(x,\tau)$.

- $f(x,x)$ is the price of a forward start call option (i.e., a future atm call) which we priced in Section 5.1. The result was $C_x(x,\tau) = g(\tau)x$ with

$$g(\tau) = \mathcal{N}(a\sqrt{\tau}) - e^{-r\tau}\mathcal{N}(a'\sqrt{\tau}); \quad [a,a'] = \frac{r \pm \frac{1}{2}\sigma^2}{\sigma}.$$

- We have used the result

$$\overset{*}{B}{}_y^+(x,\tau)\mathbb{I}(x<y) = \mathcal{I}_y[B_y^+(x,\tau)\mathbb{I}(x>y)].$$

⬚

**End-out, Fixed Strike, Lookback Call**

The lookback monitoring window is now $[T_1, T_2]$, the payoff at time $T_2$ is $(z-k)^+$, where $z = \max\limits_{T_1 \le s \le T_2} X_s$, and we seek to price the option for all $t < T_1$. Over the range $[T_1, T_2]$ we have a standard fixed strike, lookback call option with price given by Equation (8.12), $M(x, z \vee k, t) - ke^{-r\tau}$. We evaluate this price at time $T_1$, where $z = x$ (the asset price and its maximum are equal at the start of monitoring), to obtain

$$V_c(x, T_1) = M(x, x \vee k, t) - ke^{-r\tau}; \qquad (\tau = T_2 - T_1).$$

Using our expression (8.6) for $M(x, z, t)$, we obtain after some simplification,

$$V_c(x, T_1) = \begin{cases} (1+\beta)A_k^+(x,\tau) - k[B_k^+(x,\tau) + \beta\overset{*}{B}{}_k^-(x,\tau)] & \text{if } x < k \\ g(\tau)x - ke^{-r\tau} & \text{if } x > k \end{cases} \quad (8.19)$$

where, in the notation of (5.1),

$$g(\tau) = (1+\beta)\mathcal{N}(a\sqrt{\tau}) + (1-\beta)e^{-r\tau}\mathcal{N}(-a'\sqrt{\tau}).$$

It remains only to calculate the pv of this time $T_1$ payoff. Write the $T_1$ payoff in the equivalent form

$$V_c(x, T_1) = F(x,\tau)\mathbb{I}(x<k) + [g(\tau)x - ke^{-r\tau}]\mathbb{I}(x>k)$$

where, using $Q_k^+(x,\tau;k) = A_k^+(x,\tau) - kB_k^+(x,\tau)$,

$$F(x,\tau) = Q_k^+(x,\tau;k) + \beta[A_k^+(x,\tau) - \overset{*}{B}{}_k^-(x,\tau)].$$

The final result is,

$$\boxed{V_c(x,t) = \begin{cases} Q_{kk}^{-+}(x,t;k) + \beta[A_{kk}^{-+}(x,t) - \overset{*}{B}{}_{kk}^{+-}(x,t)] \\ +g(\tau)A_k^+(x,\tau_1) - ke^{-r\tau}B_k^+(x,\tau_1) \end{cases}} \quad (8.20)$$

where $\tau_1 = (T_1 - t)$ is the time remaining to the expiry date $T_1$. Unlike the other lookback options, the end-out, lookback price has no dependence on the running max/min ($z$ or $y$). Obviously, this must be the case, because at time $t < T_1$, no max/min has yet been observed.

The case of an end-out, partial-time lookback put option is not presented here, but its calculation follows a very similar analysis to the one above.

---

## 8.7 Extreme Spread Options

Consider the time horizon $[0, T_2]$ divided into two contiguous sections: $S_1 = [0, T_1]$ and $S_2 = [T_1, T_2]$. Extreme spread options are derivatives whose payoffs at time $T_2$ depend on the difference of extremal prices in the two time domains $S_1$ and $S_2$. These complex derivatives were introduced in the 1998 PhD thesis [3] of Bermin and reported in the book by Haug [29]. Their pricing (using our methods) was published in Bermin et al. [4]. We have already set up all the tools necessary to price these derivatives, so despite their complexity, their prices can be determined without much ado.

To fix the notation, let us define the following extremal variables.

$$Y_{ij} = \min_{T_i \leq s \leq T_j} X_s; \qquad Z_{ij} = \min_{T_i \leq s \leq T_j} X_s$$

for $(i, j) = [0, 1, 2]$ and where it is understood that $T_0 = 0$. We shall also denote by $y$ and $z$ the currently observed minimum and maximum asset prices, i.e., $y = Y_{0t}$ and $z = Z_{0t}$. The following theorem determines the current prices of contracts that pay, at expiry $T_2$, the relevant extremal asset values.

### THEOREM 8.3

Let $pv_t(V)$ denote the present value at time $t$, for $t < T_1 < T_2$, of a contract that pays $V$ at time $T_2$. Then

$$
\begin{array}{|l|}
\hline
pv_t(Y_{01}) = e^{-r\tau}\, m(x, y, \tau_1) \\
pv_t(Y_{12}) = m(x, x, \tau) = g'(\tau)x \\
pv_t(Y_{02}) = m(x, y, \tau_2) \\
pv_t(Z_{01}) = e^{-r\tau}\, M(x, z, \tau_1) \\
pv_t(Z_{12}) = M(x, x, \tau) = g(\tau)x \\
pv_t(Z_{02}) = M(x, z, \tau_2) \\
\hline
\end{array}
\qquad (8.21)
$$

where $\tau_i = (T_i - t)$, $\tau = (T_2 - T_1)$, and

$$[g(\tau), g'(\tau)] = (1 + \beta)\mathcal{N}(\pm a\sqrt{\tau}) + (1 - \beta)e^{-r\tau}\mathcal{N}(\mp a'\sqrt{\tau})$$

with $[a, a'] = \frac{r}{\sigma} \pm \frac{1}{2}\sigma$.

**REMARK 8.6**    Note that in the above theorem we have written the generic prices $m, M$ in terms of relative prices $(\tau, \tau_i)$ rather than absolute time $t$. Observe further that $g(\tau)$ in the theorem is the same as introduced to price the end-out, fixed strike, lookback call in the previous section.    ▯

**PROOF**    Let $y_1 = Y_{01}$, which is constant for all times in $S_2$. Hence, at time $T_1$, the value of the contract paying $Y_{01}$ at time $T_2$ is simply, $V(x_1, y_1, T_1) = e^{-r\tau} y_1$. This contract has pv, $e^{-r\tau} m(x, y, \tau_1)$.

The contract that pays $Y_{12}$ at time $T_2$, has value $m(x, y, \tau_2)$ for all times in $S_2$. At time $T_1$, its value will be $m(x_1, x_1, \tau) = g'(\tau)x_1$, since $y = x$ at the start of monitoring. The pv of this contract, $g'(\tau)$ units of the underlying, is simply $g'(\tau)x$.

The case for $Y_{02}$ is obvious and the proofs for the corresponding maximum contracts, follow similarly.    ▯

The payoff table for the four principal extreme spread options, is given below.

|      | Normal Extreme Spread | Reverse Extreme Spread |
|------|-----------------------|------------------------|
| Call | $(Z_{12} - Z_{01})^+$ | $(Y_{12} - Y_{01})^+$  |
| Put  | $(Y_{01} - Y_{12})^+$ | $(Z_{01} - Z_{12})^+$  |

The calculations that follow show how these extreme spread options are priced. First, note the identities,

$$(Y_{01} - Y_{12})^+ = Y_{01} - \min(Y_{01}, Y_{12}) = Y_{01} - Y_{02}$$
$$(Y_{12} - Y_{01})^+ = Y_{12} - \min(Y_{01}, Y_{12}) = Y_{12} - Y_{02}$$
$$(Z_{12} - Z_{01})^+ = \max(Z_{01}, Z_{12}) - Z_{01} = Z_{02} - Z_{01}$$
$$(Z_{01} - Z_{12})^+ = \max(Z_{01}, Z_{12}) - Z_{12} = Z_{02} - Z_{12}.$$

Putting together the above decompositions and the prices derived in theorem 8.3, gives the required prices of all the extreme spread options.

| Option | $T_2$ Payoff | Price for $t < T_1$ |
|---|---|---|
| Normal Call | $(Z_{12} - Z_{01})^+$ | $M(x, z, \tau_2) - e^{-r\tau}M(x, z, \tau_1)$ |
| Reverse Call | $(Y_{12} - Y_{01})^+$ | $g'(\tau)x - m(x, y, \tau_2)$ |
| Normal Put | $(Y_{01} - Y_{12})^+$ | $e^{-r\tau}m(x, y, \tau_1) - m(x, y, \tau_2)$ |
| Reverse Put | $(Z_{01} - Z_{12})^+$ | $M(x, z, \tau_2) - g(\tau)x$ |

Extensions, including partial price, extreme spread options, can be found in the Konstandatos [45] and [46].

---

## 8.8    Look-Barrier Options

The world of exotic options gets more and more complex, as new products are invented either for speculation or specialized forms of hedging. The term *vanilla exotics* has even crept into the derivatives jargon, to distinguish the standard exotic options (such as call and put barrier options) from the more esoteric ones. So what are the non-vanilla exotics called? Indeed, *exotic exotics*. The look-barrier derivative is an example of an exotic lookback option. It first appeared in the PhD thesis [3] of Bermin, and was reported in the book [29] by Haug. Pricing of this exotic option by our methods was published in Bermin et al. [4].

The look-barrier option is a hybrid option having both barrier and lookback features. The time interval $[t, T_2]$ is divided into two segments: $S_1 = [t, T_1]$ and $S_2 = [T_1, T_2]$ with $t < T_1 < T_2$. The first segment $S_1$ is a barrier window, while the second segment $S_2$ is a lookback window. There are several species of look-barrier options, depending on the type of barrier (D/O, U/O, D/I, U/I); and the type of lookback (floating/fixed strike, call/put).

The methods we have developed for pricing barrier and lookback options are ideal for such options. Despite their complexity, they are just grist to the mill for our pricing tools. We demonstrate this claim by pricing just one of the look-barriers, namely the D/O, fixed strike, look-barrier call option.

Since this look-barrier call is just a normal fixed strike, lookback option over the time interval $S_2$, we can use Equation (8.12) to find its price at time $T_1$. This yields, as we found in Equation (8.19),

$$V_c(x, T_1) = \begin{cases} Q_k^+(x, \tau; k) + \beta[A_k^+(x, \tau) - \overset{*}{B}_k^-(x, \tau)] & \text{if } x < k \\ g(\tau)x - ke^{-r\tau} & \text{if } x > k \end{cases}$$

where all the parameters have been previously defined, and we have replaced the term $[A_k^+ - kB_k^+](x, \tau)$ by the equivalent call price notation, $Q_k^+(x, \tau; k)$.

This expression, now becomes the payoff function $f(x)$ of the D/O barrier option for time interval $S_1$. And to price this D/O barrier option, we need only find its equivalent European payoff given by

$$V_c^{eq}(x, T_1) = f(x)\mathbb{I}(x > h) - \mathcal{I}_h\{f(x)\mathbb{I}(x > h)\}$$

where $h$ is the barrier price and $\mathcal{I}_h$ denotes the image wrt $x = h$.

We must tread carefully at this point, since the expression for $f(x) = V_c(x, T_1)$ above, also contains an image, but wrt $x = k$, the strike price. Thus we have a double image in this problem: first wrt $k$, then wrt $h$. We have met such double images in the previous chapter, from which we recall the formula (7.25),

$$\mathcal{I}_h \mathcal{I}_k\{U(x, t)\} = (k/h)^\alpha U(k^2 x/h^2, t)$$

for any function $U(x, t)$.

Now let us define, $E(x, \tau) = Q_k^+(x, \tau; k) + \beta[A_k^+(x, \tau) - \overset{*}{B}_k^-(x, \tau)]$ and $F(x, \tau) = g(\tau)x - ke^{-r\tau}$, so that the barrier payoff $f(x)$ has the structure,

$$f(x) = E(x, \tau)\mathbb{I}(x < k) + F(x, \tau)\mathbb{I}(x > k).$$

Then it is easy to demonstrate that

$$f(x)\mathbb{I}(x > h) = \begin{cases} F(x, \tau)\mathbb{I}(x > h) & \text{if } h > k \\ E(x, \tau)[\mathbb{I}(x > h) - \mathbb{I}(x > k)] + F(x, \tau)\mathbb{I}(x > k) & \text{if } h < k. \end{cases}$$

The price of the look-barrier call option will then be the pv of this payoff, i.e., $V_h(x, t) = \text{pv}\{f(x)\mathbb{I}(x > h)\}$, minus, its image wrt $h$, i.e., $\mathcal{I}_h\{V_h(x, t)\}$, yielding

$$V_c(x, t) = V_h(x, t) - \mathcal{I}_h\{V_h(x, t)\}.$$

We proceed to calculate $V_h(x, t)$. First,

$$F(x, \tau)\mathbb{I}(x > h) = [g(\tau)x - ke^{-r\tau}]\mathbb{I}(x > h)$$

which has pv,

$$V_1(x, t) = g(\tau)A_h^+(x, \tau_1) - ke^{-r\tau}B_h^+(x, \tau_1). \tag{8.22}$$

Second, $E(x, \tau)\mathbb{I}(x > h) = [Q_k^+(x, \tau; k) + \beta[A_k^+(x, \tau) - \overset{*}{B}_k^-(x, \tau)]\mathbb{I}(x > h)$ has pv,

$$V_2(x, t) = Q_{hk}^{++}(x, t; k) + \beta[A_{hk}^{++}(x, t) - \overset{*}{B}_{\bar{h}k}^{--}(x, t)] \tag{8.23}$$

where $\bar{h} = k^2/h$, since $\mathcal{I}_k[U(x)]\mathbb{I}(x>h) = \mathcal{I}_k[U(x)\mathbb{I}(x<\bar{h})]$. Third, the pv of $E(x,\tau)\mathbb{I}(x>k)$, is similarly given by

$$V_3(x,t) = Q_{kk}^{++}(x,t;k) + \beta[A_{kk}^{++}(x,t) - \overset{*}{B}_{kk}^{--}(x,t)]. \tag{8.24}$$

Fourth, the pv of $F(x,\tau)\mathbb{I}(x>k)$ is

$$V_4(x,t) = g(\tau)A_k^+(x,\tau_1) - ke^{-r\tau}B_k^+(x,\tau_1). \tag{8.25}$$

One further result is needed before we put all these expressions together to get the final result. This is the following homogeneity property of second-order bond binaries:

$$B_{ab}^{--}(\lambda x, t) = B_{\frac{a}{\lambda}\frac{b}{\lambda}}^{--}(x,t) \tag{8.26}$$

as is readily verified from the formula (5.9). From this, we obtain the double images,

$$\mathcal{I}_h\overset{*}{B}_{\bar{h}k}^{--}(x,t) = (k/h)^\alpha\, B_{\bar{h}k}^{--}(k^2x/h^2,t) = (k/h)^\alpha\, B_{h\bar{k}}^{--}(x,t)$$

$$\mathcal{I}_h\overset{*}{B}_{kk}^{--}(x,t) = (k/h)^\alpha\, B_{kk}^{--}(k^2x/h^2,t) = (k/h)^\alpha\, B_{\bar{k}\bar{k}}^{--}(x,t)$$

where we have now defined, $\bar{k} = h^2/k$ (recall $\bar{h} = k^2/h$).

Collecting all the terms for $V_i(x,t)$ given in Equations (8.22), (8.23) and (8.24) and subtracting images wrt $h$, we get the final expression for the D/O, fixed strike, look-barrier call option, below.

<u>For $h > k$</u>

$$\left.\begin{array}{rl} V_c(x,t) = & g(\tau)[A_h^+(x,\tau_1) - \mathcal{I}_hA_h^+(x,\tau_1)] \\[2mm] & -ke^{-r\tau}[B_h^+(x,\tau_1) - \mathcal{I}_hB_h^+(x,\tau_1)] \end{array}\right\} \tag{8.27}$$

For $h < k$

$$
\begin{aligned}
V_c(x,t) = \quad & g(\tau)[A_k^+(x,\tau_1) - \mathcal{I}_h A_k^+(x,\tau_1)] \\
& -ke^{-r\tau}[B_k^+(x,\tau_1) - \mathcal{I}_h B_k^+(x,\tau_1)] \\
& +[Q_{hk}^{++}(x,t;k) - \mathcal{I}_h Q_{hk}^{++}(x,t;k)] \\
& +\beta[A_{hk}^{++}(x,t) - \mathcal{I}_h A_{hk}^{++}(x,t)] \\
& -\beta[\overset{*}{B}_{\bar{h}k}^{--}(x,t) - (k/h)^\alpha B_{h\bar{k}}^{--}(x,t)] \\
& -[Q_{kk}^{++}(x,t;k) - \mathcal{I}_h Q_{kk}^{++}(x,t;k)] \\
& -\beta[A_{kk}^{++}(x,t) - \mathcal{I}_h A_{kk}^{++}(x,t)] \\
& +\beta[\overset{*}{B}_{kk}^{--}(x,t) - (k/h)^\alpha B_{\bar{k}\bar{k}}^{--}(x,t)]
\end{aligned} \tag{8.28}
$$

Even in our heavily reduced notation, this expression is quite daunting. To see what it looks like in terms of its basic components, containing only univariate and bivariate Gaussian distribution functions, see Haug [29]. It is not a pretty sight!

Prices, using the above method, for all the other look-barrier calls and puts, within the family of look-barriers, can be found in the quoted references.

---

## 8.9   Summary

In this chapter we have also used powerful, specialized methods for pricing lookback options in the BS framework. We showed that lookback options can be expressed in terms of integrated barrier options. Indeed, we were able to find equivalent European payoffs for lookback options, just as we did for barrier options, and these were valid for essentially arbitrary payoffs of the form $f(x,y)$ (resp. $F(x,z)$).

Thus, we were also able to embed path-dependent lookback options into an equivalent European pricing problem. To find the equivalent payoff, all we had to do was integrate the barrier option equivalent payoff wrt the running minimum (resp running maximum). In practice, these involved little more than integrals of elementary power functions.

For the six standard lookback calls and puts, we showed that all their prices can be derived in terms of just two generic options. These are the option that pays the running minimum $m(x, y, t)$, and the one that pays the running maximum $M(x, z, t)$. Our equivalent payoff method makes it a simple matter to price these basic contracts.

Our method readily handles extensions like partial time lookback options and partial price lookback options. We also showed how to price a number of exotic lookback options, including extreme spread options, and a hybrid class of mixed lookback and barrier options, called look-barrier options.

The next chapter is our final one on path-dependent options. Here we follow traditional methods to some degree, in pricing the family of geometric mean Asian options: the only ones with closed analytic prices in the BS framework. However, some new insights, particularly using the pde approach do emerge.

---

## Exercise Problems

Where appropriate assume standard Black–Scholes dynamics and standard notation.

1. The purpose of this question is to prove the assertion that lookback options are integrated versions of barrier options, using the FTAP. The starting point will therefore be the representation

$$V_{\min}(x, y, t) = e^{-r\tau} \mathbb{E}_Q\{f(X_T, Y_{0,T})|\mathcal{F}_t\}$$

for the price of a minimum type lookback option, where $\tau = (T - t)$ and $\mathcal{F}_t$ is defined by $x = X_t$; $y = Y_{0,t}$. Show by formally differentiating this expression wrt the parameter $y$, that the the result is equivalent to a certain D/O barrier option.
Repeat the process for a maximum type lookback option.

2. Consider a minimum type lookback option with expiry payoff $f(x, y) = f(x)\mathbb{I}(y > b)$. Find the associated equivalent European payoff and show that it reduces to that of a D/O barrier option, with barrier level $b$.
*Hint*: The identity $\frac{d}{d\xi}\mathbb{I}(\xi > b) = \delta(\xi - b)$, where $\delta(\xi)$ is the Dirac delta function is needed.

3. Verify by direct differentiation, that the boundary conditions

$$\frac{\partial}{\partial y}m(x, y, t) = 0 \text{ at } x = y; \qquad \frac{\partial}{\partial z}M(x, z, t) = 0 \text{ at } x = z$$

are indeed satisfied.

4. Equation (8.5) breaks down in the special case of zero interest rates, since $\beta = \sigma^2/2r$. Show that the correct expression for $m(x,y,t)$ when $r = 0$ is given by

$$m(x,y,t) = y\mathcal{N}(d') + x\mathcal{N}(-d) + \sigma\sqrt{\tau}\,x[d\mathcal{N}(-d) - \phi(d)]$$

where $[d,d'](x,y,\tau) = \frac{\log(x/y)}{\sigma\sqrt{\tau}} \pm \frac{1}{2}\sigma\sqrt{\tau}$, and $(\phi, \mathcal{N})$ are the usual pdf and cdf of a standard $N(0,1)$ variate.

Derive the corresponding expression for $M(x,z,t)$ when $r = 0$.

5. Consider the turbo-lookback option with expiry $T$ payoff $V(x,y,T) = y^k$ for some real $k$. Show that the equivalent European payoff is given the expression

$$V^{eq}(x,y,T) = y^k\mathbb{I}(x>y) + \left(\frac{2k+\alpha}{k+\alpha}\right)x^k\mathbb{I}(x<y) - \left(\frac{k}{k+\alpha}\right)y^k\overset{*}{\mathbb{I}}(x>y)$$

where the asterisk denotes the image wrt $x = y$.

Hence price the lookback derivative in terms of turbo binaries (see Section 4.8) and their images.

6. Consider the turbo-lookback option with expiry $T$ payoff $V(x,z,T) = z^k$ for some real $k$. Show that the equivalent European payoff is given the expression

$$V^{eq}(x,z,T) = z^k\mathbb{I}(x<z) + \left(\frac{2k+\alpha}{k+\alpha}\right)x^k\mathbb{I}(x>z) - \left(\frac{k}{k+\alpha}\right)z^k\overset{*}{\mathbb{I}}(x<z)$$

where the asterisk denotes the image wrt $x = z$. Hence price the lookback option, as above, in terms of turbo binaries.

7. Consider a minimum lookback option with payoff $V(x,y,T) = f(y)$, independent of $x$. Show that the equivalent European payoff can be expressed in the form

$$V^{eq}(x,y,T) = f(y)\mathbb{I}(x>y) - h(y)\overset{*}{\mathbb{I}}(x>y) + [f(x) + h(x)]\mathbb{I}(x<y)$$

where

$$h(x) = f(x) - \alpha x^{-\alpha}g(x) \quad \text{and} \quad g(x) = \int^x \xi^{\alpha-1}f(\xi)\,d\xi.$$

Verify that this formula agrees with the equivalent European payoff for the case $f(y) = y^k$, considered in Q5 above.

8. Find the equivalent European payoffs for the minimum and maximum type lookback options with respective payoffs: (i) $V(x,y,T) = (x/y)^\gamma$

and (ii) $V(x, z, T) = (x/z)^\gamma$, where $\gamma$ is some real parameter.

Hence determine their current values and verify that they reduce to the ZCB price when $\gamma = 0$.

9. Verify Equation (8.17)

$$V_p(x, z, t) = P_{\mu z}(x, \tau) + \mu^{\alpha+2}\beta[A_{z'}^+(x, \tau) - z'\overset{*}{B}_{z'}^-(x, \tau)]$$

for the pv of a partial price, lookback put option. Here $z' = z/\mu$ and the payoff is $V_p(x, z, T) = (\mu z - x)^+$ with $\mu < 1$.

10. Consider a simultaneous partial-price, partial-time lookback call option with expiry payoff $V_c(x, y, T_2) = (x - \lambda y)^+$, $(\lambda \geq 1)$ and lookback window $[0, T_1]$ with $T_1 < T_2$. Show that its value for any time $t$ in $(0, T_1)$, is given by

$$V_c(x, y, t) = Q_{y,\lambda y}^{++}(x, t; \lambda y) + g(\tau)A_y^-(x, \tau_1)$$
$$+\beta\lambda^{\alpha+2}[y'\overset{*}{B}_{y,y'}^{++}(x, t) - A_{y,y'}^{--}(x, t)]$$

in terms of second-order, $Q$ options and asset and bond binaries, where $\tau_1 = (T_1 - t)$, $y' = y/\lambda$ and $g(\tau)$ is a function of $\tau = (T_2 - T_1)$ only, to be determined. The image is wrt $y'$.

Verify that this price reduces to the case of a start-out partial time lookback call when $\lambda \to 1$.

11. Let $B_\xi^\pm(x, \tau)$ denote prices of up and down type bond binaries, with exercise price $\xi$ and time $\tau = (T - t)$ remaining to expiry, and let $\overset{*}{B}_\xi^\pm(x, \tau)$ denote the corresponding image price relative to $x = \xi$. Show that

$$\int_{y_1}^{y_2} \left[B_\xi^+(x, \tau) - \overset{*}{B}_\xi^+(x, \tau)\right] d\xi = m(x, y_2, \tau) - m(x, y_1, \tau)$$

$$\int_{z_1}^{z_2} \left[B_\xi^-(x, \tau) - \overset{*}{B}_\xi^-(x, \tau)\right] d\xi = M(x, z_2, \tau) - M(x, z_1, \tau)$$

where $m(x, y, \tau)$ and $M(x, z, \tau)$ denote the prices of generic minimum and maximum lookback options.

12. (a) Consider a min-max lookback option whose expiry $T$ payoff depends on both the minimum asset price and maximum asset price over $[0, T]$, i.e., $V(x, y, z, T) = f(y, z)$ for some function $f(x, y, z)$. Show that this option is an integrated version of a double knock-out barrier option.

(b) (*Hard*): Hence show that the price of the lookback option with expiry payoff $V_T = (z - ky)^+$, where $k > 1$, is given (in the notation of this chapter) by

$$V(x, y, z, t) = M(x, z, t) - km(x, y, t) + \mathbb{I}(z < ky) \sum_{n=-\infty}^{\infty} k^{n\alpha} \cdot$$

$$\{[m(k^{2n+1}x, ky, t) - m(k^{2n}x, ky, t)]$$
$$- [m(k^{2n+1}x, z, t) - m(k^{2n}x, z, t)]\}$$

and $\alpha = (2r/\sigma^2 - 1)$.

*Hint*: The result of Q11 above should help.

—ooOoo—

# Chapter 9

# Asian Options

## 9.1 Introduction

Asian options were first introduced in the 1990s decade, partly to discourage the possibility of market manipulation. For example, a call option which is currently itm, just before expiry, could be forced otm at expiry, by a sudden down selling of the underlying asset. This would be virtually impossible to do for a fixed strike Asian option. There was an additional benefit, in that Asian options were often much cheaper than corresponding standard options, thus providing a cost effective method of risk management.

Similar to lookbacks, Asian options are path dependent and the standard ones come in floating and fixed strike versions. A floating strike Asian call option is similar to a European call, except that the strike price is replaced by an average of the asset price over some specified time window. A fixed strike Asian option replaces the asset price at expiry by a similar average. The great variety of Asian options currently traded are due to the many different forms of averaging and monitoring windows adopted. Averaging may be arithmetic, geometric, discrete, continuous or other; monitoring windows may be complete or partial.

It transpires that closed-form, analytic prices for Asian options in the BS world, generally only exist for geometric averaging. The reason for this is not difficult to understand. When the underlying asset price is log-normal, so is its geometric mean (discrete or continuous). However, arithmetic or other means will not be log-normal. This outcome was to some extent a disappointment to the options community, because arithmetic mean averaging is the preferred market place mode for Asian options.

Some institutions are content to approximate the arithmetic mean by the geometric mean, and as Figure 9.1 shows, this may not be too far off the mark.

The first closed-form analysis of geometric mean Asian options was by Kemna and Vorst [42] in 1990. A lot of research can also be found in the literature for pricing arithmetic mean Asian options, mostly by computational

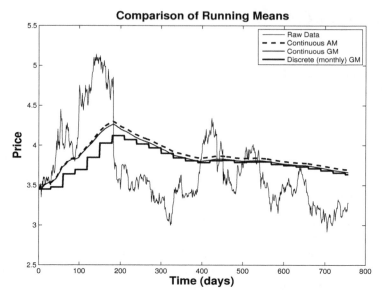

**FIGURE 9.1:** Comparison of the running means for a given underlying asset process $x = X_t$. Upper curve (dashed) is continuous arithmetic mean; middle curve (dotted) is continuous geometric mean; lower curve(solid) discrete monthly geometric mean.

methods. In a somewhat offbeat paper, Linetsky [52] presented a formal solution for an arithmetic mean Asian call option, in terms of an infinite spectral sum of parabolic cylinder functions. In the spirit of the rest of this book, for which pricing methodology and exact solutions are paramount, we shall concentrate in this chapter exclusively on geometric mean Asian options.

## 9.2 Pricing Framework

We follow here the approach presented in Chapter 13 of Wilmott et al. in [77], but skip over most of the detail. This leads to the derivation of a modified BS-pde for a fairly general class of Asian options.

As usual, for BS dynamics, assume the underlying asset price $x = X_t$, satisfies the sde

$$dx = x(\mu dt + \sigma dB_t)$$

under the real world measure. Assume the continuous time monitoring window occupies the range $[0, T]$, and define a new state variable $y = Y_t$ through

$$y = \int_0^t A(X_s, s)\, ds. \tag{9.1}$$

It is useful to think of $A(x, t)$ as an averaging kernel, and $y$ as a running asset price average over $[0, t]$. Asian options are then defined to be any derivative whose payoff at time $T$ depends on $Y_T$, and possibly also on $X_T$; i.e., $f(X_T, Y_T)$ for some payoff function $f(x, y)$. For a small time increment $dt$, we have

$$dy = A(x, t)\, dt$$

which, despite having no Brownian jump, is implicitly stochastic through its dependence on $x = X_t$.

The price $V(x, y, t)$ of an Asian option, will have a corresponding increment (by Itô's Lemma), given by

$$\begin{aligned} dV &= (V_t + \tfrac{1}{2}\sigma^2 x^2 V_{xx})dy + V_x dx + V_y dy \\ &= (V_t + \mu x V_x + A(x, t)dy + \tfrac{1}{2}\sigma^2 x^2 V_{xx})dt + \sigma x\, dB_t. \end{aligned}$$

Now we can set up a hedge portfolio of one Asian option and $h$ units short of the underlying asset, exactly as for standard European options. This portfolio can be made riskless with the unique choice $h = V_x$, leading to the following BS-pde for Asian options

$$\boxed{V_t = rV - rx V_x - A(x, t)V_y - \tfrac{1}{2}\sigma^2 x^2 V_{xx}; \qquad V(x, y, T) = f(x, y)} \tag{9.2}$$

**REMARK 9.1**  This TV-pde differs from the standard BS-pde (1.19) only by the addition of the term $A(x, t)V_y$. But this extra term makes a significant difference to the solutions of the pde. While the standard BS-pde has only one state variable $x$, it is one-dimensional; the pde for Asian options has two state variables $(x, y)$, and is therefore two-dimensional.

For floating strike Asian options on the (continuous) arithmetic mean, we take $A(x, t) = x/T$ and $f(x, y) = (x - y)^+$ for calls and $f(x, y) = (y - x)^+$ for puts. Note that for this choice of averaging kernel,

$$y(T) = \frac{1}{T}\int_0^T X_s ds$$

which is just the continuous arithmetic mean asset price over the interval $[0, T]$. We shall be interested in solving (9.2) for all $0 < t < T$, so that at time $t$, as for lookback options, we will already have observed the running mean $y(t)$ up to time $t$.

One cannot hope to solve Equation (9.2) for arbitrary $A(x, t)$. Indeed, as we have already remarked, it cannot be solved in simple analytical terms, even for $A(x, t) = x/T$. ∎

As for standard options, there is an FTAP for Asian options, which is given by the following theorem.

**THEOREM 9.1**
*Let $V(x, y, t)$ denote the price of an Asian option, with expiry $T$ payoff function $f(x, y)$, as described above. Then,*

$$\boxed{V(x, y, t) = e^{-r(T-t)} \, \mathbb{E}_Q\{f(X_T, Y_T)|(X_t = x, Y_t = y)\}} \qquad (9.3)$$

*where in the BS economy, $X_T = xe^{(r - \frac{1}{2}\sigma^2)\tau + \sigma\sqrt{\tau}Z}$, $Y_T = \int_0^T A(X_s, s)ds$ and $\tau = (T - t)$.*

The proof is relegated to the Exercise Problems (see Q1) for this chapter.

---

## 9.3 Geometric Mean Asian Options

Let $X_t$ denote the underlying asset price at time $t > 0$. Then the discrete and continuous geometric means of $X_t$ are given by the following definitions.

**DEFINITION 9.1**

1. *Let $X_i$, for $i = 1, 2, \ldots, n$ denote asset prices at fixed times $T_i$. Then the discrete geometric mean (GM) of the $X_i$ is defined to be*

$$G_n = \left[\prod_{i=1}^n X_i\right]^{1/n} = \sqrt[n]{X_1 X_2 \cdots X_n}. \qquad (9.4)$$

2. *The continuous GM over the time interval $[0, T]$ is defined by*

$$G_T = \exp\left[\frac{1}{T}\int_0^T \log X_s \, ds\right]. \qquad (9.5)$$

While it may not appear that these two definitions are consistent, they in fact are, as can be seen by writing them in the equivalent forms,

$$\log G_n = \frac{1}{n}\sum_{i=1}^n \log X_i \quad \text{and} \quad \log G_T = \frac{1}{T}\int_0^T \log X_s \, ds.$$

Both also show that the log of the GM is just the arithmetic mean (AM) of the log-prices.

GM Asian options therefore have expiry payoffs which can generally be expressed in the form: $V_T = f(X_T, G_T)$ for some function $f$. That is, the payoff depends on the GM, $G_T$, over range $[0, T]$ and possibly, also on the asset price $X_T$, at expiry $T$. The standard GM Asian calls and puts have payoffs given in the following table.

| Standard GM Mean Asian Payoffs | | |
| --- | --- | --- |
| | Fixed Strike | Floating Strike |
| Call | $(G_T - k)^+$ | $(X_T - G_T)^+$ |
| Put | $(k - G_T)^+$ | $(G_T - X_T)^+$ |

We mentioned in the previous section that, as for most other derivatives, Asian options can be priced by either the pde method or using the FTAP (i.e., discounted expectations under the EMM). The two methods are very different in their details, and as they are very instructive, we present both methods in what follows.

We assume, as above, that time zero is the start of GM monitoring and the expiry date $T$ coincides with the end of GM monitoring. We shall price all options at some intermediate time $t$, such that $0 < t < T$. Thus the GM over $[0, t]$, will already have been observed, and we shall denote this GM by $g = G_t$. If we want the start of monitoring to coincide with the present time, we need only set the current time $t = 0$ in the prices we obtain.

## 9.4 FTAP Method for GM Asian Options

Under the EMM, $Q$ we have the representation $X_s = xe^{(r - \frac{1}{2}\sigma^2)(s-t) + \sigma B_{s-t}}$, of the underlying asset price, for all $s > t$, where $B_t$ denotes Brownian motion. The expression (9.5) for the continuous GM, can be split into two parts: one over $[0, t]$ for which the GM has already been observed, and the other over $(t, T]$, which is the as yet unobserved, random part.

$$\log G_T = \frac{1}{T} \int_0^t \log X_s ds + \frac{1}{T} \int_t^T \log X_s ds$$

$$= \frac{t}{T} \log G_t + \frac{1}{T} \int_t^T [\log x + (r - \tfrac{1}{2}\sigma^2)(s - t) + \sigma B_{s-t}] ds$$

$$= \frac{t}{T} \log G_t + (1 - \frac{t}{T}) \log x + \tfrac{1}{2}(r - \tfrac{1}{2}\sigma^2)\frac{\tau^2}{T} + \frac{\sigma}{T} \int_0^\tau B_u du$$

where $\tau = (T - t)$ and we have changed variables in the integral through $u = (s - t)$. We therefore see that, when $X_s$ follows gBm, the continuous GM over $[0, T]$, conditional on being at time $t$ with $0 < t < T$, is given by the expression

$$G_T = x \left( \frac{G_t}{x} \right)^{t/T} \exp \left\{ \tfrac{1}{2}(r - \tfrac{1}{2}\sigma^2)\frac{\tau^2}{T} + \frac{\sigma}{T} \int_0^\tau B_u du \right\}.$$

To make further progress, we need the properties of the stochastic integral appearing in the above formula for $G_T$. The following lemma provides the necessary results.

**LEMMA 9.1**
Let $I_\tau = \int_0^\tau B_u du$. Then $I_\tau$ is a zero mean Gaussian rv with

$$\text{var}\{I_\tau\} = \tfrac{1}{3}\tau^3 \quad \text{and} \quad \text{corr}\{B_\tau, I_\tau\} = \tfrac{1}{2}\sqrt{3}.$$

**PROOF**  See Q2 Exercise Problems at the end of this chapter.  ⌑

Let $B_\tau = \sqrt{\tau} Z$, where $Z$ is a standard Gaussian rv. Then, from this lemma, our expression above for $G_T$ can be written as

$$G_T = x \left( \frac{g}{x} \right)^{t/T} \exp \left\{ \tfrac{1}{2}(r - \tfrac{1}{2}\sigma^2)\frac{\tau^2}{T} + \sigma \frac{\tau^{3/2}}{\sqrt{3T}} Z' \right\}$$

where $Z' \sim N(0,1)$ is also a standard Gaussian rv, correlated to $Z$, with correlation coefficient $\rho = \tfrac{1}{2}\sqrt{3}$. Since $G_T$ is essentially the exponential of a Gaussian rv, it is also log-normally distributed.

**Alternative Expression for $G_T$**
When pricing derivatives dependent on the GM, $G_T$, we shall find it convenient to express the above formula for $G_T$ in a more suggestive form. Let

$$s = \frac{\tau}{T} = 1 - \frac{t}{T}; \quad \bar{\sigma}(s) = \frac{\sigma s}{\sqrt{3}}; \quad y = x \left( \frac{g}{x} \right)^{1-s} \tag{9.6}$$

and define $\bar{q}(s)$ by

$$\bar{q}(s) = r - \tfrac{1}{2}(r - \tfrac{1}{2}\sigma^2)s - \tfrac{1}{6}\sigma^2 s^2. \tag{9.7}$$

Note that $s$ denotes at time $t$, the fraction of GM monitoring time still remaining to expiry $T$. Then, under the EMM, $Q$, we have the following representations for $X_T$ and $G_T$:

$$\boxed{\begin{aligned} X_T &= xe^{(r-\frac{1}{2}\sigma^2)\tau + \sigma\sqrt{\tau}Z} \\ G_T &= ye^{(r-\bar{q}-\frac{1}{2}\bar{\sigma}^2)\tau + \bar{\sigma}\sqrt{\tau}Z'} \end{aligned}} \tag{9.8}$$

where $(Z, Z') \sim N(0, 1; \frac{1}{2}\sqrt{3})$.

Written in this form, the GM, $G_T$, looks exactly like another asset of current price $y$; rms time-dependent volatility $\bar{\sigma}(s)$; and paying a mean, time-dependent dividend yield $\bar{q}(s)$. Let us call this hypothetical asset, $Y$. The reader is referred back to Section 2.9 to see the effect of time-dependent parameters in derivative pricing.

Note further, that if we are currently at the start of monitoring for the GM, then $t = 0$ and $s = 1$, so the expression for $G_T$ reduces to the simpler result,

$$G_T = x\, e^{(r - q_0 - \frac{1}{2}\sigma_0^2)T + \sigma_0 \sqrt{T} Z'} \tag{9.9}$$

with

$$q_0 = \tfrac{1}{2}(r + \tfrac{1}{2}\sigma_0^2) \quad \text{and} \quad \sigma_0 = \sigma/\sqrt{3}. \tag{9.10}$$

The expression for $\sigma_0$ is sometimes called the "root-3 volatility rule" by traders in Asian options.

It now follows from the FTAP, that the pv of any continuous GM Asian option with payoff function $F(X_T, G_T)$, is given by,

$$V(x, G_t, t) = e^{-r\tau}\, \mathbb{E}_Q\{F(X_T, G_T) | \mathcal{F}_t\} \tag{9.11}$$

with $X_T$ and $G_T$ given by Equation (9.8).

The next example illustrates the application of these formulae to a simple GM Asian option.

### Example 9.1

Consider the generic GM Asian option, which pays at expiry $T$, simply the continuous GM of the underlying asset price, over $[0, T]$, i.e., $f(X_T, G_T) = G_T$, independent of $X_T$.

From what we have just discussed, this derivative has pv given by

$$V(x, g, t) = e^{-r\tau}\, \mathbb{E}_Q\{G_T | \mathcal{F}_t\} = y e^{-\bar{q}(s)\tau} = x(g/x)^{1-s} e^{-\bar{q}(s)\tau}$$

For the case $t = 0$, we have $g = x$ and $V(x, g, t)$ reduces to $V_0(x, T) = x e^{-\frac{1}{2}(r + \frac{1}{2}\sigma_0^2)T}$. □

### 9.4.1  GM Asian Calls and Puts

**Fixed Strike GMA Call/Put**

For a fixed strike GM Asian call option of strike price $k$ and expiry date $T$, using the FTAP method above, we need only calculate the expectation,

$$V_c(x, g, t) = e^{-r\tau}\, \mathbb{E}_Q\{(G_T - k)^+ | \mathcal{F}_t\}.$$

But this is just the price of a standard European call option on the hypothetical asset $Y$. Hence, we can write down its price immediately, as the pair of equations

$$
\boxed{
\begin{aligned}
V_c(x, g, t) &= y e^{-\bar{q}(s)\tau} \mathcal{N}(d_1) - k e^{-r\tau} \mathcal{N}(d_2) \\
[d_1, d_2] &= \frac{\log(y/k) + [r - \bar{q}(s) \pm \frac{1}{2}\bar{\sigma}^2(s)]\tau}{\bar{\sigma}(s)\sqrt{\tau}}
\end{aligned}
}
\tag{9.12}
$$

If the current time coincides with the start of averaging ($t = 0$; $s = 1$), this reduces to the expression

$$
V_0^c(x, T) = x e^{-q_0 T} \mathcal{N}(d_0 + \sigma_0 \sqrt{T}) - k e^{-rT} \mathcal{N}(d_0)
\tag{9.13}
$$

with

$$
d_0(x, T) = \frac{\log(x/k) + \frac{1}{2}(r - \frac{1}{2}\sigma^2)T}{\sigma_0 \sqrt{T}}
$$

which is an oft quoted formula for GM Asian call options (e.g., see Zhang [78]).

The corresponding formulae for GM Asian put options, for $t > 0$,

$$
\boxed{
V_p(x, g, t) = -y e^{-\bar{q}(s)\tau} \mathcal{N}(-d_1) + k e^{-r\tau} \mathcal{N}(-d_2)
}
\tag{9.14}
$$

and for $t = 0$, the simpler result,

$$
V_0^p(x, T) = -x e^{-q_0 T} \mathcal{N}(-d_0 - \sigma_0 \sqrt{T}) + k e^{-rT} \mathcal{N}(-d_0).
$$

### Floating Strike GMA Call/Put

To price the corresponding floating strike GMA call, by the FTAP method, we start from

$$
V_c(x, g, t) = e^{-r\tau} \mathbb{E}_Q \{ (X_T - G_T)^+ | \mathcal{F}_t \}.
$$

We immediately see that this expectation is going to be more complicated than the one we calculated above for the fixed strike GMA call. But, if we think about this for a moment, we can convince ourselves that it is nothing more than the price of an exchange option for asset $X$, and hypothetical asset $Y$. We have already priced this option in Section 6.2, where we derived the Margrabe formula. We repeat the formula here for convenience, and at the same time include dividend yields $q_1$ and $q_2$ for the two assets.

$$
\begin{aligned}
V_{ex}(x, y, t) &= x e^{-q_1\tau} \mathcal{N}(d_1) - y e^{-q_2\tau} \mathcal{N}(d_2) \\
d_{1,2}(x, y, t) &= [\log(x/y) + (q_2 - q_1 \pm \tfrac{1}{2}\nu^2)\tau]/\nu\sqrt{\tau} \\
\nu^2 &= \sigma_1^2 + \sigma_2^2 - 2\rho\sigma_1\sigma_1.
\end{aligned}
$$

For the floating strike GM Asian call, we make the identifications:

$$
\begin{aligned}
q_1 &= 0; & q_2 &= \bar{q}(s); & \rho &= \tfrac{1}{2}\sqrt{3}; \\
\sigma_1 &= \sigma; & \sigma_2 &= \bar{\sigma}(s); & y &= x(g/x)^s.
\end{aligned}
$$

This leads to the formula

$$
\boxed{
\begin{aligned}
V_c(x, g, t) &= x\,\mathcal{N}(d_1) - ye^{-\bar{q}(s)\tau}\,\mathcal{N}(d_2) \\
d_{1,2} &= \frac{\log(x/y) + [\bar{q}(s) \pm \frac{1}{2}\nu^2(s)]\tau}{\nu(s)\sqrt{\tau}} \\
\nu^2(s) &= \sigma^2(1 - s + \frac{1}{3}s^2)
\end{aligned}
}
\tag{9.15}
$$

For the case $t = 0$, $(s = 1, y = x)$, this reduces to

$$
V_0^c(x, T) = x[\mathcal{N}(d_1) - e^{-q_0 T}\mathcal{N}(d_2)]; \qquad d_{1,2} = [q_0/\sigma_0 \pm \tfrac{1}{2}\sigma_0]\sqrt{T}
$$

which is reminiscent of the price of a forward-start call option considered in Section 5.1.

The corresponding prices for a floating strike GM Asian put option are, for $t > 0$,

$$
\boxed{V_p(x, g, t) = -x\,\mathcal{N}(-d_1) + ye^{-\bar{q}(s)\tau}\,\mathcal{N}(-d_2)}
\tag{9.16}
$$

and, for $t = 0$,

$$
V_0^p(x, T) = x[-\mathcal{N}(-d_1) + e^{-q_0 T}\mathcal{N}(-d_2)].
$$

---

## 9.5 PDE Method for GM Asian Options

We now turn our attention to pricing GM Asian options using the BS-pde (9.2). The first thing to note is that for the continuous GM, we take the averaging kernel to be $A(x, t) = \log x$, independent of $t$. So the pde to solve in $(0 < t < T, \; x > 0, y > 0)$ is,

$$
V_t = rV - rxV_x - (\log x)V_y - \tfrac{1}{2}\sigma^2 x^2 V_{xx}
\tag{9.17}
$$

where $V = V(x, y, t)$ and $y = Y_t = \int_0^t \log X_s ds$. The GM at expiry is then $G_T = e^{Y_T/T}$. Hence, the payoff function for the pde is,

$$
V(x, y, T) = f(x, G) = f(x, e^{y/T}).
\tag{9.18}
$$

### 9.5.1 Fixed Strike GM Asian Calls

For a fixed strike call, the payoff function $f(G) = (e^{y/T} - k)^+$ is independent of $x$. We can use this simple observation to solve the pde (9.17) by a variable reduction technique. The details are motivated by similar considerations in Wilmott et al. [77]. This clever device, reduces the problem to solving a 1D-pde, rather than the 2D-pde described by (9.17). The next proposition shows how this can be done.

## PROPOSITION 9.1

*Define a new variable $\xi > 0$, by*

$$\xi = e^{y/T} x^{1-t/T}. \tag{9.19}$$

*Then, $V(x, y, t)$ transforms to $U(\xi, t)$ and the 2D BS-pde (9.17) reduces to the 1D-pde*

$$U_t = rU + [r - q(s)]\xi U_\xi - \tfrac{1}{2}\nu^2(s)\xi^2 U_{\xi\xi} \tag{9.20}$$

*where, with $s = 1 - t/T = \tau/T$,*

$$q(s) = r(1 - s) + \tfrac{1}{2}\sigma^2 s(1 - s) \quad and \quad \nu(s) = \sigma s \tag{9.21}$$

*and the corresponding payoff at expiry is $U(\xi, T) = f(\xi)$.*

## REMARK 9.2

1. It is important to realize that proposition 9.1 only applies to derivatives with payoffs $f(G_T)$, depending on $G_T$ alone. If the payoff also depends on $X_T$, the method breaks down. Hence the reduction will work fine for fixed strike GM Asian calls and puts with payoffs, $(G_T - k)^+$ and $(k - G_T)^+$. But, it will not work (at least not directly) for floating strike versions with payoffs, $(X_T - G_T)^+$ and $(G_T - X_T)^+$.

2. Note that the reduction variable $\xi$ is equal to the GM, $G_T = e^{y/T}$ at the expiry date $T$.

3. We have intentionally written the transformed pde (9.20) in the form, of an equivalent 1D, BS-pde on a dividend paying asset of price $\xi$, having continuous time-varying dividend yield $q(s)$, and a deterministic, time-varying volatility $\nu(s)$. Naturally, this identification will present a quick and immediate way to solve the pde and hence price the option.

□

**PROOF** The proof involves little more than an application of the Chain Rule for partial differentiation. We find, with $\log \xi = y/T + s \log x$,

$$V_t = U_t + U_\xi \xi_t \quad = U_t - (\xi/T)(\log x) U_\xi$$
$$xV_x = x(U_x + U_\xi \xi_x) = xU_x + s\xi U_\xi$$
$$(\log x)V_y = (\log x)U_\xi \xi_y = (\xi/T)(\log x) U_\xi$$

and $x^2 V_{xx} = s^2 U_{\xi\xi} - s(1 - s)\xi U_\xi$. Substituting into the BS-pde (9.17) and collecting the terms leads to the results stated in the proposition. Note that the terms in $(\log x)\xi U_\xi$ cancel. □

As an illustration of this method, consider the same example 9.1 we solved previously by the FTAP. One would have to conclude that the pde method in this case is much the simpler.

**Example 9.2**

We wish to solve (9.20) with payoff $U(\xi, T) = \xi$. From the third remark above, the solution of this problem for $0 < t < T$, can be written down by inspection as

$$U(\xi, t) = \xi e^{-\bar{q}\tau}$$

where according to Section 2.9, we replace the time varying dividend yield $q(s)$, by its time average over $[t, T]$. This yields,

$$\bar{q}(s) = \tau^{-1} \int_t^T q(1 - u/T) du = s^{-1} \int_0^s q(v) dv$$

$$= s^{-1} \int_0^s \left[ r(1 - v) + \tfrac{1}{2}\sigma^2 v(1 - v) \right] dv$$

$$= r - \tfrac{1}{2}(r - \tfrac{1}{2}\sigma^2)s - \tfrac{1}{6}\sigma^2 s^2.$$

This is precisely the same expression for $\bar{q}(s)$ we obtained in Equation (9.7).

Substituting for $\xi = e^{y/T} x^{1-t/T}$ and $g = e^{y/t}$, then leads to the same result as in example 9.1. That is, $V(x, g, t) = x(g/x)^{t/T} e^{-\bar{q}(s)\tau}$. ☐

To price the GM Asian call option by the pde method, we simply have to find the time averaged dividend yield, as in the above example, and also the rms volatility, and substitute these for the fixed parameters in the standard BS call price formula, on the hypothetical asset $Y$. The rms volatility comes from

$$\bar{\sigma}^2(s) = \frac{\sigma^2}{\tau} \int_t^T (1 - u/T)^2 \, du = \frac{\sigma^2}{s} \int_0^s v^2 \, dv = \tfrac{1}{3}\sigma^2 s^2$$

which also agrees with the rms volatility defined in Equation (9.6). With these values of $\bar{q}(s)$ and $\bar{\sigma}(s)$ replacing the fixed $q$ and $\sigma$ in the standard BS call price, we immediately arrive at the same formula (9.12), for the price of the fixed strike GM Asian call option.

## 9.5.2 Floating Strike GM Asian Calls

We mentioned above that the variable reduction method, through the transformation $\xi = e^{y/T} x^{t/T}$, will not work for GM Asian options, when the payoff depends on $X_T$ as well as the GM, $G_T$. However, as we saw in proposition 9.1, the transformation had another very important feature: it eliminated the troublesome $(\log x)V_y$ term in the pde. Hence, it is worthwhile sticking with the transformation. But now, instead of getting the reduction $V(x, y, t) \to U(\xi, t)$, we get $V(x, y, t) \to U(x, \xi, t)$. That is, we no longer reduce from a 2D-pde to a 1D-pde. Under these circumstances, the counterpart to proposition 9.1 becomes the following.

### PROPOSITION 9.2

*Under the transformation, $\xi = e^{y/T} x^s$, with $s = (1 - t/T)$, $V(x, y, t)$ transforms to $U(x, \xi, t)$ and the 2D, BS-pde (9.17) transforms to*

$$U_t = rU - rxU - s[r - \tfrac{1}{2}\sigma^2(1-s)]\xi U_\xi - \tfrac{1}{2}\sigma^2[x^2 U_{xx} + s^2 \xi^2 U_{\xi\xi} - 2sx\xi U_{x\xi}] \quad (9.22)$$

*and the corresponding payoff at expiry is $U(x, \xi, T) = f(x, \xi)$.*

We leave the details of the proof, which is similar to that for proposition 9.1, to the reader.

In the general case, we are forced into solving this 2D, BS-pde with time varying parameters. However, for many payoff functions in practice, such as calls and puts for example, we can apply a different reduction to a 1D-pde, using homogeneity arguments. Indeed, that is precisely what we did in the pde approach to pricing the exchange option in Section 6.2. Hopefully the reader will have recognized that the pde (9.22) with payoff $f(x, \xi) = (x - \xi)^+$ is the pricing equation for an exchange option, albeit with time-varying parameters.

So for a floating strike call (or put), where the payoff $U(x, \xi, T) = (x - \xi)^+$ is homogeneous in $x$ and $\xi$ of degree one, the price $U(x, \xi, t)$ will also be homogeneous of degree one. That is, if we scale up the asset price $x$ and the variable $\xi$ by a factor $\lambda$, then the price will also scale up by the same factor. So by Euler's Equation,

$$xU_x + \xi U_\xi = U$$

and some simple calculations, which parallel those in Section 6.2, we will find that the 2D-pde (9.22) reduces to the 1D, BS-pde

$$U_t = \rho(s)U - \rho(s)xU_x - \tfrac{1}{2}\nu^2(s)x^2 U_{xx} \quad (9.23)$$

where $\rho(s)$ and $\nu(s)$ are a time-varying interest rate and volatility, defined by

$$\rho(s) = [r + \tfrac{1}{2}\sigma^2 s](1 - s); \qquad \nu(s) = \sigma(1 - s). \quad (9.24)$$

Note that in the payoff $U(x, \xi, T) = (x - \xi)^+$, the variable $\xi$, just becomes a parameter representing the strike price of a European call option. The new pde for $U(x, \xi, t)$ has become independent of $\xi$.

The price of the floating strike call option can now be written down as the standard BS call price with risk-free rate $r$ replaced by the average $\bar{\rho}(s)$

over $[t, T]$, and the volatility $\sigma$ replaced by the rms average $\bar{\nu}(s)$ over the same interval. We have skipped some of the details in the simple polynomial integrations below, but with $s = (1 - t/T)$, we find,

$$
\begin{aligned}
\bar{\rho}(s) &= \frac{1}{\tau} \int_t^T \rho(1 - u/T) du \\
&= \frac{1}{s} \int_0^s [r + \tfrac{1}{2}\sigma^2 v](1 - v) dv; \qquad (v = u/T) \\
&= r - \tfrac{1}{2}(r - \tfrac{1}{2}\sigma^2)s - \tfrac{1}{6}\sigma^2 s^2
\end{aligned}
$$

and, similarly

$$
\bar{\nu}^2(s) = \frac{\sigma^2}{\tau} \int_t^T (1 - u/T)^2 du = \sigma^2(1 - s + \tfrac{1}{3}s^2).
$$

Hence,

$$
U(x, \xi, t) = x \mathcal{N}(d_1) - \xi e^{-\bar{\rho}(s)\tau} \mathcal{N}(d_2)
$$

where

$$
d_{1,2} = \frac{\log(x/\xi) + [\bar{\rho}(s) \pm \tfrac{1}{2}\bar{\nu}^2(s)]\tau}{\bar{\nu}(s)\sqrt{\tau}}.
$$

It is now a straightforward matter to show that this expression for the floating strike GM Asian call price, is indeed the same as in Equation (9.15), once the variable identifications are made.

## 9.6 Discrete GM Asian Options

To keep things simple, we shall suppose in this section that the discrete GM will be computed from the $n$ asset prices $X_i$, $(i = 1, 2, \ldots, n)$ evaluated at times $t_i > 0$. Assume further, that the current time is $t = 0$ and the expiry date is $t_n = T$. In these circumstances, the GM at expiry, $G_n$ say, is

$$
G_n = \sqrt[n]{X_1 X_2 \cdots X_n}.
$$

As usual, the asset price $X_t$ for $t > 0$ is assumed to follow gBm with drift rate $r$ (under the EMM $Q$) and volatility $\sigma$. The initial asset price at $t = 0$, is taken to be $x = X_0$. The distribution of $G_n$ is given by the following lemma.

**LEMMA 9.2**
*With $G_n$ defined above,*

$$
\log(G_n/x) \stackrel{d}{=} N\left\{(r - \tfrac{1}{2}\sigma^2)\bar{T}_n, \ \sigma^2 \hat{T}_n\right\} \tag{9.25}
$$

*where*

$$\bar{T}_n = \frac{1}{n}\sum_{i=1}^n t_i; \qquad \hat{T}_n = \frac{1}{n^2}\sum_{i=1}^n\sum_{j=1}^n \min(t_i, t_j). \qquad (9.26)$$

Naturally, Lemma 9.2 demonstrates that the discrete GM of a set of log-normal rv's is itself log-normal.

**PROOF** For gBm under the EMM, $Q$, we have $X_i = xe^{(r-\frac{1}{2}\sigma^2)t_i+\sigma B_i}$, where $B_i = B(t_i)$ is a Brownian motion under $Q$, evaluated at time $t = t_i$. This yields,

$$G_n = x\prod_{i=1}^n e^{(r-\frac{1}{2}\sigma^2)t_i/n+\sigma B_i/n}$$

and hence,

$$\log(G_n/x) = \frac{1}{n}\sum_{i=1}^n(r-\tfrac{1}{2}\sigma^2)t_i + \frac{\sigma}{n}\sum_{i=1}^n B_i \overset{d}{=} N(\mu_n, \sigma_n^2)$$

where

$$\mu_n = (r-\tfrac{1}{2}\sigma)^2 \cdot \frac{1}{n}\sum_{i=1}^n t_i = (r-\tfrac{1}{2}\sigma^2)\bar{T}_n$$

and

$$\sigma_n^2 = \mathrm{var}\left(\frac{\sigma}{n}\sum_{i=1}^n B_i\right) = \frac{\sigma^2}{n^2}\mathbb{E}\left(\sum_{i=1}^n\sum_{j=1}^n B_i B_j\right)$$

$$= \frac{\sigma^2}{n^2}\sum_{i=1}^n\sum_{j=1}^n \mathbb{E}(B_i B_j) = \frac{\sigma^2}{n^2}\sum_{i=1}^n\sum_{j=1}^n \min(t_i, t_j) = \sigma^2\hat{T}_n.$$

The last equality comes from (2.5) for the covariance of Brownian motion at two distinct times. This completes the proof. □

### Price of Discrete GM Asian Fixed Strike Call Option

Lemma 9.2 permits an immediate calculation of this option, as it gives the representation,

$$G_n = xe^{(r-\frac{1}{2}\sigma^2)\bar{T}_n+\sigma\sqrt{\hat{T}_n}Z}; \quad Z \sim N(0,1).$$

Hence for the call option of strike price $k$ and payoff $(G_n - k)^+$, we obtain (see Q4 in the Exercise Problems for details)

$$\boxed{\begin{array}{c} C_n(x, T) = xe^{-q_n T}N(d + \sigma\sqrt{\hat{T}_n}) - ke^{-rT}N(d) \\ d = \dfrac{\log(x/k)+(r-\frac{1}{2}\sigma^2)\bar{T}_n}{\sigma\sqrt{\hat{T}_n}} \\ q_n = r - (r-\tfrac{1}{2}\sigma^2)\bar{T}_n/T - \tfrac{1}{2}\sigma^2\hat{T}_n/T \end{array}} \qquad (9.27)$$

## Case of Uniform Spacing

Suppose now that the times $t_i$ are uniformly spaced. Then, $t_i = i\Delta$, for all $i$, and $\Delta$ will be the constant time interval between consecutive monitoring dates. In this case, we find,

$$\bar{T}_n = \frac{\Delta}{n} \sum_{i=1}^{n} i = \tfrac{1}{2}(n+1)\Delta; \qquad T = n\Delta.$$

$$\hat{T}_n = \frac{\Delta}{n^2} \sum_{i=1}^{n} \sum_{j=1}^{n} \min(i,j) = \frac{\Delta}{n^2} \sum_{k=1}^{n} k^2 = \frac{(n+1)(2n+1)\,\Delta}{6n}.$$

## The Continuous Limit

It is a relatively simple matter to get the continuous GM Asian prices, from the discrete ones. All we need to do is take the limits, $\Delta \to 0$ and $n \to \infty$ in such a way that, $n\Delta \to T$. In that case, both $\bar{T}_n$ and $\hat{T}_n$ have the finite limits,

$$\bar{T}_n \to \tfrac{1}{2}T \qquad \text{and} \qquad \hat{T}_n \to \tfrac{1}{3}T.$$

We leave it as an exercise for the reader to show that the corresponding fixed strike call option price in Equation (9.27) has a limit, which reproduces that in Equation (9.13) exactly.

---

## 9.7 Summary

Asian options are generally more problematic than either barrier or lookback options, when it comes to pricing. Different methods of averaging lead to very different analyses, and averaging by the geometric mean is the only one with a closed form price in a Black–Scholes economy. Even then, the corresponding BS-pde has an extra term, which not only is potentially difficult to handle, but also raises the dimension of the pde from first to second order.

We demonstrated how both the FTAP and pde methods could be used to price GM Asian options. In the case of the FTAP (or EMM) method, the trick was to express the GM at expiry in a mathematical form which resembles a second asset, also following gBm. Then, pricing could proceed by calculating a 2-asset discounted expectation. This latter expectation could either be recognized as a standard 2-asset contract (e.g., an exchange option), or could be just another candidate for the 2D, Gaussian Shift Theorem.

The pde method employed clever transformations to reduce the dimension of the pde back to first order. Both methods require the pricing of equivalent standard options with deterministic, time-varying parameters. Fortunately, there is a simple prescription, derived in Section 2.9, which handles such problems.

The next chapter defines and develops the theory of M-binaries. These are general multi-asset, multi-period binary options, which include all previously analyzed binaries in this book, as special cases. These complex objects, require a special notation to describe them, else their pricing would descend into a nightmare of mathematical symbols. This theory also paves the way to price many more complex exotic options, including general multi-asset barrier options.

---

## Exercise Problems

1. Verify by direct substitution, that

$$v(x, y, t) = e^{-r(T-t)} \, \mathbb{E}_Q\{f(X_T, Y_T)|(X_t = x, Y_t = y)\}$$

   formally satisfies the pde

$$V_t = rV - rxV_x - A(x,t)V_y - \tfrac{1}{2}\sigma^2 x^2 V_{xx}; \qquad V(x,y,T) = f(x,y)$$

   where $Y_T = \int_0^T A(X_s, s)\, ds$.

2. Let $I_\tau = \int_0^\tau B_u du$. Prove lemma 9.1 that $I_\tau$ is Gaussian with zero mean, and

$$\text{var}\{I_\tau\} = \tfrac{1}{3}\tau^3; \qquad \text{corr}\{B_\tau, I_\tau\} = \tfrac{1}{2}\sqrt{3}.$$

3. The BS-pde for the price $V(x, y, \tau)$ of a GM Asian option is

$$V_\tau = -rV + rxV_x + (\log x)V_y + \tfrac{1}{2}\sigma^2 V_{xx}; \qquad (\tau = T - t)$$

   where $y = \int_0^t \log X_s ds$ and the payoff is $V(x, y, 0) = f(G_T)$, and $G_T$ is the continuous GM of $X_s$ over $[0, T]$. Assuming a separable solution of the pde of the form

$$\log\{V(x, y, \tau)\} = \alpha(\tau)\log x + \beta(\tau)y + \gamma(\tau),$$

   obtain ode's for $\alpha(\tau)$, $\beta(\tau)$ and $\gamma(\tau)$. Solve these ode's for the case $f(G_T) = G_T$, and hence obtain the corresponding derivative price in terms of $x$, the current asset price and $g$, the currently observed GM of the asset price.

4. Derive the expression

$$C_n(x, T) = xe^{-q_n T} \mathcal{N}(d + \sigma\sqrt{\hat{T}_n}) - ke^{-rT} \mathcal{N}(d)$$

$$d = \frac{\log(x/k) + (r - \frac{1}{2}\sigma^2)\bar{T}_n}{\sigma\sqrt{\hat{T}_n}}$$

$$q_n = r - (r - \tfrac{1}{2}\sigma^2)\bar{T}_n/T - \tfrac{1}{2}\sigma^2 \hat{T}_n/T$$

for the price (see Eq(9.27)) of a discrete GM call option.

5. Consider a derivative which pays at time $T$ the continuous arithmetic mean (AM) of an asset price over $[0, T]$. That is,

$$V_T = A_T = \frac{1}{T} \int_0^T X_s \, ds.$$

Show that in the BS framework, the time $t < T$ value of this derivative is given by

$$V(x, A_t, t) = \left[ \frac{1 - e^{-r\tau}}{rT} \right] x + \frac{t}{T} e^{-r\tau} A_t$$

where $A_t$ is the currently observed AM. Obtain this result by two distinct methods:

(a) by solving the BS-pde, assuming a separable solution of the form

$$V(x, y, t) = \alpha(t)x + \beta(t)y; \qquad y = \int_0^t X_s \, ds$$

(b) by discounted expectations under the EMM (i.e., the FTAP method).

6. Derive put-call parity relations for both fixed strike and floating strike GM Asian options.

7. Using the representation (9.8) for $X_T$ and $G_T$, determine the time $t$ price of the following GM binary options:

(a) the bond binary with payoff $B_T = \mathbb{I}(G_T > k)$; and

(b) the asset binary with payoff $A_T = X_T \mathbb{I}(G_T > k)$

for a given exercise price $k > 0$.

8. In this problem you are asked to price a forward-start GM Asian call option. The option is to be valued at $t = 0$, assuming that the holder receives a European call option at time $T_1$, which expires at time $T_2 > T_1$ and has a strike price set equal to the continuous GM over $[0, T_1]$. Show that the price can be expressed in the form

$$V_0(x; T_1, T_2) = x[\mathcal{N}(d_1) - e^{-(r\tau + q_0 T_1)} \mathcal{N}(d_2)]$$

where

$$d_1 = \frac{(r + \frac{1}{2}\sigma^2)(\tau + \frac{1}{2}T_1)}{\sigma\sqrt{\tau + \frac{1}{3}T_1}}; \qquad d_2 = \frac{(r - \frac{1}{2}\sigma^2)\tau + \frac{1}{2}(r - \frac{1}{2}\sigma_0^2)T_1}{\sigma\sqrt{\tau + \frac{1}{3}T_1}}$$

with $\tau = (T_2 - T_1)$ and other parameters as defined in the text.

*Hints*: The bivariate GST in the form

$$\mathbb{E}\{e^{cZ}F(Z, Z')\} = e^{\frac{1}{2}c^2}\mathbb{E}\{F(Z + c, Z' + \rho c)\}$$

where $\rho = \text{corr}(Z, Z')$; and the second part of Q9 in Chapter 3 Exercise Problems, should be helpful.

Verify that this formula reduces to the price of a floating strike GM Asian call option when $\tau = 0$.

9. The discrete generalized GM of a set of prices $X_i$ for $(i = 1, 2, \ldots, n)$ is given by

$$G_n = \prod_{i=1}^{n} X_i^{\alpha_i} = X_1^{\alpha_1} X_2^{\alpha_2} \cdots X_n^{\alpha_n}; \qquad \alpha_i > 0, \; \sum_i \alpha_i = 1.$$

Assuming BS dynamics for the asset $X$, show that

$$\log(G_n/x) \overset{d}{=} N\left\{(r - \tfrac{1}{2}\sigma^2)\bar{T}_n, \; \sigma^2\hat{T}_n\right\}$$

as in Lemma 9.2, but with different definitions for $\bar{T}_n$ and $\hat{T}_n$ to be determined.

Hence price a fixed strike GM Asian call option, with the above expression for the GM.

*Note:* Zhang [78] refers to this GM as a *flexible* GM.

—ooOoo—

# Chapter 10

## Exotic Multi-Options

### 10.1 Introduction

Chapters 5 and 6 of this book dealt with dual expiry and two asset rainbow options respectively. It should be obvious that these option categories have multi-period and multi-asset extensions. That is, we might want to consider options with more than two expiry dates; or options with more than just two underlying assets. This chapter considers a class of derivatives which are simultaneously multi-period and multi-asset. In particular, in this chapter, we shall be mainly concerned with a single object, which we call an M-binary. This binary is the basic building block for all multi-period, multi-asset derivatives, which have a closed-form price in the Black–Scholes economy. The M-binary is itself a multi-asset, multi-period exotic option, which includes as special cases all the binary options we have met in this book.

A few examples of derivatives that can be decomposed into portfolios of M-binaries include:

- discretely monitored barrier and lookback options

- performance driven executive stock options (ESO's)

- multiple strike reset options

- multi-asset discrete geometric mean Asian options

- options that depend on the maximum, minimum or $n$th best of a group of assets

- multi-asset exchange options

And there are many more, limited perhaps only by one's imagination.

**REMARK 10.1**   The results of this chapter stem largely from the Honours Thesis of Skipper [70]. This was an outstanding piece of work, considering Skipper was only an undergraduate at the time. The main ideas of this chapter can be found in Skipper and Buchen [72].   ☐

Having the generality described above, the M-binary is a complex object. Its payoff depends on $N$ assets and $M$ discrete monitoring periods, and hence on up to $(M \times N)$ variables. It is natural to use matrix and vector notation to describe these objects, and indeed we shall do so. However, even in this framework, the notational burden becomes enormous. Thus, in order to simplify the analysis as much as possible and arrive at the most succinct representation for the M-binary price, we have resorted to some non-standard matrix and vector notation. In particular, we shall meet quantities like $\boldsymbol{X}^A$, where $\boldsymbol{X}$ is a vector of prices, and $A$ is a matrix. No such quantity exists in the normal matrix algebra, but we shall give it a precise meaning in describing the M-binary payoff. Our special notation is described in the next section. We ask the reader to be patient and to persevere with the details of this section, with the promise that considerable benefit will come in the final analysis.

---

## 10.2   Matrix and Vector Notation

In what follows, we shall meet several different classes of mathematical and statistical quantities. Since it is easy to lose sight of which class a quantity belongs to, let us introduce the classes or categories in the list below.

$$\mathcal{I}_n = \text{The index set } \{1, 2, \ldots, n\}$$
$$\mathcal{V}_n \ (\mathcal{V}_n') = \text{The set of all } n-\text{dimensional column (row) vectors}$$
$$\mathcal{A}_{mn} = \text{The set of all } (m \times n) \text{ matrices}$$
$$\mathcal{D}_n = \text{The set of all } (n \times n) \text{ diagonal matrices}$$
$$\mathcal{C}_n \ (\mathcal{C}_n^*) = \text{The set of all } (n \times n) \text{ covariance (correlation) matrices}$$

We denote, as in Chapter 3, the transpose of a vector by a dash: e.g., $\boldsymbol{X}'$ and $A'$. In addition, to indicate that $\boldsymbol{X} \in \mathcal{V}_n$ is a Gaussian random vector with zero mean and correlation matrix matrix $R \in \mathcal{C}^*$, we write $\boldsymbol{X} \sim \mathcal{G}_n(R)$.

As an aid to understanding, we shall illustrate the special notations introduced by way of simple examples.

### The Payoff Vector

Suppose there are $N$ assets $A_i$; $i \in \mathcal{I}_N$, and $M$ monitoring times $T_j$; $j \in \mathcal{I}_M$. Let $S_{ij}$ denote the price of asset $A_i$ at time $T_j$. Then the M-binary payoff will depend on (not necessarily all) the $S_{ij}$. For each asset $A_i$, the payoff will depend on at least one $T_j$, and for each monitoring time $T_j$, there will be at least one $A_i$. Since not all components of $S_{ij}$ necessarily appear in the payoff function, we assemble all those that do, into a single vector, $\boldsymbol{X}$.

The order of these components is completely arbitrary, but each element $X_k$ corresponds to a unique pair $(i,j) \in \mathcal{I}_N \times \mathcal{I}_M$. If we are concerned with the particular asset and time indices for a given element $X_k$, we may write $X_k = X_{(ij)}$ where $X_{(ij)} = S_{ij}$, and refer to $X_{(ij)}$ as a *dual-index vector*. We refer to this vector of asset prices as the *payoff* vector and its dimension $n$ as the *payoff dimension*. From the above discussion, it should be clear that $\boldsymbol{X} \in \mathcal{V}_n$ and $N \leq n \leq MN$.

### Example 10.1
As an illustration of a payoff vector, consider a multi-asset, multi-period call option with payoff at expiry $T$ given by

$$V_T = (M - k)^+; \qquad M = \max\{S_1(T_1),\ S_1(T_2),\ S_2(T_1),\ S_3(T_2)\}$$

where $T_1 < T_2 \leq T$. This call option is a 3-asset, 2-period derivative with payoff dimension $n = 4$. One possible representation of the payoff vector is

$$\boldsymbol{X} = [X_1,\ X_2,\ X_3,\ X_4]' = [X_{(11)},\ X_{(12)},\ X_{(21)},\ X_{(32)}]'$$

but any other ordering of the components is permissable. In our vector notation, we would then write the payoff of this call option in the form: $V_T(\boldsymbol{X}) = (\max\{\boldsymbol{X}\} - k)^+$. Note that in this example only four of the possible six components of $S_{ij}$ appear in the option payoff. ☐

### Vector and Matrix Functions
Let $\boldsymbol{x} \in \mathcal{V}_n$ be some positive column vector of dimension $n$. Then we shall freely use expressions like $\boldsymbol{x}^2$, $\sqrt{\boldsymbol{x}}$, $e^{\boldsymbol{x}}$, $\log \boldsymbol{x}$ etc. to denote new vectors, of the same dimension, obtained by component-wise evaluation. Examples include,

$$\log \boldsymbol{x} = [\log x_1,\ \log x_2,\ \cdots,\log x_n]'$$
$$\sqrt{\boldsymbol{x}} = [\sqrt{x_1},\ \sqrt{x_2},\ \cdots,\sqrt{x_n}]'$$
$$\boldsymbol{x}/\boldsymbol{y} = [x_1/y_1,\ x_2/y_2,\cdots,x_n/y_n]'.$$

**DEFINITION 10.1**    Let $A \in \mathcal{A}_{mn}$ be an $(m \times n)$ matrix, and $\boldsymbol{x} \in \mathcal{V}_n$ a positive $n-$vector. Then we define vector $\boldsymbol{x}^A \in \mathcal{V}_m$ by

$$\boldsymbol{x}^A = \exp(A \log \boldsymbol{x}). \tag{10.1}$$

Note that the rhs of the expression in (10.1) is well defined, since $\log \boldsymbol{x} \in \mathcal{V}_n \Rightarrow A \log \boldsymbol{x} \in \mathcal{V}_m$. Indeed, we can also legitimately write $\log \boldsymbol{x}^A = A \log \boldsymbol{x}$ in this formalism.

An important special case of (10.1) is for $A = \boldsymbol{\alpha} \in \mathcal{A}_{1n}$, i.e., a row vector of dimension $n$. As a harmless abuse of this notation, we shall also allow a column vector here. The vector power function becomes, the scalar quantity,

$$\boldsymbol{x}^{\boldsymbol{\alpha}} = x_1^{\alpha_1} x_2^{\alpha_2} \cdots x_n^{\alpha_n}. \tag{10.2}$$

The parameter $\alpha$ is called the *payoff index vector* and plays a significant role in the M-binary algebra. For example, if $\alpha = \frac{1}{n}\mathbf{1}$, where $\mathbf{1} = [1, 1, \ldots, 1]' \in V_n$, then $x^\alpha = \sqrt[n]{x_1 x_2 \cdots x_n}$ denotes the discrete geometric mean of the $x_i$.

**Example 10.2**
Evaluate $x^A$ when

$$
x = \begin{bmatrix} x \\ y \\ z \end{bmatrix}; \qquad A = \begin{bmatrix} 1 & 0 & -1 \\ \frac{1}{2} & \frac{1}{2} & 0 \end{bmatrix}.
$$

**Solution:**

$$
x^A = \exp(A \log x) = \exp \begin{bmatrix} \log x - \log z \\ \frac{1}{2}\log x + \frac{1}{2}\log y \end{bmatrix} = \begin{bmatrix} xz^{-1} \\ \sqrt{xy} \end{bmatrix}.
$$

$\square$

It should be clear from the above example, that each row of $x^A$ is determined as a scalar similar to Equation (10.2), with the index set $\alpha$ obtained from each corresponding row of matrix $A$.

**Multi-variate Indicator Function**
We start with a definition.

**DEFINITION 10.2**  *Let $y$, $a \in V_m$. Define the multi-variate indicator function of dimension $m$ for $(y > a)$ as*

$$
\mathbb{I}_m(y > a) = \mathbb{I}(y_1 > a_1))\mathbb{I}(y_2 > a_2) \cdots \mathbb{I}(y_m > a_m) = \prod_{i=1}^{m} \mathbb{I}(y_i > a_i). \qquad (10.3)
$$

Thus the $m$ dimensional indicator function is the product of $m$ one-dimensional indicator functions.

Since $\mathbb{I}(y < a) = \mathbb{I}(-y > -a)$, it is a simple matter to change any "less-than" indicator functions into corresponding "greater-than" indicator functions. Let $S \in \mathcal{D}_m$ be a diagonal matrix of dimension $m$, with all entries equal to $\pm 1$. Then the indicator $\mathbb{I}_m(Sy > Sa)$ has $i$-component $\mathbb{I}(y_i > a_i)$ if $S_{ii} = 1$, and has $i$-component $\mathbb{I}(y_i < a_i)$ if $S_{ii} = -1$. Thus it is possible, with the appropriate choice of matrix $S$, to have any product of both "less-than" and "greater-than" indicators. We shall call matrix $S$, the *exercise indicator matrix*.

We say that an M-binary has *exercise dimension* $m$, if its payoff depends on some $m$ dimensional indicator function. In practice, we require the important constraint $1 \leq m \leq n$, where $n$ is the payoff dimension. If this were not the case, then the exercise condition may be over-determined. In other cases, it should always be possible to decompose the exercise condition into a sum of lower order exercise conditions, using the Inclusion-Exclusion Principle. This is precisely what we did in Section 6.5, when solving the ICIAM problem. There we had an $m = 3$ exercise dimension, and an $n = 2$ payoff dimension. In that problem, we showed how it was possible to decompose the dimension-3 exercise condition, into the sum of three dimension-2 exercise conditions.

**Example 10.3**
Consider the previous example 10.2 for $x^A$ and let

$$a = \begin{bmatrix} 1 \\ k \end{bmatrix}; \quad S = \begin{bmatrix} 1 & 0 \\ 0 & -1 \end{bmatrix}.$$

Interpret the exercise condition $\mathbb{I}_m(Sx^A > Sa)$.
**Solution:**

$$\mathbb{I}_m(Sx^A > Sa) = \mathbb{I}(xz^{-1} > 1) \cdot \mathbb{I}(\sqrt{xy} < k).$$

Thus the exercise condition will be the two-dimensional indicator: $\mathbb{I}(x > z)\mathbb{I}(\sqrt{xy} < k)$, which in words translates to: payoff occurs iff $x > z$ and the geometric mean (GM), $\sqrt{xy} < k$. $\quad\quad\quad\quad\quad$ ▯

**Multi-Variate Normals**
It transpires that our general $\mathbb{M}$-binary will be priced in terms of a multi-variate normal distribution function. This function was defined in Equation (3.26) where the notation $\mathcal{N}_m(d;\ R)$ was introduced. Here, $d$ is a vector of real numbers in $\mathcal{V}_m$, and $R \in \mathcal{C}_m^*$ is an $m$ dimensional correlation matrix. Recall, that this function represents the probability $\mathbb{P}\{Z < d\}$, where $Z \sim \mathcal{G}_m(R)$ is a standard Gaussian $m$-vector with covariance[1] matrix $R$, i.e., $Z \sim N(0, 1; R)$.

The multi-variate normal can also be expressed in terms of a multi-dimensional indicator function. This is

$$\mathcal{N}_m(d; R) = \mathbb{E}\{\mathbb{I}_m(Z < d)\} \tag{10.4}$$

where $Z \sim \mathcal{G}_m(R)$. We shall find this representation useful, when we come to the proof of the $\mathbb{M}$-binary pricing valuation formula.

It will also be useful to note the following from Equation (3.33). If $\Gamma \in \mathcal{C}_m$ is a covariance matrix, then its diagonal elements contain the variances. Let $D \in \mathcal{D}_m$ denote the square-root of this diagonal matrix. Then, the corresponding correlation matrix will be $R = D^{-1}\Gamma D^{-1} \in \mathcal{C}_m^*$.

---

[1]Note that covariance and correlation are synonymous for zero mean, unit variance rv's.

This completes our discussion of notation used to describe the M-binary. As mentioned previously, and reinforced here, the extra effort spent in this section will pay big dividends when we come to describe, price, and apply the M-binary in the sections that follow.

**Black–Scholes Asset Prices**

Let $t < T_1$ denote the current time and suppose the $N$ assets $A_i$ have observed prices, $S_i(t) = x_i$; $i \in \mathcal{I}_N$. We assume that under the EMM, the future prices $S_i(u)$ follow correlated gBm, characterized by the system of sde's,

$$dS_i(u) = S_i(u)[(r - q_i)\, du + \sigma_i\, dB_i(u)]; \quad u > t, \ S_i(t) = x_i.$$

As usual, $r$ will denote the risk-free rate. Further, $q_i$ and $\sigma_i$ are the dividend yield and volatility of asset $A_i$. We shall assume these parameters are constants. There would be no difficulty if they were also deterministic functions of time, but this would needlessly add to the already daunting expanse of notation. The Brownian motions $B_i(t)$, $B_{i'}(t)$ are assumed to have constant correlation coefficient $\rho_{ii'}$.

The solution of the above sde is given by $S_i(u) = x_i e^{(r - q_i - \frac{1}{2}\sigma_i^2)(u-t) + \sigma_i B_i(u-t)}$. The payoff vector described earlier in this section, can be now constructed by evaluating $S_i(u)$ at the corresponding monitoring times $T_j$, $j \in \mathcal{I}_M$. That is,

$$X_{(ij)} = S_i(T_j) = x_i e^{(r - q_i - \frac{1}{2}\sigma_i^2)\tau_j + \sigma_i B_i(\tau_j)} \tag{10.5}$$

where $\tau_j = (T_j - t)$. Since $t < T_1$, all the $\tau_j$ are positive and represent the times remaining to each of the monitoring instances $T_j$.

---

## 10.3 The M-Binary Payoff

Several preliminaries are needed before we give a formal description of the M-binary payoff. We start by defining four sets of parameters.

1. The *asset parameter set* $\mathbb{A}$ for the M-binary is defined to be the set

$$\mathbb{A} = [x_i, \ q_i, \ \sigma_i, \ \rho_{ii'}]; \quad (i, i') \in \mathcal{I}_N \tag{10.6}$$

where for each asset $A_i$, parameters $(x_i, q_i, \sigma_i)$ are the current asset price, dividend yield, and volatility respectively. The final parameter $\rho_{ii'}$ is the correlation coefficient of the log-returns of assets $A_i$ and $A_{i'}$.

2. The *tenor set* $\mathbb{T}$ for an M-binary is the set containing the monitoring times $T_j, j \in \mathcal{I}_M$ and the expiry date $T$.

$$\mathbb{T} = [T_1, \ T_2, \ \cdots, \ T_M, T]. \tag{10.7}$$

The expiry date $T$ is taken as the date at which the payoff of the M-binary is actually made. We assume that times are ordered such that

$$t < T_1 < T_2 < \cdots < T_M \leq T$$

where $t$ is the current time. Often, the final monitoring date $T_M$ will coincide with the expiry date $T$, but they may also be different.

3. For an $N$-asset, $M$-period M-binary of payoff dimension $n$ and exercise dimension $m$, the *dimension set* for the M-binary is defined by

$$\mathbb{D} = [N, \ M, \ n, \ m]. \tag{10.8}$$

Recall $N \leq n \leq MN$ and $1 \leq m \leq n$.

4. The *payoff parameter set* $\mathbb{P}$ for an M-binary is the vector/matrix set

$$\mathbb{P} = [\boldsymbol{\alpha}, \ \boldsymbol{a}, \ S, \ A] \tag{10.9}$$

where $\boldsymbol{\alpha} \in \mathcal{V}_n$ is the *payoff index vector*; $\boldsymbol{a} \in \mathcal{V}_m$ is the *exercise price vector*; $S \in \mathcal{D}_m$ is the *exercise indicator matrix* with all components equal to $\pm 1$; and $A \in \mathcal{A}_{mn}$ is the *exercise condition matrix*. The first parameter $\boldsymbol{\alpha}$ determines what is actually paid at expiry $T$, while the other three parameters determine the conditions under which exercise at expiry takes place.

**DEFINITION 10.3**    *An* M-*binary on assets* $A_i$ *with tenor set* $\mathbb{T}$, *dimension set* $\mathbb{D}$, *payoff parameter set* $\mathbb{P}$, *and payoff vector* $\boldsymbol{X}$ *is a multi-asset, multi-period binary option with expiry* $T$ *payoff function*

$$\boxed{V(\boldsymbol{X}, T) = \boldsymbol{X}^{\boldsymbol{\alpha}} \, \mathbb{I}_m(S\boldsymbol{X}^A > S\boldsymbol{a})} \tag{10.10}$$

*This* M-*binary might suitably be called a* $[\mathbb{T}, \mathbb{D}, \mathbb{P}]-$ M-*binary.*

**REMARK 10.2**    This definition for the M-binary payoff is a succinct way of writing down what would otherwise be a very complex function. It is only possible to do so because of the special vector and matrix notation, we introduced earlier.                                                                                 ⬚

If exercised, the actual payoff is seen to be $\boldsymbol{X}^{\boldsymbol{\alpha}}$, which is just a product of various powers of the asset prices $X_k$. This payoff, as we shall see, includes the possibility of: a bond or cash payoff of \$1; single asset payoffs; geometric means of all or some of the assets; asset products and quotients; and many others.

The nature of the exercise condition, $S\boldsymbol{X}^A > S\boldsymbol{a}$ is governed by the matrix $A$. This too, allows considerable flexibility in the range of possible exercise scenarios. It includes all the standard exercise conditions such as $X_k \gtrless a_k$, but also comparison conditions like $X_k \gtrless X_\ell$; and also, as we shall see, many other types.

This M-binary payoff is the most general one possible, in the class of power binaries, that has a closed analytical price in the BS framework.

The following examples are provided to illustrate the types of payoffs captured by the M-binary.

### Example 10.4

1. $\boldsymbol{\alpha} = \boldsymbol{0} \in \mathcal{I}_n$, the zero vector, gives a \$1 bond payoff $\boldsymbol{X}^\alpha = 1$.

2. Let $\boldsymbol{1}_k \in \mathcal{V}_n$ denote the vector whose $k$th element is 1 and all others are zero. Then $\boldsymbol{\alpha} = \boldsymbol{1}_k$ gives a payoff equal to the $k$th asset price (i.e., the $k$th element of the payoff vector $\boldsymbol{X}$). In other words, $\boldsymbol{X}^\alpha = X_k$.

3. Let $\boldsymbol{1} \in \mathcal{V}_n$ denote the vector of all ones. Then, $\boldsymbol{\alpha} = \frac{1}{n}\boldsymbol{1}$ gives a payoff which is the discrete GM of the $X_k$, i.e., $\boldsymbol{X}^\alpha = G_n = \sqrt[n]{X_1 X_2 \cdots X_n}$.

4. $\boldsymbol{\alpha} = (\boldsymbol{1}_k - \boldsymbol{1}_\ell)$ gives a payoff equal to the quotient or ratio of $X_k$ and $X_\ell$, i.e., $\boldsymbol{X}^\alpha = X_k/X_\ell$. The choice $\boldsymbol{\alpha} = \boldsymbol{1}_k + \boldsymbol{1}_\ell$ gives the product $X_k X_\ell$.

$\square$

In the third example above, note that the GM may be over prices for the same asset at different times (as for GM Asian options), or for different assets at the same time, or even a combination of both. More complex payoffs can be constructed as portfolios of M-binaries. For example, a derivative that pays the maximum of a set of prices can be expressed as such a portfolio. Indeed, we shall derive this portfolio in Section 10.7 of this chapter.

The next few examples illustrate the types of exercise conditions captured by the M-binary.

### Example 10.5

1. The case $A = I_n \in \mathcal{A}_{nn}$, the $n$ dimensional identity matrix, gives the exercise condition

$$\mathbb{I}_n(\boldsymbol{X}^A > \boldsymbol{a}) = \mathbb{I}_n(\boldsymbol{X} > \boldsymbol{a}) = \mathbb{I}(X_1 > a_1)\mathbb{I}(X_2 > a_2) \cdots \mathbb{I}(X_n > a_n).$$

Any, or all of these conditions can easily be reversed by including an appropriate diagonal, exercise indicator matrix $S$.

2. The choice $A = \frac{1}{n}\mathbf{1} \in \mathcal{A}_{1n} = \mathcal{I}_n$ gives the discrete GM condition:
$\mathbb{I}_n(\boldsymbol{X}^A > \boldsymbol{a}) = \mathbb{I}(G_n > a)$.

3. In the case $n = 2$, $m = 1$, the choice $A = [-1,\, 1] \in \mathcal{A}_{12}$, gives

$$\mathbb{I}_m(\boldsymbol{X}^A > \boldsymbol{a}) = \mathbb{I}(X_1^{-1}X_2 > a) = \mathbb{I}(X_2 > aX_1).$$

4. For a more complex example, the derivative with expiry payoff

$$V_T(\boldsymbol{X}) = \sqrt{X_1 X_2}\ \mathbb{I}(\sqrt{X_1 X_3} > X_2)\mathbb{I}(X_4 < a).$$

is an M-binary with $n = 4$, $m = 2$, payoff vector $\boldsymbol{X} = [X_1, X_2, X_3, X_4]'$ and payoff parameter set $\mathbb{P} = [\boldsymbol{\alpha},\, \boldsymbol{a},\, S,\, A]$ where

$$\boldsymbol{\alpha} = \begin{bmatrix} \frac{1}{2} \\ \frac{1}{2} \\ 0 \\ 0 \end{bmatrix}; \quad \boldsymbol{a} = \begin{bmatrix} 1 \\ a \end{bmatrix}; \quad S = \begin{bmatrix} 1 & 0 \\ 0 & -1 \end{bmatrix}; \quad A = \begin{bmatrix} \frac{1}{2} & -1 & \frac{1}{2} & 0 \\ 0 & 0 & 0 & 1 \end{bmatrix}.$$

$\square$

---

## 10.4   Valuation of the M-Binary

The present value of the M-binary is stated in the following theorem.

**THEOREM 10.1**
*Let assets $A_i$ have parameter set $\mathbb{A}$. Then the pv, for all $t < \min(T_i)$, of the M-binary with payoff function of Equation (10.10), is given by*

$$\boxed{V(\boldsymbol{x}, t) = \boldsymbol{x}^{\boldsymbol{\alpha}}\, e^{\theta(t)}\, \mathcal{N}_m[S\boldsymbol{d}(\boldsymbol{x}, t);\ SCS]} \tag{10.11}$$

*where $\boldsymbol{x} = pv(\boldsymbol{X})$, $\tau = (T - t)$, and*

$$\boxed{\begin{aligned} \theta(t) &= -r\tau + \boldsymbol{\alpha}'\boldsymbol{\mu} + \tfrac{1}{2}\boldsymbol{\alpha}\Gamma\boldsymbol{\alpha}; \\ \boldsymbol{d}(\boldsymbol{x}, t) &= D^{-1}[\log(\boldsymbol{x}^A/\boldsymbol{a}) + A(\boldsymbol{\mu} + \Gamma\boldsymbol{\alpha})] \quad \in \mathcal{V}_m \end{aligned}} \tag{10.12}$$

*where*

$$
\begin{array}{ll}
\boldsymbol{\mu}(t) = (r - q_i - \tfrac{1}{2}\sigma_i^2)\tau_j & \in \mathcal{V}_n \\[4pt]
\Sigma(t) = \mathrm{diag}(\sigma_i\sqrt{\tau_j}) & \in \mathcal{D}_n \\[4pt]
C(t) = D^{-1}(A\Gamma A')D^{-1} & \in \mathcal{C}_m^* \\[6pt]
\Gamma(t) = \Sigma R \Sigma & \in \mathcal{C}_n \\[6pt]
D^2(t) = \mathrm{diag}(A\Gamma A') & \in \mathcal{D}_m \\[6pt]
R(t) = \dfrac{\rho_{ii'}\,\min(\tau_j,\tau_{j'})}{\sqrt{\tau_j \tau_{j'}}} & \in \mathcal{C}_n^* \\[8pt]
\tau_j = T_j - t &
\end{array}
\tag{10.13}
$$

The remarks below are intended to help explain the various terms appearing in this valuation formula.

### REMARK 10.3

1. All the parameters such as $\theta(t)$, $\boldsymbol{\mu}(t)$ etc. are time dependent, but independent of the asset prices in the vector $\boldsymbol{x}$. Only the parameter $\boldsymbol{d}(\boldsymbol{x},t)$ has $\boldsymbol{x}$ dependence.

2. Care must be taken in correctly interpreting the vector $\boldsymbol{x}$. It represents the pv of the payoff vector $\boldsymbol{X}$, and not the pv the original $N$ asset prices $x_i$. The two are quite different, as the following example illustrates.

   Suppose our M-binary has $N = M = 2$ with $n = 3$ and the payoff vector is $\boldsymbol{X} = [X(T_1), Y(T_1), X(T_2)]'$. Let $[x, y]$ denote the pv of the two assets $X, Y$, where $x = X(t), y = Y(t)$. Then the pv of the payoff vector is $\boldsymbol{x} = [x, y, x]$.

3. Observe that $\boldsymbol{\mu} = \mu_k = \mu_{(ij)}$ is a dual-index vector, as is the diagonal of matrix $\Sigma$. Further, $R = R_{kk'} = R_{(ij)(i'j')}$ is a dual-index correlation matrix. Recall that the index pair $k = (ij)$ corresponds to the value $X_k = S_i(T_j)$ of asset $A_i$ at time $T_j$. Similarly, $(k, k')$ corresponds to the two-fold pairs $[(ij), (i'j')]$ for prices $S_i(T_j)$ and $S_{i'}(T_{j'})$.

4. The pv of the M-binary depends on the $m$th order normal distribution function. This may seems surprising at first, since $m$ denotes the exercise dimension, and not the asset or payoff dimension. But a little thought, should convince the reader that the result makes sense. For example, a fixed strike call option on the discrete GM, $G_n$, of any set of asset prices, will have payoff $(G_n - k)^+ = (G_n - k)\mathbb{I}(G_n > k)$. This payoff involves only first order exercise conditions and therefore has a price depending only on the univariate normal.

5. Matrix $\Gamma$ is essentially the covariance matrix for the Gaussian vector $\log \boldsymbol{X}$. Matrix $C$ is the correlation matrix for the transformed Gaussian vector $A \log \boldsymbol{X} = \log \boldsymbol{X}^A$, and the diagonal matrix $D$ is the corresponding normalization matrix.

6. A significant simplification of the M-binary pricing formula (10.11) occurs if the exercise condition matrix $A$ is equal to the $n$ dimensional identity matrix $I_n$. In this case the exercise condition, $\mathbb{I}_n(S\boldsymbol{X} > S\boldsymbol{a})$, is determined by whether asset prices at certain dates are above or below fixed exercise prices, and we get the reductions: $D = \Sigma$ and $C = R$. That is, the correlation matrix $SCS$ in the normal distribution function, will essentially be the same as that related to the asset price vector $\boldsymbol{X}$.

$\Box$

The proof of theorem 10.1 is presented next. It basically depends on four results, all of which we have met previously in this book. They are: the FTAP, the multi-variate representation of $N$ assets following correlated gBm's, the multi-variate GST and linear transformations of Gaussian vectors.

**PROOF**    The valuation formula for the M-binary is the multi-variate version of the FTAP, which may be stated as

$$V(\boldsymbol{x}, t) = e^{-r\tau}\, \mathbb{E}_Q\{V(\boldsymbol{X}, T) | \boldsymbol{X}(t) = \boldsymbol{x}\}.$$

Under multi-variate BS dynamics and the EMM, $Q$, we saw in Equation (10.5), that the price of asset $A_i$ at time $T_j$ was

$$X_{(ij)} = S_i(T_j) = x_i e^{(r - q_i - \frac{1}{2}\sigma_i^2)\tau_j + \sigma_i B_i(\tau_j)}.$$

Now observe that

$$\mathbb{E}_Q\{B_i(\tau_j)\, B_{i'}(\tau_{j'})\} = \rho_{ii'}\left[\frac{\min(\tau_j, \tau_{j'})}{\sqrt{\tau_j \tau_{j'}}}\right] = R_{(ii')(jj')}$$

where $\rho_{ii'}$ is the correlation coefficient (ccf) of the log-asset prices for assets $A_i$ and $A_{i'}$, at the *same* time $\tau_j$; and the term in square brackets is the ccf of the *same* asset $A_i$ at the two distinct times $T_j$ and $T_{j'}$. The expression above, gives the ccf for Brownian motions of different assets *and* different times.

Since the Brownian motion, $B_i(\tau_j)$ can be written equivalently as $\sqrt{\tau_j}\, Z_{(ij)}$ for some Gaussian rv $Z_{(ij)}$, it follows that we can write the log of the payoff vector $X_k = X_{(ij)}$, in vector/matrix notation, as

$$\log \boldsymbol{X} \overset{d}{=} \log \boldsymbol{x} + \boldsymbol{\mu}(t) + \Sigma(t)\, \boldsymbol{Z}$$

where $\boldsymbol{Z} = Z_{(ij)} \sim \mathcal{G}_n(R)$. With this representation for $\log \boldsymbol{X}$, we easily find,

$$\boldsymbol{X}^{\boldsymbol{\alpha}} = \boldsymbol{x}^{\boldsymbol{\alpha}} \exp[\boldsymbol{\alpha}'\boldsymbol{\mu} + \boldsymbol{c}'\boldsymbol{Z}]; \qquad \boldsymbol{c} = \Sigma\,\boldsymbol{\alpha}.$$

Now consider the exercise condition, $S\boldsymbol{X}^A > S\boldsymbol{a}$. This is equivalent to

$$S A[\log \boldsymbol{x} + \boldsymbol{\mu} + \Sigma \boldsymbol{Z}] > S \log \boldsymbol{a}$$
$$\text{or} \quad B\boldsymbol{Z} > -S\boldsymbol{u}; \qquad \boldsymbol{u} = \log(\boldsymbol{x}^A/\boldsymbol{a}) + A\boldsymbol{\mu},$$

where $B = SA\Sigma \in \mathcal{A}_{mn}$. However, we know from Section 3.11 on linear transformations of Gaussian rv's, that $B\boldsymbol{Z}$ will be Gaussian with zero mean and covariance

$$\Lambda = BRB' = SA(\Sigma R\Sigma)A'S = S(A\Gamma(t)\,A')S \quad \in \mathcal{C}_m$$

provided $B$ has rank $m$. Putting all this together leads to,

$$V(\boldsymbol{x},t) = e^{-r\tau}\,\mathbb{E}_Q\{\boldsymbol{X}^{\boldsymbol{\alpha}}\mathbb{I}_m(S\boldsymbol{X}^A > S\boldsymbol{a})\}$$
$$= \boldsymbol{x}^{\boldsymbol{\alpha}}\,e^{-r\tau+\boldsymbol{\alpha}'\boldsymbol{\mu}}\,\mathbb{E}_Q\{e^{\boldsymbol{c}'\boldsymbol{Z}}\,\mathbb{I}_m(B\boldsymbol{Z} > -S\boldsymbol{u})\}.$$

Next use the multi-variate GST described by (3.27) to evaluate the above expectation to get

$$V(\boldsymbol{x},t) = \boldsymbol{x}^{\boldsymbol{\alpha}}\,e^{-r\tau+\boldsymbol{\alpha}'\boldsymbol{\mu}+\frac{1}{2}\boldsymbol{c}'Rc}\,\mathbb{E}_Q\{\mathbb{I}_m(B(\boldsymbol{Z}+Rc) > -S\boldsymbol{u})\}$$
$$= \boldsymbol{x}^{\boldsymbol{\alpha}}\,e^{\theta(t)}\,\mathbb{E}_Q\{\mathbb{I}_m(B\boldsymbol{Z} > -S\boldsymbol{u} - BR\Sigma\boldsymbol{\alpha})\}$$
$$= \boldsymbol{x}^{\boldsymbol{\alpha}}\,e^{\theta(t)}\,\mathbb{E}_Q\{\mathbb{I}_m(B\boldsymbol{Z} < S\boldsymbol{u} + BR\Sigma\boldsymbol{\alpha})\}$$
$$= \boldsymbol{x}^{\boldsymbol{\alpha}}\,e^{\theta(t)}\,\mathbb{E}_Q\{\mathbb{I}_m(B\boldsymbol{Z} > S[\log(\boldsymbol{x}^A/\boldsymbol{a}) + A\boldsymbol{\mu} + A\Gamma\boldsymbol{\alpha}])\}$$
$$= \boldsymbol{x}^{\boldsymbol{\alpha}}\,e^{\theta(t)}\,\mathbb{E}_Q\{\mathbb{I}_m(B\boldsymbol{Z} < S D\boldsymbol{d})\}.$$

We have taken the liberty to replace $\boldsymbol{Z}$ by $-\boldsymbol{Z}$ in the third line, which negates all elements of $\boldsymbol{Z}$, and hence does not change its correlation matrix. The final step in the proof takes note of the fact that if $B\boldsymbol{Z}$ is a zero mean Gaussian rv with covariance matrix $\Lambda$, then $D^{-1}(B\boldsymbol{Z})$ with $D^2 = \text{diag}(\Lambda)$, is a standard Gaussian rv (i.e., all elements have unit variance) with correlation matrix $D^{-1}\Lambda D^{-1}$. Therefore, since $D$ has only positive elements,

$$\mathbb{E}_Q\{\mathbb{I}_m(B\boldsymbol{Z} < S D\boldsymbol{d})\} = \mathbb{E}_Q\{\mathbb{I}_m(D^{-1}B\boldsymbol{Z} < S\boldsymbol{d})\}$$

and the rv, $\boldsymbol{Y} = D^{-1}B\boldsymbol{Z} \sim \mathcal{G}_m(K)$ , with

$$K = D^{-1}\Lambda D^{-1} = D^{-1}S(A\Gamma A')SD^{-1} = SD^{-1}(A\Gamma A')D^{-1}S = SCS.$$

Note that, $D^{-1}S = SD^{-1}$ as both matrices are diagonal. Thus,

$$V(\boldsymbol{x},t) = \boldsymbol{x}^{\boldsymbol{\alpha}}\,e^{\theta(t)}\,\mathbb{E}_Q\{\mathbb{I}_m(\boldsymbol{Y} < S\boldsymbol{d})\} = \boldsymbol{x}^{\boldsymbol{\alpha}}\,e^{\theta(t)}\,\mathcal{N}(S\boldsymbol{d};\,SCS).$$

This completes the proof. ⬚

**REMARK 10.4**    The proof above requires that the matrix $B = SA\Sigma$ should be of rank $m$. This will certainly be the case if $A$ has rank $m$, because $S$ and $\Sigma$ are non-singular diagonal matrices. Since, $A \in \mathcal{A}_{mn}$, it will have rank $m \leq n$ iff all its rows are linearly independent. What does this mean for the exercise condition $\boldsymbol{X}^A > \boldsymbol{a}$ ?

Let $(\boldsymbol{\alpha}, \boldsymbol{\beta}, \boldsymbol{\gamma})$ be three linearly dependent rows of matrix $A$, corresponding to exercise conditions

$$
\begin{aligned}
X &= x_1^{\alpha_1}\, x_2^{\alpha_2} \cdots x_n^{\alpha_n} > a \\
Y &= x_1^{\beta_1}\, x_2^{\beta_2} \cdots x_n^{\beta_n} > b \\
Z &= x_1^{\gamma_1}\, x_2^{\gamma_2} \cdots x_n^{\gamma_n} \gtrless c.
\end{aligned}
$$

Suppose $\boldsymbol{\gamma} = A\boldsymbol{\alpha} + B\boldsymbol{\beta}$ for some constants, $(A, B)$. Then this implies $Z = X^A \cdot Y^B > a^A b^B = d$, say. But the third condition is $Z \gtrless c$. We see that linear dependence in the rows of $A$, leads to conditions like: $\mathbb{I}(Z > d)\mathbb{I}(Z > c)$ or $\mathbb{I}(Z > d)\mathbb{I}(Z < c)$. But, these conditions are always reducible: the first to $\mathbb{I}(Z > \max(c, d))$, and the second to $[\mathbb{I}(Z > d) - \mathbb{I}(Z > c)]\mathbb{I}(c > d)$.

Thus, in order to eliminate such reducible inequalities, we need only assume that the exercise condition matrix $A$ has no linearly dependent rows. In that case, it will have full rank $m$, and so will matrix $B = SA\Sigma$. ⬚

---

## 10.5    Previous Results Revisited

In this section we revisit some of our previous results which can be obtained as direct applications of the M-binary formula.

**Two-Asset Rainbow Binaries**
We considered four such binaries in Section 6.1. Here we shall price the third of these as an example of the power of the M-binary formula. Specifically, we seek the pv of the 2-asset, 1-period binary option with $T$ payoff

$$
V_3(x, y, T) = x^p y^q\, \mathbb{I}(x > h)\mathbb{I}(x < y).
$$

This M-binary has dimension set $\mathbb{D} = [2, 1, 2, 2]$ and payoff parameter set

$$
\boldsymbol{\alpha} = \begin{bmatrix} p \\ q \end{bmatrix}; \quad
A = \begin{bmatrix} 1 & 0 \\ 1 & -1 \end{bmatrix}; \quad
\boldsymbol{a} = \begin{bmatrix} h \\ 1 \end{bmatrix}; \quad
S = \begin{bmatrix} 1 & 0 \\ 0 & -1 \end{bmatrix}.
$$

The following calculations then follow:

$$\mu = \begin{bmatrix} r - \frac{1}{2}\sigma_1^2 \\ r - \frac{1}{2}\sigma_2^2 \end{bmatrix}\tau; \quad \Sigma = \begin{bmatrix} \sigma_1 & 0 \\ 0 & \sigma_2 \end{bmatrix}\tau; \quad \Gamma = \begin{bmatrix} \sigma_1^2 & \rho\sigma_1\sigma_2 \\ \rho\sigma_1\sigma_2 & \sigma_2^2 \end{bmatrix}\tau$$

$$A\Gamma A' = \begin{bmatrix} \sigma_1^2 & \sigma_1^2 - \rho\sigma_1\sigma_2 \\ \sigma_1^2 - \rho\sigma_1\sigma_2 & \sigma^2 \end{bmatrix}\tau; \quad D = \begin{bmatrix} \sigma_1 & 0 \\ 0 & \sigma \end{bmatrix}\sqrt{\tau}$$

$$\sigma^2 = \sigma_1^2 + \sigma_2^2 - 2\rho\sigma_1\sigma_2; \quad C = \begin{bmatrix} 1 & \rho_1 \\ \rho_1 & 1 \end{bmatrix}; \quad \rho_1 = \frac{\sigma_1 - \rho\sigma_2}{\sigma}$$

$$X = \begin{bmatrix} X_T \\ Y_T \end{bmatrix}; \quad x = \begin{bmatrix} x \\ y \end{bmatrix}; \quad \log(x^A/a) = \begin{bmatrix} \log(x/h) \\ \log(x/y) \end{bmatrix}$$

$$d = \begin{bmatrix} \{\log(x/h) + (r - \frac{1}{2}\sigma_1^2 + p\sigma_1^2 + q\rho\sigma_2\sigma_2)\tau\}/\sigma_1\sqrt{\tau} \\ \{\log(x/y) + (\frac{1}{2}(\sigma_1^2 - \sigma_2^2) + p\sigma_1^2 - q\sigma_2^2 + (q-p)\rho\sigma_1\sigma_2)\tau\}/\sigma\sqrt{\tau} \end{bmatrix}$$

$$\alpha'\Gamma\alpha = p^2\sigma_1^2 + q^2\sigma_2^2 + 2pq\rho\sigma_1\sigma_2$$

$$\theta = [-r + p(r - \frac{1}{2}\sigma_1^2) + q(r - \frac{1}{2}\sigma_2^2) + \frac{1}{2}(p^2\sigma_1^2 + q^2\sigma_2^2 + 2pq\rho\sigma_1\sigma_2)]\tau.$$

Writing $d = [d_1, d]'$, the M-binary valuation formula (10.11) then yields the price of the binary as

$$V_3(x, y, t) = x^p y^q\, e^{\theta(t)}\, \mathcal{N}(d_1, -d; -\rho_1)$$

which naturally agrees with Equation (6.4) derived earlier.

## A Multi-Asset, Multi-Period ESO

Consider the performance measured executive stock option of Section 6.6. Here we shall make a slight extension, where the performance criterion is monitored at time $T_1 < T = T_2$ rather than at the expiry date of the option. The payoff of this ESO is therefore

$$V(X, T) = (X_2 - k)^+\, \mathbb{I}(X_1 > Y_1)$$

which corresponds to a call option on asset $X$, of strike price $k$, paid at expiry $T = T_2$, provided the performance criterion, $X_1 > Y_1$ is satisfied at time $T_1 < T$, and $Y$ is a benchmark index.

This payoff can be decomposed into a portfolio of 2-asset, 2-period M-binaries. In particular, $V(X, T) = V_1(X, T) - kV_2(X, T)$ with

$$V_1(X, T) = X_2\mathbb{I}(X_1 > Y_1)\mathbb{I}(X_2 > k) \quad \text{and} \quad V_2(X, T) = \mathbb{I}(X_1 > Y_1)\mathbb{I}(X_2 > k).$$

Thus, $V_1$ is an asset M-binary and $V_2$ is a bond M-binary. For these M-binaries, with dimension set $\mathbb{D} = [2, 2, 3, 2]$ we compute the following variables and parameters.

For $V_1(X, T) = X_2\mathbb{I}(X_1 > Y_1)\mathbb{I}(X_2 > k)$

$$X = \begin{bmatrix} X_1 \\ Y_1 \\ X_2 \end{bmatrix}; \quad x = \begin{bmatrix} x \\ y \\ x \end{bmatrix}; \quad \alpha = \begin{bmatrix} 0 \\ 0 \\ 1 \end{bmatrix}; \quad a = \begin{bmatrix} 1 \\ k \end{bmatrix}$$

and

$$S = \begin{bmatrix} 1 & 0 \\ 0 & 1 \end{bmatrix}; \quad A = \begin{bmatrix} 1 & -1 & 0 \\ 0 & 0 & 1 \end{bmatrix}.$$

Furthermore, with $\tau_1 = T_1 - t$ and $\tau_2 = T_2 - t$, we obtain:

$$\Sigma = \begin{bmatrix} \sigma_1\sqrt{\tau_1} & 0 & 0 \\ 0 & \sigma_2\sqrt{\tau_1} & 0 \\ 0 & 0 & \sigma_1\sqrt{\tau_2} \end{bmatrix}$$

$$R = \begin{bmatrix} 1 & \rho & \tau_{12} \\ \rho & 1 & \rho\tau_{12} \\ \tau_{12} & \rho\tau_{12} & 1 \end{bmatrix}; \quad \tau_{12} = \sqrt{\tau_1/\tau_2}$$

$$\Gamma = \Sigma R \Sigma = \begin{bmatrix} \sigma_1^2\tau_1 & \rho\sigma_1\sigma_2\tau_1 & \sigma_1^2\tau_1 \\ \rho\sigma_1\sigma_2\tau_1 & \sigma_2^2\tau_1 & \rho\sigma_1\sigma_2\tau_1 \\ \sigma_1^2\tau_1 & \rho\sigma_1\sigma_2\tau_1 & \sigma_1^2\tau_2 \end{bmatrix}$$

$$\boldsymbol{\mu} = \begin{bmatrix} (r - \tfrac{1}{2}\sigma_1^2)\tau_1 \\ (r - \tfrac{1}{2}\sigma_2^2)\tau_1 \\ (r - \tfrac{1}{2}\sigma_1^2)\tau_2 \end{bmatrix}$$

$$\theta = -r\tau_2 + (r - \tfrac{1}{2}\sigma_1^2)\tau_2 + \tfrac{1}{2}\sigma_1^2\tau_2 = 0.$$

Next we compute:

$$A\Gamma = \begin{bmatrix} (\sigma_1^2 - \rho\sigma_1\sigma_2)\tau_1 & (\rho\sigma_1\sigma_2 - \sigma_2^2)\tau_1 & (\sigma_1^2 - \rho\sigma_1\sigma_2)\tau_1 \\ \sigma_1^2\tau_1 & \rho\sigma_1\sigma_2\tau_1 & \sigma_1^2\tau_2 \end{bmatrix}$$

$$A\Gamma A' = \begin{bmatrix} \sigma^2\tau_1 & (\sigma_1^2 - \rho\sigma_1\sigma_2)\tau_1 \\ (\sigma_1^2 - \rho\sigma_1\sigma_2)\tau_1 & \sigma_1^2\tau_2 \end{bmatrix}$$

where $\sigma^2 = \sigma_1^2 + \sigma_2^2 - 2\rho\sigma_1\sigma_2$. Then,

$$D = \sqrt{\text{diag}(A\Gamma A')} = \begin{bmatrix} \sigma\sqrt{\tau_1} & 0 \\ 0 & \sigma_1\sqrt{\tau_2} \end{bmatrix}$$

$$C = \begin{pmatrix} 1 & \rho^* \\ \rho^* & 1 \end{pmatrix}; \quad \rho^* = \left( \frac{\sigma_1 - \rho\sigma_2}{\sigma} \right)\sqrt{\tau_1/\tau_2}$$

$$\boldsymbol{x}^{\alpha} = x^0 y^0 x^1 = x$$

$$A\boldsymbol{\mu} = \begin{bmatrix} \tfrac{1}{2}(\sigma_2^2 - \sigma_1^2)\tau_1 \\ (r - \tfrac{1}{2}\sigma_1^2)\tau_2 \end{bmatrix}; \quad A\Gamma\boldsymbol{\alpha} = \begin{bmatrix} (\sigma_1^2 - \rho\sigma_1\sigma_2)\tau_1 \\ \sigma_1^2\tau_2 \end{bmatrix}.$$

Next we calculate $\boldsymbol{d} = \begin{bmatrix} d_1 \\ d_2 \end{bmatrix}$ as

$$\boldsymbol{d} = \begin{bmatrix} \sigma\sqrt{\tau_1} & 0 \\ 0 & \sigma_1\sqrt{\tau_2} \end{bmatrix}^{-1} \left( \log\begin{bmatrix} x/y \\ x/k \end{bmatrix} + \begin{bmatrix} \tfrac{1}{2}(\sigma_2^2 - \sigma_1^2)\tau_1 \\ (r - \tfrac{1}{2}\sigma_1^2)\tau_2 \end{bmatrix} + \begin{bmatrix} (\sigma_1^2 - \rho\sigma_1\sigma_2)\tau_1 \\ \sigma_1^2\tau_2 \end{bmatrix} \right)$$

$$= \begin{bmatrix} \{\log(x/y) + \tfrac{1}{2}\sigma^2\tau_1\}/\sigma\sqrt{\tau_1} \\ \{\log(x/k) + (r + \tfrac{1}{2}\sigma_1^2)\tau_2\}/\sigma_1\sqrt{\tau_2} \end{bmatrix}$$

Hence, finally, we can write $V_1(x, y, t) = x \mathcal{N}(d_1, d_2; \rho^*)$.

For $V_2(\boldsymbol{X}, T) = \mathbb{I}(X_1 > Y_1)\mathbb{I}(X_2 > k)$

The only parameter change compared to $V_1$ is $\boldsymbol{\alpha} = [0, 0, 0]'$. This leads to the following expressions:

$$\theta = -r\tau_2$$
$$\boldsymbol{x}^{\alpha} = x^0 y^0 x^0 = 1$$
$$\boldsymbol{d} = \begin{bmatrix} \{\log(x/y) + \frac{1}{2}(\sigma_2^2 - \sigma_1^2)\tau_1\}/\sigma\sqrt{\tau_1} \\ \{\log(x/k) + (r - \frac{1}{2}\sigma_1^2)\tau_2\}/\sigma_1\sqrt{\tau_2} \end{bmatrix} = \begin{bmatrix} d_1' \\ d_2' \end{bmatrix}; \qquad \text{(say)}$$

Note that $d_1' = d_1 - \rho^*\sigma_1\sqrt{\tau_2}$ and $d_2' = d_2 - \sigma_1\sqrt{\tau_2}$. This yields the following expression $V_2(x, y, t) = e^{-r\tau_2}\mathcal{N}(d_1', d_2'; \rho^*)$, for the bond M-binary. Hence putting the two expressions together we get the ESO present value as

$$\boxed{V(x, y, t) = x \mathcal{N}(d_1, d_2; \rho^*) - ke^{-r\tau_2}\mathcal{N}(d_1', d_2'; \rho^*)} \qquad (10.14)$$

where

$$\begin{cases} d_1 = [\log(x/y) + \frac{1}{2}\sigma^2\tau_1]/\sigma\sqrt{\tau_1} \\ d_1' = d_1 - \rho^*\sigma_1\sqrt{\tau_2} \\ d_2 = [\log(x/k) + (r + \frac{1}{2}\sigma_1^2)\tau_2]/\sigma_1\sqrt{\tau_2} \\ d_2' = d_2 - \sigma_1\sqrt{\tau_2} \\ \rho^* = (\frac{\sigma_1 - \rho\sigma_2}{\sigma})\sqrt{\tau_1/\tau_2} \\ \sigma^2 = \sigma_1^2 + \sigma_2^2 - 2\rho\sigma_1\sigma_2. \end{cases}$$

## 10.6  Multi-Asset, 1-Period Asset and Bond Binaries

These are the multi-asset extensions of simple asset and bond binaries consider in Section 4.2. The payoffs of these binaries at time $T$ are given by

$$A_k(\boldsymbol{X}, T; S, \boldsymbol{a}) = X_k\mathbb{I}(S\boldsymbol{X} > S\boldsymbol{a}); \qquad B(\boldsymbol{X}, T; S, \boldsymbol{a}) = \mathbb{I}(S\boldsymbol{X} > S\boldsymbol{a}).$$

Here $\boldsymbol{X}$ denotes $n$ asset prices evaluated at time $T$. The asset binary $A_k$ pays one unit of asset $k$ at time $T$, while the bond binary, as usual, pays one dollar (or one unit of cash). The exercise condition, $S\boldsymbol{X} > S\boldsymbol{A}$, shows that the payoff is made iff the asset prices at time $T$ are above or below (depending on $S$) the given exercise prices contained in the vector $\boldsymbol{a}$.

Hence for both these binaries, we have the common parameters (see final comment in Remark 10.3), $A = I_n$; $C = R$. Further, for the asset binary we have $\boldsymbol{\alpha} = \boldsymbol{1}_k$ and for the bond binary $\boldsymbol{\alpha} = \boldsymbol{0}$. These lead to the following formulae, for all $t < T$,

$$\boxed{A_k(\boldsymbol{x}, t; S, \boldsymbol{a}) = x_k e^{-q_k\tau}\mathcal{N}_n(S\boldsymbol{d}; SRS)} \qquad (10.15)$$

for the asset binary, and

$$\boxed{B(\boldsymbol{x}, t; S, \boldsymbol{a}) = e^{-r\tau} \mathcal{N}_n(S\boldsymbol{d}'; SRS)} \tag{10.16}$$

for the bond binary, where $\tau = (T = t)$, $R = \rho_{ij}$ is the correlation matrix, and

$$d_i' = \frac{\log(x_i/a_i) + (r - q_i - \frac{1}{2}\sigma_i^2)\tau}{\sigma_i \sqrt{\tau}}; \qquad d_i = d_i' + \rho_{ik}\sigma_k\sqrt{\tau}.$$

---

## 10.7 Quality Options

We define a *quality option* as a multi-variate option whose payoff depends on the maximum or minimum asset price within a basket of assets. Calls and puts on the minimum and maximum of two assets were considered in Section 6.3. Here the extension is to more than two assets.

The following lemma is integral to the pricing of such best and worst derivatives.

**LEMMA 10.1**
*Define $A \in \mathcal{A}_{nn}$ and $a \in \mathcal{V}_n$ by their elements,*

$$A_{ij} = \begin{cases} 1 & \text{if } j = p; \\ -1 & \text{if } i = j \neq p; \\ 0 & \text{otherwise}; \end{cases} \qquad a_i = \begin{cases} k & \text{if } i = p \\ 1 & \text{if } i \neq p \end{cases} \tag{10.17}$$

*where $p \in \mathcal{I}_n$ is a fixed integer in the range $1 \leq p \leq n$ and $k$ is a positive constant. Then the exercise condition $\mathbb{I}_n(s\boldsymbol{X}^A > s\boldsymbol{a})$ is satisfied for $s = 1$ if $\max(\boldsymbol{X}, k) = X_p$, and for $s = -1$ if $\min(\boldsymbol{X}, k) = X_p$.*

**PROOF** After taking logs, we find for row $i$ of the condition corresponding to $s = +1$, $A\log \boldsymbol{X} > \log \boldsymbol{a}$

$$\begin{aligned} \text{for } i \neq p; & \qquad -\log X_i + \log X_p > \quad 0 \\ \text{for } i = p; & \qquad \log X_p > \quad \log k. \end{aligned}$$

Thus $X_p > X_i$ and $X_p > k$ implying $X_p = \max(\boldsymbol{X}, k)$. A similar argument applies for the case $A\log \boldsymbol{X} < \log(a)$ when $s = -1$. □

**DEFINITION 10.4** *We say that a binary option is a $W_p^s$–binary if it pays at expiry $T$, one unit of asset $p$, provided $X_p$: (a) is the maximum of a* *basket of assets and a fixed amount of cash $k$ if $(s = +)$; or (b) the minimum of a basket of assets and a fixed amount of cash $k$ if $(s = -)$.*

According to Lemma 10.1, the payoff of this binary can be written in the form

$$W_p^s(\boldsymbol{X}, T; k) = X_p \, \mathbb{I}_n(s\boldsymbol{X}^A > sk\boldsymbol{1}_p) \tag{10.18}$$

where matrix $A \in \mathcal{A}_{nn}$ is defined in (10.17). To save on notational complexity, we have dropped the explicit dependence of $A$ on the integer $p \in \mathcal{I}_n$.

The next theorem determines the price of this M-binary.

## THEOREM 10.2

The present value, for all $t < T$, of the $W_p^s$−binary is

$$\boxed{W_p^s(\boldsymbol{x}, t; k) = x_p e^{-q_p \tau} \, \mathcal{N}(s\boldsymbol{d}; C)} \tag{10.19}$$

where

$$d_i = \begin{cases} \dfrac{\log(x_p/k) + (r - q_p + \frac{1}{2}\sigma_p^2)\tau}{\sigma_p \sqrt{\tau}} & \text{if } i = p \\[3mm] \dfrac{\log(x_p/x_i) - (q_p - \frac{1}{2}\sigma_{ip}^2)\tau}{\sigma_{ip}\sqrt{\tau}} & \text{if } i \neq p \end{cases}$$

$$C_{ij} = \begin{cases} 1 & \text{if } i = j \\[2mm] \dfrac{\sigma_p - \rho_{ip}\sigma_i}{\sigma_{ip}} & \text{if } i \neq p = j \\[2mm] \dfrac{\sigma_{ip}^2 + \sigma_{jp}^2 - \sigma_{ij}^2}{2\sigma_{ip}\sigma_{jp}} & \text{if } i \neq p \neq j \\[2mm] C_{ji} & \text{otherwise} \end{cases}$$

$$\sigma_{ij}^2 = \sigma_i^2 + \sigma_j^2 - 2\rho_{ij}\sigma_i\sigma_j \qquad \text{for all } (i, j) \in \mathcal{I}_n.$$

**REMARK 10.5**   The elements $C_{ij}$ of the correlation matrix $C$ agree with those of Johnson [40] and Rich and Chance [62] who considered pricing of multi-asset options in the BS framework. ⬜

**PROOF**   Since the order of elements in the payoff matrix $\boldsymbol{X}$ is irrelevant, we can either take $p = 1$ or move the $p$th element $X_p$ to the first position. In either case, we can then partition vectors and matrices accordingly: quantities relating to $X_p$, and those relating to $X_i$, with $i \neq p$. Hence, take

$$A = \begin{bmatrix} 1 & \boldsymbol{0}' \\ \boldsymbol{1} & -I \end{bmatrix}; \qquad \boldsymbol{a} = \begin{bmatrix} k \\ \boldsymbol{1} \end{bmatrix}; \qquad \Gamma = \tau \begin{bmatrix} \sigma_p^2 & \boldsymbol{g}' \\ \boldsymbol{g} & G \end{bmatrix}$$

where all the terms have appropriate dimensions, $g_i = \rho_{ip}\sigma_i\sigma_p$, and $G$ contains terms like $\rho_{ij}\sigma_i\sigma_j$ with $(i, j) \neq p$. This leads to

$$A\Gamma A' = \tau \begin{bmatrix} \sigma_p^2 & \sigma_p^2 \boldsymbol{1}' - \boldsymbol{g}' \\ \sigma_p^2 \boldsymbol{1} - \boldsymbol{g} & H \end{bmatrix}$$

where $H = \sigma_p^2 11' - g1' - 1g' + G$. Thus, $\text{diag}(H) = \sigma_{ip}$, with $i \neq p$, and

$$D = \sqrt{\text{diag}(A\Gamma A')} = \text{diag}(\sigma_p, \sigma_{2p}, \sigma_{3p}, \cdots, \sigma_{np}).$$

This yields the expression for $C = D^{-1}(A\Gamma A')D^{-1}$. Using,

$$A(\boldsymbol{\mu} + \Gamma\alpha) = \begin{bmatrix} r + \frac{1}{2}\sigma_p^2 \\ \frac{1}{2}\sigma_{ip}^2 \end{bmatrix}; \qquad (i \neq p)$$

with $\boldsymbol{\alpha} = [1, \mathbf{0}]'$, then leads to the results stated in the theorem. $\qquad \square$

### Best and Worst Options

The generalization of best and worst options to more than two assets appears in Rich and Chance [62]. While their martingale method appears extremely complicated, our M-binary approach gives the result in a much simpler fashion. For the best-option in a basket of $n$ assets and cash $k$, we calculate as follows:

$$W_{\max}(\boldsymbol{X}, T) = \max(\boldsymbol{X}, k)$$
$$= \sum_{p=1}^{n} W_p^+(\boldsymbol{X}, T; k) + k\mathbb{I}_n(\boldsymbol{X} < \boldsymbol{k}); \qquad (\boldsymbol{k} = k\mathbf{1}_n).$$

The present value of the option is therefore,

$$\boxed{W_{\max}(\boldsymbol{x}, t; k) = \sum_{p=1}^{n} W_p^+(\boldsymbol{x}, t; k) + kB(\boldsymbol{x}, t; -I_n, \boldsymbol{k})} \qquad (10.20)$$

where $W_p^+(\boldsymbol{x}, t; k)$ is given by Equation 10.18) with $s = +$ and $B(\boldsymbol{x}, t; I_n, \boldsymbol{k})$ is the multi-asset bond binary of Equation (10.16).

Similarly, for the worst-option in the basket, with payoff $W_{\min}(\boldsymbol{X}, T; k) = \min(\boldsymbol{X}, k)$, the present value is

$$\boxed{W_{\min}(\boldsymbol{x}, t; k) = \sum_{p=1}^{n} W_p^-(\boldsymbol{x}, t; k) + kB(\boldsymbol{x}, t; I_n, \boldsymbol{k})} \qquad (10.21)$$

### Calls and Puts on the Max/Min

Johnson [40] presents the generalization of calls and puts on the maximum and minimum of two assets (the Stulz formulae [73] derived in Section 6.3) to the multi-asset case. Our analysis, using previous results of this section, are immediate.

For a call option, of strike price $k$, on the maximum of $n$ assets, we have the payoff:

$$C_{\max}(\boldsymbol{X}, T; k) = (\max \boldsymbol{X} - k)^+ = \max(\boldsymbol{X}, k) - k.$$

Hence, the present value, for all $t < T$, is

$$\boxed{C_{\max}(\boldsymbol{x}, t; k) = W_{\max}(\boldsymbol{x}, t; k) - ke^{-r\tau}}. \tag{10.22}$$

For a put option, of strike price $k$, on the minimum of $n$ assets, we have the payoff:

$$P_{\min}(\boldsymbol{X}, T; k) = (k - \min \boldsymbol{X})^+ = k - \min(\boldsymbol{X}, k).$$

Here the present value, for all $t < T$, is

$$\boxed{P_{\min}(\boldsymbol{x}, t; k) = ke^{-r\tau} - W_{\min}(\boldsymbol{x}, t; k)} \tag{10.23}$$

The calculations of calls on the minimum and puts on the maximum are deferred to the Exercise Problems.

---

## 10.8 Compound Exchange Option

Margrabe's formula for pricing a simple exchange option was derived in Section 6.2. The compound exchange option is an interesting extension of this derivative, and was considered in Carr (1988) [12]. The valuation of this option is a fine example of the power of the M-binary formula.

Like other compound options considered in Section 5.5, the compound exchange option is a dual-expiry option with payoff depending on two future dates $T_1 < T_2$. At time $T_1$, the holder has the right to receive an exchange option maturing at time $T_2$, for $k$ units of an asset $Y$. The exchange option gives its holder the right, at time $T_2$, to swap one unit of asset $X$ for one unit of asset $Y$. The compound exchange option is therefore also a sequential exchange option, with asset $Y$ being the delivery asset for both exchanges. Carr gives several examples of situations where such compound exchange options might find practical application.

The payoff of the compound exchange option at time $T_1$ is given by:

$$V(x_1, y_1, T_1) = [V_{ex}(x_1, y_1, \tau) - ky_1]^+$$

where $V_{ex}(x, y, \tau)$ denotes Margrabe's Formula (6.8) for the value of an exchange option with time $\tau = (T_2 - T_1)$ remaining to expiry. That is,

$$V(x, y, T_1) = x\mathcal{N}(d) - y\mathcal{N}(d')$$

$$(d, d') = \frac{\log(x/y) \pm \frac{1}{2}\sigma^2\tau}{\sigma\sqrt{\tau}}; \quad \sigma^2 = \sigma_1^2 + \sigma_2^2 - 2\rho\sigma_1\sigma_2.$$

Let $z = c$ solve the equation,

$$z\mathcal{N}(d) - \mathcal{N}(d') = k; \qquad z = x/y. \tag{10.24}$$

This transcendental equation for $z = x/y$, will have a unique solution since the lhs is monotonic increasing in $z$. With $c$ so defined, the time $T_1$ payoff can now be written as,

$$V(x_1, y_1, T_1) = [V_{ex}(x_1, y_1, \tau) - ky_1]\,\mathbb{I}(x_1 > cy_1).$$

The corresponding time $T_2$ payoff will then be,

$$V(x_2, y_2, T_2) = [(x_2 - y_2)^+ - ky_1]\,\mathbb{I}(x_1 > cy_1). \tag{10.25}$$

This payoff is seen to be a portfolio of three M-binaries: $V = V_1 - V_2 - kV_3$, with

$$V_1 = x_2\mathbb{I}(x_1 > cy_1)\mathbb{I}(x_2 > y_2); \quad V_2 = y_2\mathbb{I}(x_1 > cy_1)\mathbb{I}(x_2 > y_2); \quad V_3 = y_1\mathbb{I}(x_1 > cy_1).$$

We consider each of these in turn.

### $V_1(\boldsymbol{X}, T_2) = x_2\mathbb{I}(x_1 > cy_1)\mathbb{I}(x_2 > y_2)$

This is a 2-asset, 2-period M-binary with dimension set $\mathbb{D} = [2, 2, 4, 2]$. Define,

$$\boldsymbol{X} = \begin{bmatrix} x_1 \\ y_1 \\ x_2 \\ y_2 \end{bmatrix}; \quad \boldsymbol{x} = \begin{bmatrix} x \\ y \\ x \\ y \end{bmatrix}; \quad \boldsymbol{\alpha} = \begin{bmatrix} 0 \\ 0 \\ 1 \\ 0 \end{bmatrix}; \quad A = \begin{bmatrix} 1 & -1 & 0 & 0 \\ 0 & 0 & 1 & -1 \end{bmatrix}; \quad \boldsymbol{a} = \begin{bmatrix} c \\ 1 \end{bmatrix}.$$

Since the exercise condition involves only "greater-than" constraints, we may ignore the exercise condition matrix $S$. Let $\tau_i = (T_i - t)$ and $s = \sqrt{\tau_1/\tau_2}$. Then $s$ denotes the ccf for the same assets at the two times $T_1$ and $T_2$. The correlation matrix $R$ for payoff vector $\boldsymbol{X}$, and corresponding diagonal volatility matrix $\Sigma$, will be

$$R = \begin{bmatrix} 1 & \rho & s & \rho s \\ \rho & 1 & \rho s & s \\ s & \rho s & 1 & \rho \\ \rho s & s & \rho & 1 \end{bmatrix}; \quad \Sigma = \begin{bmatrix} \sigma_1\sqrt{\tau_1} & & & \\ & \sigma_2\sqrt{\tau_1} & & \\ & & \sigma_1\sqrt{\tau_2} & \\ & & & \sigma_2\sqrt{\tau_2} \end{bmatrix}.$$

This leads, after some elementary matrix operations, to

$$\Gamma = \Sigma R\Sigma = \begin{bmatrix} \sigma_1^2\tau_1 & \rho\sigma_1\sigma_2\tau_1 & \sigma_1^2\tau_1 & \rho\sigma_1\sigma_2\tau_1 \\ \rho\sigma_1\sigma_2\tau_1 & \sigma_2^2\tau_1 & \rho\sigma_1\sigma_2\tau_1 & \sigma_2^2\tau_1 \\ \sigma_1^2\tau_1 & \rho\sigma_1\sigma_2\tau_1 & \sigma_1^2\tau_2 & \rho\sigma_1\sigma_2\tau_2 \\ \rho\sigma_1\sigma_2\tau_1 & \sigma_2^2\tau_1 & \rho\sigma_1\sigma_2\tau_2 & \sigma_2^2\tau_2 \end{bmatrix}; \quad A\Gamma A' = \sigma^2\begin{bmatrix} \tau_1 & \tau_1 \\ \tau_1 & \tau_2 \end{bmatrix}$$

where $\sigma^2 = \sigma_1^2 + \sigma_2^2 - 2\rho\sigma_1\sigma_2$. Next, we compute

$$D = \sigma \begin{bmatrix} \sqrt{\tau_1} & 0 \\ 0 & \sqrt{\tau_2} \end{bmatrix}; \quad C = \begin{bmatrix} 1 & \sqrt{\tau_1/\tau_2} \\ \sqrt{\tau_1/\tau_2} & 1 \end{bmatrix}; \quad \boldsymbol{\mu} = \begin{bmatrix} (r - \frac{1}{2}\sigma_1^2)\tau_1 \\ (r - \frac{1}{2}\sigma_2^2)\tau_1 \\ (r - \frac{1}{2}\sigma_1^2)\tau_2 \\ (r - \frac{1}{2}\sigma_2^2)\tau_2 \end{bmatrix}$$

and

$$\boldsymbol{x}^{\alpha} = x; \quad \boldsymbol{x}^A = \begin{bmatrix} x/y \\ x/y \end{bmatrix}; \quad A(\boldsymbol{\mu} + \Gamma\boldsymbol{\alpha}) = \frac{1}{2}\sigma^2 \begin{bmatrix} \tau_1 \\ \tau_2 \end{bmatrix}; \quad \boldsymbol{d} = \begin{bmatrix} d_1 \\ d_2 \end{bmatrix}$$

where

$$d_1(x, y, \tau_1) = \frac{\log(x/cy) + \frac{1}{2}\sigma^2\tau_1}{\sigma\sqrt{\tau_1}}; \quad d_2(x, y, \tau_2) = \frac{\log(x/y) + \frac{1}{2}\sigma^2\tau_2}{\sigma\sqrt{\tau_2}}.$$

Finally, with $\boldsymbol{\alpha}'\boldsymbol{\mu} = (r - \frac{1}{2}\sigma_1^2)\tau_2$ and $\boldsymbol{\alpha}'\Gamma\boldsymbol{\alpha} = \frac{1}{2}\sigma_1^2\tau_2$, we find

$$\theta = -r\tau_2 + (r - \frac{1}{2}\sigma_1^2)\tau_2 + \frac{1}{2}\sigma_1^2\tau_2 = 0$$

and the M-binary formula (10.11) gives

$$V_1(x, y, t) = x\mathcal{N}(d_1, d_2; s); \quad s = \sqrt{\tau_1/\tau_2} \tag{10.26}$$

in terms of a bivariate normal.

$$V_2(\boldsymbol{X}, T_2) = y_2\mathbb{I}(x_1 > cy_1)\mathbb{I}(x_2 > y_2)$$

Here the only parameter change is to $\boldsymbol{\alpha} = [0, 0, 0, 1]'$, leading to changed values for:

$$\boldsymbol{x}^{\alpha} = y; \quad A(\boldsymbol{\mu} + \Gamma\boldsymbol{\alpha}) = -\frac{1}{2}\sigma^2 \begin{bmatrix} \tau_1 \\ \tau_2 \end{bmatrix}; \quad \boldsymbol{d} = \begin{bmatrix} d_1' \\ d_2' \end{bmatrix}$$

where

$$d_1'(x, y, \tau_1) = \frac{\log(x/cy) - \frac{1}{2}\sigma^2\tau_1}{\sigma\sqrt{\tau_1}}; \quad d_2'(x, y, \tau_2) = \frac{\log(x/y) - \frac{1}{2}\sigma^2\tau_2}{\sigma\sqrt{\tau_2}}.$$

Hence, as $\theta(t)$ is still zero, we obtain from the M-binary formula

$$V_2(x, y, t) = y\mathcal{N}(d_1', d_2'; s); \quad s = \sqrt{\tau_1/\tau_2}. \tag{10.27}$$

$$V_3(\boldsymbol{X}, T_2) = y_1\mathbb{I}(x_1 > cy_1)$$

Finally, the pv (details omitted) of this 2-asset first order binary is

$$V_3(x, y, t) = y\mathcal{N}(d_1') \tag{10.28}$$

in terms of a univariate normal.

Putting the three expressions (10.26), (10.27) and (10.28) together leads to the result

$$V(x, y, t) = x\mathcal{N}(d_1, d_2; \sqrt{\tfrac{\tau_1}{\tau_2}}) - y\mathcal{N}(d_1', d_2'; \sqrt{\tfrac{\tau_1}{\tau_2}}) - ky\mathcal{N}(d_1') \qquad (10.29)$$

for the price of the compound exchange option for all $t < T_1$. All the parameters in this expression have been previously defined in the above analysis. Furthermore, this price agrees exactly with the price given in the Carr reference [12], although its derivation is not provided in that paper.

## 10.9 Multi-Asset Barrier Options

We follow closely the work of Skipper [71], who has derived a very general result, including a Method of Images for multi-asset barrier options. Care must be taken in how such an option is defined. Let $X_i$ denote a set of $n$ assets whose prices follow correlated gBm, in the usual way. The option payoff at the expiry date $T$, will be some function $f(\boldsymbol{X})$. However, this payoff will only be made subject to a barrier condition. In the multi-variate case, this barrier condition will be governed by a single variable $y = \boldsymbol{x}^{\boldsymbol{\alpha}}$. In other words, the barrier variable will, in the most general case, be some product of powers of the asset prices $x_i$: $y = x_1^{\alpha_1} x_2^{\alpha_2} \cdots x_n^{\alpha_n}$.

This variable includes: any of the single assets, e.g., $y = x_k$, by taking $\boldsymbol{\alpha} = \mathbf{1}_k$; or more complex cases such as the geometric mean of two or more assets, e.g., $y = \sqrt{x_k x_\ell}$, by taking $\boldsymbol{\alpha} = \tfrac{1}{2}(\mathbf{1}_k + \mathbf{1}_\ell)$; or even ratios of two assets, e.g., $y = x_k/x_\ell$, by taking $\boldsymbol{\alpha} = (\mathbf{1}_k - \mathbf{1}_\ell)$.

Once the barrier variable $y$ is defined, the four standard barrier options can then be associated with it. These are the two knock-out barrier options (D/O, U/O) and the two knock-in options (D/I, U/I). The down barrier options depend on whether the barrier $y = b$ is hit from above; the up barriers on whether $y = b$ is hit from below. As might be expected, the four barrier options are connected by parity relations, so in practice, only one of them needs to be priced. We arbitrarily, choose the D/O as the benchmark barrier option.

The key idea in pricing the multi-asset D/O barrier option, will of course depend on the appropriate image function relative to our barrier variable. The theorem below, provides the details.

### THEOREM 10.3 (Skipper 2003)

*The image $\overset{*}{V}(\boldsymbol{x},t)$ of the multi-variate function $V(\boldsymbol{x},t)$ wrt the barrier vari-*
*able $y = \boldsymbol{x}^{\alpha} = b$ and the multi-asset BS-pde, is given by*

$$\boxed{\overset{*}{V}(\boldsymbol{x},t) = \left(\frac{b}{y}\right)^{\kappa} V(\hat{\boldsymbol{x}},t); \qquad \hat{x}_i = x_i(b/y)^{c_i}} \tag{10.30}$$

*where*

$$\kappa = \frac{2\mu'\alpha}{\lambda}; \qquad c = \frac{2\Gamma\alpha}{\lambda}; \qquad \lambda = \alpha'\Gamma\alpha \tag{10.31}$$

*and $\mu_i = (r - \frac{1}{2}\sigma_i^2)$; $\Gamma_{ij} = \rho_{ij}\sigma_i\sigma_j$ is the covariance matrix associated with $\boldsymbol{X}$.*

**REMARK 10.6**  It is convenient here to think of $\overset{*}{V}$ as the image function, and $\hat{\boldsymbol{x}}$ as the image point. Observe that the multi-variate image function
reduces to the 1D image function $\overset{*}{V} = (b/x)^{\kappa}V(b^2/x,t)$, with $\kappa = 2r/\sigma^2 - 1$,
when $n = 1$, $y = x$, $\alpha = 1$ and $\Gamma = \sigma^2$. In this case, the image point is
$\hat{x} = b^2/x$. Note further, that in the general case, $\lambda = \alpha'\Gamma\alpha = \sigma_y^2$ is the
instantaneous variance associated with the rv $Y = \boldsymbol{X}^{\alpha}$.

In many cases, the function $V(\boldsymbol{x},t)$ will depend explicitly on the barrier
variable $y$. In that case, since $\alpha'c = 2$, the image point corresponding to $y$ is
given by

$$\hat{y} = \hat{\boldsymbol{x}}^{\alpha} = \boldsymbol{x}^{\alpha}(b/y)^{\alpha'c} = b^2/y. \tag{10.32}$$

⬚

### COROLLARY 10.1

*The multi-asset image function $\mathcal{I}_b(y)[V(\boldsymbol{x},t)] = \overset{*}{V}(\boldsymbol{x},t)$, (10.30) satisfies*
*the same four properties stated in theorem 7.1 for one-dimensional image*
*functions. Thus*

1. *$\mathcal{I}_b^2(y)[V(\boldsymbol{x},t)] = V(\boldsymbol{x},t)$. That is, $\mathcal{I}_b(y)$ is an involution.*

2. *$\overset{*}{V}(\boldsymbol{x},t)$ satisfies the multi-asset BS-pde whenever $V(\boldsymbol{x},t)$ does.*

3. *$\overset{*}{V}(\boldsymbol{x},t) = V(\boldsymbol{x},t)$ when $y = b$. That is, $V$ and its image $\overset{*}{V}$ agree on the barrier.*

4. *If $y \gtrless b$ is the active domain for $V(\boldsymbol{x},t)$, then the active domain for $\overset{*}{V}(\boldsymbol{x},t)$ is the complementary domain $y \lessgtr b$.*

The proof of theorem 10.3, which is presented next, adopts a matrix approach. This proof is furthermore, just an extension of the 1D approach
of transforming the BS-pde to the heat equation, and noting that the image solution of the heat equation $U_t = kU_{\xi\xi}$ relative to $\xi = 0$ is simply

$\overset{*}{U}(\xi,t) = U(-\xi,t)$. The image function $\overset{*}{U}(\xi,t)$ also satisfies the four image properties above.

**PROOF** Our first task is to transform the multi-variate BS-pde (3.31), repeated here for convenience,

$$V_t = rV - r\sum_{i=1}^n x_i V_i - \tfrac{1}{2}\sum_{i=1}^n \sum_{j=1}^n \rho_{ij}\sigma_i\sigma_j x_i x_j V_{ij}$$

to a new pde with constant coefficients and only a second order derivative in the independent price variable. To achieve this, define

$$\boldsymbol{u} = \log \boldsymbol{x} - \log \boldsymbol{a}; \qquad \tau = (T-t); \qquad V(\boldsymbol{x},t) = e^{-\boldsymbol{\beta}'\boldsymbol{u}-\omega\tau}F(\boldsymbol{u},\tau).$$

The parameters $(\boldsymbol{\beta}, \omega)$ are to be chosen to satisfy the requirements mentioned above, while $\boldsymbol{a}$ will be chosen to translate the barrier to $\boldsymbol{\alpha}'\boldsymbol{u} = 0$.

1. Let $\epsilon = e^{-\boldsymbol{\beta}'\boldsymbol{u}-\omega\tau}$. Then, the relevant partial differentiations are,

$$\begin{aligned}
V_t &= [\omega F - F_\tau]\,\epsilon \\
x_i V_i &= [-\beta_i F + F_i]\,\epsilon \\
x_i x_j V_{ij} &= [\beta_i\beta_j F - \beta_i F_j - \beta_j F_i + F_{ij} + (\beta_i F - F_i)\delta_{ij}]\,\epsilon
\end{aligned}$$

where subscripts on $F$ denote partial derivatives wrt $u_i$ and $u_j$ and $\delta_{ij}$ is the Kronecker delta, equal to 1 if $i = j$; 0 otherwise.

The BS-pde then becomes,

$$-\omega F + F_\tau = -rF + \sum_i (r - \tfrac{1}{2}\sigma_i^2)(F_i - \beta_i F) + \tfrac{1}{2}\sum_{i,j}(F_{ij} - 2\beta_j F_i + \beta_i\beta_j F)\Gamma_{ij}.$$

In deriving this expression, we have used,

$$\sum_{i,j} A_i \Gamma_{ij}\delta_{ij} = \sum_i A_i\sigma_i^2 \quad \text{and} \quad \sum_{i,j}[A_{ij} + A_{ji}]\Gamma_{ij} = 2\sum_{i,j} A_{ij}\Gamma_{ij}$$

for any $A_i$ and $A_{ij}$, since $\Gamma_{ii} = \sigma_i^2$ and $\Gamma_{ij}$ is symmetric (i.e., equal to $\Gamma_{ji}$).

Now let us choose $\beta_i$ and $\omega$ to kill the coefficients of $F$ and $F_i$ to get,

$$\omega = r + \sum_i (r - \tfrac{1}{2}\sigma_i^2)\beta_i - \tfrac{1}{2}\sum_{i,j}\beta_i\beta_j\Gamma_{ij} = r + \boldsymbol{\beta}'\boldsymbol{\mu} - \tfrac{1}{2}\boldsymbol{\beta}'\Gamma\boldsymbol{\beta}$$

and

$$0 = \sum_i [(r - \tfrac{1}{2}\sigma_i^2) - \Gamma_{ij}\beta_j]F_i = [\boldsymbol{\mu} - \Gamma\boldsymbol{\beta}]'(\nabla F).$$

So, assuming $\Gamma$ is positive definite, we take

$$\boxed{\beta = \Gamma^{-1}\mu \qquad \text{and} \qquad \omega = r + \tfrac{1}{2}\mu'\Gamma^{-1}\mu} \tag{10.33}$$

This achieves the first task, in transforming the multi-variate BS-pde, to

$$F_\tau(\mathbf{u},\tau) = \tfrac{1}{2}\sum_{i,j}\Gamma_{ij}F_{ij}(\mathbf{u},\tau) \tag{10.34}$$

in the domain ($\mathbf{u} \in \mathbb{R}^n$; $\tau > 0$).

2. Now the barrier $y = x^\alpha = b$ is equivalent to $\log y = \alpha'\log x = \log b$, or $\alpha'\mathbf{u} = \log b - \alpha'\log a$, and this can be set equal to zero by the choosing $a$ so that $b = a^\alpha$ (using our special vector notation as in (10.2)). That is, for this choice of $a$, $y = b$ transforms to $\alpha'\mathbf{u} = 0$.

3. Next seek a transformation $\mathcal{T} \in \mathcal{A}_{nn}$ such that $\hat{\mathbf{u}} = \mathcal{T}\mathbf{u}$ is the image point in $\mathbf{u}$−space. This transformation must have the image properties: (i) $F(\mathcal{T}\mathbf{u},\tau)$ satisfies the pde (10.34); and (ii) $\hat{\mathbf{u}} = \mathbf{u}$ on the barrier $\alpha'\mathbf{u} = 0$.

Let $D_i = \partial u_i$ denote the gradient operator wrt $u_i$; and $\hat{D}_i = \partial\hat{u}_i$, that wrt $\hat{u}_i$. Then, since $\mathbf{DF} = \mathcal{T}'\hat{\mathbf{D}}F$ and the pde (10.34) transforms to

$$F_\tau = \tfrac{1}{2}\mathbf{D}'\Gamma\mathbf{D}F = \tfrac{1}{2}\hat{\mathbf{D}}'(\mathcal{T}\Gamma\mathcal{T}')\hat{\mathbf{D}}F.$$

Hence, $F(\mathcal{T}\mathbf{u},\tau)$ will satisfy the pde, iff $\mathcal{T}\Gamma\mathcal{T}' = \Gamma$. Further, to satisfy the barrier condition, $\mathcal{T}\mathbf{u} = 0$ when $\alpha'\mathbf{u} = 0$, we now show that $\mathcal{T}$ must have the representation, $\mathcal{T} = I_n - c\alpha'$, where $I_n$ is the $(n \times n)$ identity matrix, and $c \in \mathcal{V}_n$ is a vector to be determined.

Observe first, that $\mathcal{T}\mathbf{u} = [\mathbf{u} - c\alpha'\mathbf{u}] = \mathbf{u}$ when $\alpha'\mathbf{u} = 0$, so the barrier condition is met by this choice of $\mathcal{T}$. Second, we calculate,

$$\begin{aligned}
\mathcal{T}\Gamma\mathcal{T} &= (I_n - c\alpha')\Gamma(I_n - \alpha c')\\
&= \Gamma - c\alpha'\Gamma - \Gamma\alpha c' + c(\alpha'\Gamma\alpha)c'\\
&= \Gamma - \lambda[cv' + vc' - cc']
\end{aligned}$$

writing $v = \Gamma\alpha/\lambda$, where $\lambda = \alpha'\Gamma\alpha$. Thus, $\mathcal{T}\Gamma\mathcal{T} = \Gamma$ if $c = 2v$. It follows, that

$$\boxed{\mathcal{T} = I_n - \frac{2\Gamma\alpha\alpha'}{\lambda}; \qquad \lambda = \alpha'\Gamma\alpha} \tag{10.35}$$

At this point, let us make a few observations:

a. $\mathcal{T}^2 = (I_n - 2v\alpha')(I_n - 2v\alpha') = I_n - 4v\alpha' + 4v(\alpha'v)\alpha' = I_n$ since $\alpha'v = 1$.

b. $\alpha'\hat{u} = \alpha'(I_n - 2v\alpha')u = -\alpha'u$. Thus the two sides of the barrier, $\alpha'u \gtrless 0$ correspond to the complementary domains $\alpha'\hat{u} \lessgtr 0$.

c. In the 1D case, $\alpha = 1, \Gamma = \lambda = \sigma^2$ we find $\mathcal{T} = -1$. Thus the image solution of the 1D heat equation, $F_\tau(u, \tau) = \frac{1}{2}\sigma^2 F_{uu}(u, \tau)$, will be $F(-u, \tau)$ relative to the barrier at $u = 0$.

4. The final step in the proof is to back-transform to the original variables.

$$\overset{*}{V}(\boldsymbol{x}, t) = e^{-\boldsymbol{\beta}'\boldsymbol{u} - \omega\tau} F(\hat{\boldsymbol{u}}, \tau)$$
$$= e^{-\boldsymbol{\beta}'\boldsymbol{u} - \omega\tau} \cdot e^{\boldsymbol{\beta}'\hat{\boldsymbol{u}} + \omega\tau} V(\hat{\boldsymbol{x}}, t)$$
$$= e^{\boldsymbol{\beta}'(\hat{\boldsymbol{u}} - \boldsymbol{u})} V(\hat{\boldsymbol{x}}, t).$$

Now from (10.33) and (10.35), we obtain

$$\boldsymbol{\beta}'(\hat{\boldsymbol{u}} - \boldsymbol{u}) = \boldsymbol{\beta}'(\mathcal{T}\boldsymbol{u} - \boldsymbol{u}) = -\left(\frac{2\boldsymbol{\beta}'\Gamma\alpha\alpha'\boldsymbol{u}}{\lambda}\right)$$
$$= \frac{-2(\boldsymbol{\mu}'\alpha)}{\lambda} [\alpha' \log \boldsymbol{x} - \log b] = -\kappa[\log y - \log b].$$

Hence $e^{\boldsymbol{\beta}'(\hat{\boldsymbol{u}} - \boldsymbol{u})} = (b/y)^\kappa$ and $\overset{*}{V}(\boldsymbol{x}, t) = (b/y)^\kappa V(\hat{\boldsymbol{x}}, t)$.

The expression for the image point $\hat{\boldsymbol{x}}$, is calculated from:

$$\log \hat{\boldsymbol{x}} = \hat{\boldsymbol{u}} + \log \boldsymbol{a} = \mathcal{T}\boldsymbol{u} + \log \boldsymbol{a}$$
$$= (I_n - c\alpha\alpha')(\log \boldsymbol{x} - \boldsymbol{a}) + \log \boldsymbol{a}; \qquad c = 2\Gamma\alpha/\lambda$$
$$= \log \boldsymbol{x} - c \log(y/b); \quad \text{since } \log(y/b) = \alpha'[\log \boldsymbol{x} - \log \boldsymbol{a}].$$

Hence, finally, $\hat{x}_i = x_i (b/y)^{c_i}$, in agreement with Equation (10.30). This completes the proof of theorem 10.3. $\qquad \square$

## Example 10.6

Consider the 2-asset case with asset prices $(x, y)$ and $y$ being the barrier variable. We stated the image function for this case in Equation (7.46). Let us now verify the result using theorem 10.3.

For this case, we obtain:

$$\Gamma = \begin{bmatrix} \sigma_1^2 & \rho\sigma_1\sigma_2 \\ \rho\sigma_1\sigma_2 & \sigma_2^2 \end{bmatrix}; \qquad \alpha = \begin{bmatrix} 0 \\ 1 \end{bmatrix}; \qquad \lambda = (\alpha'\Gamma\alpha) = \sigma_2^2.$$

Then

$$\mathcal{T} = (I_2 - 2\Gamma\alpha\alpha'/\lambda) = \begin{bmatrix} 1 & -2\rho\sigma_1/\sigma_2 \\ 0 & -1 \end{bmatrix},$$

and

$$\kappa = \frac{2\mu'\alpha}{\lambda} = \frac{2r}{\sigma_2^2} - 1; \qquad c = \begin{bmatrix} c_1 \\ c_2 \end{bmatrix} = \frac{2\Gamma\alpha}{\lambda} = \begin{bmatrix} 2\rho\sigma_1/\sigma_2 \\ 2 \end{bmatrix},$$

and $\hat{\boldsymbol{x}} = [\hat{x}, \hat{y}]'$, with

$$\hat{x} = x(b/y)^{c_1} = x(b/y)^{2\rho\sigma_1/\sigma_2}; \qquad \hat{y} = y(b/y)^{c_2} = b^2/y.$$

Thus, for this 2D example,

$$\overset{*}{V}(x, y, t) = (b/y)^\kappa V\left[x(b/y)^{2\rho\sigma_1/\sigma_2}, b^2/y, t\right].$$

This agrees precisely with Equation (7.46). $\qquad\qquad\qquad\qquad\qquad$ ▯

## Method of Images

Now that we have the image function relative to a barrier $y = x^\alpha = b$, we are in a position to price the four standard barrier types. The Method of Images discussed in Section 7.2 carries over to the multi-barrier case by simply selecting the appropriate image operator. Indeed, we can go even further and use the Method of Images to determine the equivalent European payoffs for the four multi-asset barrier types. This leads immediately to

$$\begin{array}{|l|} \hline V_{do}^{eq}(\boldsymbol{x}, T) = f(\boldsymbol{x})\mathbb{I}(y > b) - \overset{*}{f}(\boldsymbol{x})\mathbb{I}(y < b) \\ V_{di}^{eq}(\boldsymbol{x}, T) = [f(\boldsymbol{x}) + \overset{*}{f}(\boldsymbol{x})]\mathbb{I}(y < b) \\ V_{ui}^{eq}(\boldsymbol{x}, T) = [f(\boldsymbol{x}) + \overset{*}{f}(\boldsymbol{x})]\mathbb{I}(y > b) \\ V_{uo}^{eq}(\boldsymbol{x}, T) = f(\boldsymbol{x})\mathbb{I}(y < b) - \overset{*}{f}(\boldsymbol{x})\mathbb{I}(y > b) \\ \hline \end{array} \qquad (10.36)$$

where $f(\boldsymbol{x})$ is the multi-asset payoff function. Notice that the terms $f(\boldsymbol{x})\mathbb{I}(y > b)$ and $f(\boldsymbol{x})\mathbb{I}(y < b)$ are often examples of M-binaries and so can be priced using our M-binary pricing formula.

We give just one example of how to use the Method of Images for a multi-asset barrier option. This is an extension of the ESO priced in Section 6.6. There, the ESO paid a call option of strike price $k$, but only if the asset price at expiry $T$, was above some index. Here we shall assume the option is knocked out at any time prior to expiry, if the asset price falls below $b$ times the index.

### Example 10.7

Let $X_1, X_2$ denote the asset and index respectively, assumed to follow correlated gBm's. At time $T$ the executive gets $(X_1 - k)^+$ but only if $X_1 > bX_2$ over the interval $[t, T]$. Then the ESO is thus a D/O barrier option with equivalent payoff

$$V(x_1, x_2, T) = (x_1 - k)^+\mathbb{I}(x_1 > bx_2) - \mathcal{I}_b(y)\{\cdots\cdots\}$$

where $\mathcal{I}_b(y)$ denotes the image operator wrt $y = b$, with barrier variable $y = x_1/x_2$.

Now the first term above, $V_b$ say, is essentially the payoff of the earlier ESO, whose price is given in Equation (7.45). We need only replace there, $x$ by $x_1$ and $y$ by $bx_2$, to obtain (with a slight change in notation)

$$V_b(x_1, x_2, t) = x_1 \mathcal{N}(d_k, d_b; \rho_1) - ke^{-r\tau} \mathcal{N}(d_k', d_b'; \rho_1)$$

where (with $y = x_1/x_2$)

$$d_k(x_1, \tau) = \frac{\log(x_1/k) + (r + \frac{1}{2}\sigma_1^2)\tau}{\sigma_1\sqrt{\tau}}; \qquad d_k' = d_k - \sigma_1\sqrt{\tau}$$

$$d_b(y, \tau) = \frac{\log(y/b) + \frac{1}{2}\sigma^2\tau}{\sigma\sqrt{\tau}}; \qquad d_b' = d_b - \rho_1\sigma_1\sqrt{\tau}$$

$$\sigma^2 = \sigma_1^2 + \sigma_2^2 - 2\rho\sigma_1\sigma_2; \qquad \rho_1 = \frac{\sigma_1 - \rho\sigma_2}{\sigma}.$$

Now, for the image part of the price, we identify the parameters:

$$\Gamma = \begin{bmatrix} \sigma_1^2 & \rho\sigma_1\sigma_2 \\ \rho\sigma_1\sigma_2 & \sigma_2^2 \end{bmatrix}; \qquad \alpha = \begin{bmatrix} 1 \\ -1 \end{bmatrix}; \qquad \lambda = (\alpha'\Gamma\alpha) = \sigma^2$$

Then, we calculate,

$$\kappa = \frac{2(\alpha'\mu)}{\lambda} = \frac{2}{\sigma^2}[1, \, -1]\begin{bmatrix} r - \frac{1}{2}\sigma_1^2 \\ r - \frac{1}{2}\sigma_2^2 \end{bmatrix} = \frac{\sigma_2^2 - \sigma_1^2}{\sigma^2}$$

and

$$c = \begin{bmatrix} c_1 \\ c_2 \end{bmatrix} = \frac{2\Gamma\alpha}{\lambda} = \frac{2}{\sigma^2}\begin{bmatrix} \sigma_1^2 - \rho\sigma_1\sigma_2 \\ \rho\sigma_1\sigma_2 - \sigma_2^2 \end{bmatrix} = \begin{bmatrix} \frac{2\rho_1\sigma_1}{\sigma} \\ \frac{2\rho_2\sigma_2}{\sigma} \end{bmatrix}$$

where we have defined $\rho_2 = (\rho\sigma_1 - \sigma_2)/\sigma$. The image points, then can be computed from

$$\hat{x}_1 = x_1(b/y)^{c_1}; \qquad \hat{x}_2 = x_2(b/y)^{c_2}; \qquad \hat{y} = b^2/y.$$

For the last of these see Equation (10.32), but of course it also follows from $\hat{y} = \hat{x}_1/\hat{x}_2$, and the observation $(c_1 - c_2) = 2$. In this book we have generally avoided writing out all the terms of complicated expressions, content to leave them in a shorter, more concise notation. Let us break that tradition here, to see how the final expression looks in terms of its most basic constituents.

$$\overset{*}{V}_b(x_1, x_2, t) = (b/y)^{\kappa}\left[x_1(b/y)^{c_1}\mathcal{N}(\hat{d}_k, \hat{d}_b; \rho_1) - ke^{-r\tau}\mathcal{N}(\hat{d}_k', \hat{d}_b'; \rho_1)\right]$$

where

$$\hat{d}_k(x_1, x_2, \tau) = \frac{\log(x_1/k) + c_1 \log(b/y) + (r + \frac{1}{2}\sigma_1^2)\tau}{\sigma_1\sqrt{\tau}}; \quad \hat{d}'_k = \hat{d}_k - \sigma_1\sqrt{\tau}$$

$$\hat{d}_b(y, \tau) = \frac{\log(b/y) + \frac{1}{2}\sigma^2\tau}{\sigma\sqrt{\tau}}; \qquad\qquad \hat{d}'_b = \hat{d}_b - \rho_1\sigma_1\sqrt{\tau}.$$

The price of the barrier ESO is then $V(x_1, x_2, t) = V_b(x_1, x_2, t) - \overset{*}{V}_b(x_1, x_2, t)$, and all the parameters appearing in this formula have been previously defined above. ▯

**Internal and External Barriers**
The relationship between the barrier variable $y$ and the payoff variables $x_i$ permits a further classification of multi-asset barrier options. Most applications appearing in the literature have $y$ as a variable which is functionally (but not statistically) independent of the payoff variables. Typically, $y$ will be the price of some other asset, such as an index or related equity. When the set of asset price(s) in $y$ is different to the set describing the payoff, the barrier is called an *external* barrier. Heynen and Kat [30] and Carr [13] value options on a single payoff asset with a different asset acting as an external barrier, while Kwok et al. [48] extend the valuation to multi-asset payoffs with an external barrier.

The theory we have presented above allows for any type of barrier, not necessarily an external barrier. The assets comprising the barrier variable may partially or fully comprise the payoff function. Barriers with this feature might sensibly be called *internal* barriers. But in a sense, this distinction is irrelevant, as the theory above ignores the difference between internal and external barriers. Historically, this distinction may have been important, because pricing options with external barriers, by first passage density methods, for example, would be considerably easier than for those with internal barriers. The Method of Images presented here ignores this difference.

---

## 10.10   Summary

This chapter developed the theory of M-binaries which is the most general class of power binary with closed form analytic prices in the BS model. Particular attention is paid to developing very general exercise conditions for these binaries. A special, non-standard vector and matrix notation was adopted in order to simplify what would otherwise lead to a notational nightmare.

These binaries include all the binary options previously met in this book as special cases, and obviously include many that have not. M-binaries are the building blocks for most of the exotic options you are ever likely to meet.

A typical M-binary is described by a large number of parameters. Recall that associated with every M-binary is a 4-fold matrix asset parameter set $\mathbb{A}$ containing information about the underlying assets; a discrete tenor set $\mathbb{T}$ listing the times at which asset prices are monitored to determine the final payoff; a 4-fold dimension set $\mathbb{D}$; and a 4-fold matrix payoff set $\mathbb{P}$, which defines the payoff and exercise condition. Even for a simple 3-asset, 2-period M-binary, the total number of elements in all these sets will total 36 inputs.

With the M-binary pricing formula at hand, pricing a multi-asset, multi-period exotic option becomes one of decomposing its payoff into the appropriate portfolio of M-binaries. While this is often a fairly direct task, it can also be quite challenging.

We demonstrated how the M-binary formula could be used in several examples, including some previously worked problems and new ones as well. The final section of the chapter shows how the M-binary formulation can be extended to price multi-asset barrier options in which there is a single barrier variable of the form $y = x^\alpha$.

In a very real sense this chapter on M-binaries is probably about as far as one can push the Black–Scholes framework, at least with regard to exotic options having exact pricing solutions. And indeed this is where we bring our study of BS pricing to a close.

Further developments in exotic option pricing are presently being undertaken under various models of stochastic volatility. This topic is beyond the scope of this introductory text, but as remarked at the start of the book, current stochastic volatility models with few exceptions, do not admit analytic prices. We mention, however, the work of Agliardi [1], who presents semi-analytic pricing formulae for M-binary analogues, similar to those considered in this chapter, for the class of regular Lévy processes of exponential type. These asset price processes are a logical extension to gBm, and have the capacity to model stylized features of observed asset prices, such as jumps and leptokurtic (fat-tailed) distributions. These processes also reduce to standard gBm as a special case, so many of the formulae derived in this book could in principle be recovered from this more general model.

## Exercise Problems

1. Let $(X_i, Y_i, Z_i)$ for $(i = 1, 2)$ denote the prices of three assets $(X, Y, Z)$ at the two times $T_1$ and $T_2$. Determine the payoff vector $\boldsymbol{X}$, dimension set $\mathbb{D}$ and payoff parameter set $\mathbb{P}$ for the M-binaries with payoffs

$$(a) \qquad V_1(\boldsymbol{X}) = X_1 \mathbb{I}(\sqrt{X_2 Y_2} > Z_2) \mathbb{I}(Z_1 < c)$$

$$(b) \qquad V_2(\boldsymbol{X}) = \frac{Y_1}{X_1} \mathbb{I}(Y_1 > X_1) \mathbb{I}(Y_2 > X_2) \mathbb{I}(Z_2 < c).$$

2. Show that the price of a call option of strike price $k$, on the multi-asset variable $Y = \boldsymbol{X}^{\alpha}$, with payoff $V_T = (Y - k)^+$, is given by the expression

$$V(y, \tau) = y e^{\theta \tau} \mathcal{N}(d + \sigma_y \sqrt{\tau}) - k e^{-r\tau} \mathcal{N}(d)$$

where $\sigma_y^2 = \boldsymbol{\alpha}' \bar{\Gamma} \boldsymbol{\alpha}$ is the instantaneous variance of $Y$,

$$d(y, \tau) = \frac{\log(y/k) + (\boldsymbol{\alpha}' \boldsymbol{c}) \tau}{\sigma_y \sqrt{\tau}}; \qquad \theta = -r + \boldsymbol{\alpha}' \boldsymbol{c} + \tfrac{1}{2}\sigma_y^2$$

and $\boldsymbol{c}$ is a vector with components $c_i = (r - \tfrac{1}{2}\sigma_i^2)$.

3. Use the results of the previous question to confirm the prices of the following two asset call options:

   (a) $V_T = (X_1/X_2 - k)^+$ (quotient call of Section 6.4)
   (b) $V_T = (X_1 X_2 - k)^+$ (product call, see Section 6.7)
   (c) $V_T = (\sqrt{X_1 X_2} - k)^+$ (GM call, see Section 6.7).

4. Let $X_i$ for $i = 1, 2, \ldots, n$ denote the prices of a single asset at time the $n$ times $T_1, T_2, \ldots, T_n$. Price, for all $t < T_1$, the

   (a) bond binary with time $T$ payoff

$$V_b(X_i, T) = \mathbb{I}(X_1 < k_1) \mathbb{I}(X_2 < k_2) \cdots \mathbb{I}(X_n < k_n)$$

   (b) the asset binary with time $T$ payoff

$$V_a(X_i, T) = X_k \mathbb{I}(X_1 < k_1) \mathbb{I}(X_2 < k_2) \cdots \mathbb{I}(X_n < k_n)$$

   for some $k$ in $[1, n]$.

5. Use the M-binary valuation formula to derive the prices for the two-asset binaries $V_1, V_2$ and $V_4$ given in theorem 6.1 in the chapter on Two-Asset Rainbow Options. Note that $V_3$ was priced in this chapter.

6. Use the M-binary valuation formula to price the ICIAM bond binary stated in Section 6.5. Notice in this problem that the payoff dimension is $n = 2$ and the exercise dimension is $m = 3$, which violates the condition $m \leq n$. Hence an application of the Inclusion-Exclusion Principle will have to precede any M-binary calculations.

7. Section 10.7 of this chapter derived prices for calls on the maximum of a set of $n$ assets and puts on the the minimum. Derive the corresponding formulae for calls on the minimum and puts on the maximum.

8. Price a two asset D/O quotient call barrier option when the barrier variable is the product of the two assets. That is, the payoff at time $T$ is $(x_1/x_2 - k)^+$, provided the barrier variable $y = x_1 x_2$ remains above the barrier price $y = b$ during the life of the option.

   Use parity relations to price the corresponding D/I, U/I and U/O versions of this option.

—ooOoo—

# References

[1] Agliardi, R. The quintessential option pricing formula under lévy processes. *Applied Mathematical Letters*, 22:1626–1631, 2009.

[2] Aitchison, J. and J.A.C. Brown. *The log-normal distribution.* CUP, Cambridge, United Kingdom, 1957.

[3] Bermin, H. *Essays on lookback and barrier options: a Malliavin calculus approach.* Lund University PhD thesis, 1998.

[4] Bermin, H.P., P.W. Buchen, and O. Konstandatos. Two exotic lookback options. *Applied Mathematical Finance*, 15(4):387–402, 2008.

[5] Black, F. and M.S. Scholes. The pricing of options and corporate liabilities. *Journal of Political Economy*, 81:637–653, 1973.

[6] Broadie, M., P. Glasserman, and S.G. Kou. A continuity correction for discrete barrier options. *Mathematical Finance*, 7:325–349, 1997.

[7] Buchen, P.W. Image options and the road to barriers. *Risk Magazine*, 14(9):127–130, 2001.

[8] Buchen, P.W. The pricing of dual expiry options. *Quantitative Finance*, 4:101–108, 2004.

[9] Buchen, P.W. and O. Konstandatos. A new method of pricing lookback options. *Mathematical Finance*, 15(2):245–259, 2005.

[10] Buchen, P.W. and O. Konstandatos. A new approach to pricing double-barrier options with arbitrary payoffs and exponential barriers. *Applied Mathematical Finance*, 16(6):497–515, 2009.

[11] Cannon, J.R. *The one-dimensional heat equation.* Addison-Wesley, Reading, Massachusetts, 1984.

[12] Carr, P. The valuation of sequential exchange opportunities. *Journal of Finance*, 43(5):1235–1256, 1988.

[13] Carr, P. Two extensions to barrier option valuation. *Applied Mathematical Finance*, 2:173–284, 1995.

[14] Carr, P., K. Ellis, and V. Gupta. Static hedging of exotic options. *Journal of Finance*, 53(3):1165–1190, 1998.

[15] Chiarella, C., A. Kucera, and A. Ziogas. A survey of the integral representation of american options. *UTS Quantitative Finance Research Centre (working paper)*, 118:1–65, 2004.

[16] Conze, A. and R. Viswanathan. Path dependent options: the case of lookback options. *Journal of Finance*, 46(5):1893–1907, 1991.

[17] Cox, J., S. Ross, and M. Rubinstein. Option pricing: a simplified approach. *Journal of Financial Economics*, 7:229–263, 1979.

[18] Delbaen, F. and W. Schachermayer. A general version of the fundamental theroem of asset pricing. *Mathematische Annalen*, 300:463–520, 1994.

[19] Demeterfi, K., E. Derman, M. Kamal, and J. Zou. More than you ever wanted to know about volatility swaps. *Quantitative Strategies Research Notes (Goldman-Sachs)*, March, 1999.

[20] Dixit, A.K. and R.S. Pindyck. *Investment under uncertainty*. Princeton University Press, New Jersey, 1994.

[21] Drezner, Z. Computation of the bivariate normal integral. *Mathematics of Computation*, 32(141):277–279, 1978.

[22] Erdélyi, A., W. Magnus, F. Oberhettinger, and G. Tricomi. *Tables of Integral Transforms*, volume 1 of *Bateman Manuscript*. McGraw-Hill, New York, 1954.

[23] Freedman, D.H. A formula for economic calamity. *Scientific American*, 305(5):59–61, November 2011.

[24] Genz, A. Numerical computation of multivariate normal probabilities. *Journal of Computational and Graphical Statistics*, 1:141–149, 1992.

[25] Geske, R. The valuation of compound options. *Journal of Financial Economics*, 7:63–81, 1979.

[26] Goldman, M., H. Sosin, and M. Gatto. Path dependent options: buy at the low, sell at the high. *Journal of Finance*, 34(5):1111–1127, 1979.

[27] Grenadier, S.R. Valuing lease contracts: a real options approach. *Journal of Financial Economics*, 78(4):1173–1214, 2005.

[28] Harrison, J.M. and S.R. Pliska. Martingales and stochastic integrals in the theory of continuous trading. *Stochastic Processes and their Applications*, 11:215–260, 1981.

[29] Haug, E.G. *The complete guide to option pricing formulas*. McGraw-Hill, New York, 1998.

[30] Heynen, R. and H. Kat. Crossing barriers. *Risk Magazine*, 7(6), 1994.

[31] Heynen, R. and H. Kat. Partial barrier options. *Journal of Financial Engineering*, 3:253–274, 1994.

[32] Heynen, R. and H. Kat. Lookback options with discrete and partial monitoring of the underlying price. *Applied Mathematical Finance*, 2:273–209, 1995.

[33] Heynen, R. and H. Kat. Brick by brick. *Risk Magazine*, 9(6), 1996.

[34] Ho-Shon, K. *Real estate leases and real options*. Sydney University PhD thesis, School of Mathematics and Statistics, 2008.

[35] Hyer, T., A. Lipton, and D. Pugachevsky. Passport to success. *Risk Magazine*, 10(9):127–131, 1997.

[36] Ingersoll, J. Digital contracts: simple tools for pricing complex derivatives. *Journal of Business*, 73:67–88, 2000.

[37] Jamshidian, F. An analysis of american options. *Review of Futures Markets*, 11:72–80, 1992.

[38] Jarrow, R. (ed). *Over the rainbow: developments in exotic options and complex swaps*. Risk Publications, London, United Kingdom, 1995.

[39] Jeanblanc, M., M. Yor, and M. Chesney. *Mathematical methods for financial markets*. Springer Finance. Springer–Verlag, London, United Kingdom, 2009.

[40] Johnson, H. Options on the maximum and minimum of several assets. *Journal of Financial and Quantitative Analysis*, 22(3):277–283, 1987.

[41] Karatzas, I. and S. Shreve. *Brownian motion and stochastic calculus*. Springer-Verlag, New York, 1991.

[42] Kemna, A.G.Z. and A.C.F Vorst. A pricing method for options based upon average asset values. *Journal of Banking and Finance*, 14:113–129, 1990.

[43] Kim, I.J. The analytic valuation of american options. *Review of Financial Studies*, 3:547–572, 1990.

[44] Kloeden, P.E. and E. Platen. *Numerical solution of stochastic differential equations*. Springer-Verlag, Berlin, Germany, 1999.

[45] Konstandatos, O. *A new framework for pricing barrier and lookback options*. Sydney University PhD thesis, School of Mathematics and Statistics, 2003.

[46] Konstandatos, O. *Pricing path dependent options: a comprehensive framework*. Applications of Mathermatics. VDM-Verlag, Saarbrucken, Germany, 2008.

[47] Kunimoto, N. and M. Ikeda. Pricing options with curved boundaries. *Mathematical Finance*, 2:276–298, 1992.

[48] Kwok, Y., L. Wu, and H. Yu. Pricing multi-asset options with an external barrier. *International Journal of Theoretical and Applied Finance*, 1(4):141–183, 1998.

[49] Kyng, T. *Application of Black-Scholes exotic option pricing theory to real options and ESO's*. Sydney University PhD thesis, School of Mathematics and Statistics, 2011.

[50] Kyprianou, A., W. Schoutens, and P. Wilmott. *Exotic option pricing and advanced Lévy models*. J. Wiley & Sons, New York, 2005.

[51] Lamberton, D. and B. Lapeyre. *Introduction to stochastic calculus applied to finance*, volume 11 of *Financial Mathematics Series*. Chapman & Hall/CRC, New York, 2007.

[52] Linetsky, V. Spectral expansions for asian (average price) options. *Operations Research*, 52(6):856–867, 2004.

[53] Lipton, A. (ed). *Exotic options: the cutting-edge collection*. Risk Books, London, United Kingdom, 2003.

[54] Long, Y.L. *The multivariate normal distribution*. Springer-Verlag, New York, 1990.

[55] Longstaff, F. Pricing options with extendible maturities: analysis and applications. *Journal of Finance*, 45(3):935–957, 1990.

[56] Malloch, H. *The valuation of options on traded accounts: continuous and discrete time models*. Sydney University PhD thesis, School of Mathematics and Statistics, 2010.

[57] Margrabe, W. The value of an option to exchange one asset for another. *Journal of Finance*, 33:177–186, 1978.

[58] Merton, R. Theory of rational option pricing. *Bell Journal of Economics and Management Science*, 4(1):141–183, 1973.

[59] Neuberger, A.J. The log contract: a new instrument to hedge volatility. *Journal of Portfolio Management*, pages 74–80, 1994.

[60] Øksendal, B. *Stochastic differential equations: an introduction with applications*. Springer-Verlag, Berlin, Germany, 2003.

[61] Pfeffer, D. Sequential barrier options. *Algo Research Quarterly*, 4(3):65–73, 2001.

[62] Rich, D. and D. Chance. An alternative approach to the pricing of options on multiple assets. *Journal of Financial Engineering*, 2(3):271–285, 1993.

[63] Rich, Don R. The mathematical foundations of barrier option pricing theory. *Advances in Futures and Options Research*, 7, 1994.

[64] Rubinstein, M. Options for the undecided. *Risk Magazine*, 4(4):43, 1991.

[65] Rubinstein, M. and E. Reiner. Unscrambling the binary code. *Risk Magazine*, 4:75–83, 1991.

[66] Schönbucher, P.J. *Credit derivatives: models, pricing and implementation.* Wiley Finance Series, 2003.

[67] Schoutens, W. *Lévy processes in finance: pricing financial derivatives.* Wiley Series in Probability and Statistics, 2003.

[68] Shefrin, H. *Behavioral approach to asset pricing.* Advanced Finance Series. Academic Press, San Diego USA, London, United Kingdom, 2008.

[69] Shreve, S. and J. Vecer. Options on a traded account: vacation calls, vacation puts and passport options. *Finance and Stochastics*, 4:255–274, 2000.

[70] Skipper, M. *Technicolour rainbow exotics: unlocking the quintessential formula.* University of Sydney Honours Thesis, School of Mathematics and Statistics, 2002.

[71] Skipper, M. Pricing multi-asset barrier options. *Oxford Finance Research Centre (working paper)*, pages 1–16, 2003.

[72] Skipper, M. and P.W. Buchen. A valuation formula for multi-asset, multi-period binaries in a black-scholes economy. *ANZIAM Journal*, 50(4):475–485, 2009.

[73] Stulz, R.M. Options on the minimum and maximum of two risky assets. *Journal of Financial Economics*, 10:161–185, 1982.

[74] Thomas, B. Something to shout about. *Risk Magazine*, 6(5):56–58, 1994.

[75] Van der Hoek, J. *Binomial models in finance.* Springer Finance. Springer–Verlag, London, United Kingdom, 2006.

[76] Wiersema, U. *Brownian motion calculus.* Wiley Finance Series. John Wiley and Sons, United Kingdom, 2008.

[77] Wilmott, P., S. Howison, and J. Dewynne. *The mathematics of financial derivatives.* CUP, Cambridge, United Kingdom, 1997.

[78] Zhang, P.G. *Exotic options: a guide to second generation options.* World Scientific, Singapore, 1998.

# *Index*

Milton Keynes UK
Ingram Content Group UK Ltd.
UKHW040446071024
449327UK00020B/1036

9 780367 381721